Sustainable Manufacturing and Remanufacturing Management

Weidong Li · Sheng Wang
Editors

Sustainable Manufacturing and Remanufacturing Management

Process Planning, Optimization and Applications

Springer

Editors
Weidong Li
Faculty of Engineering,
Environment and Computing
Coventry University
Coventry
UK

Sheng Wang
Faculty of Engineering,
Environment and Computing
Coventry University
Coventry
UK

ISBN 978-3-030-08798-2 ISBN 978-3-319-73488-0 (eBook)
https://doi.org/10.1007/978-3-319-73488-0

Printed on acid-free paper

This Springer imprint is published by the registered company Springer International Publishing AG part of Springer Nature
The registered company address is: Gewerbestrasse 11, 6330 Cham, Switzerland

Preface

The rapid development of the global economies has been hindered by the unprecedented challenges of the scarcity of fossil fuel resources and ecological damage. To balance the multi-faceted dimensions of economic growth, resource utilization and environmental protection, major economies in the world have developed action plans and schemes for improving the efficiency of energy consumption. A series of product design and manufacturing regulations and directives have been proposed in detail, requiring original equipment manufacturers (OEMs) and their supply chains to adapt to total energy-efficient management.

The road map of the European Horizon 2020 Factories of the Future programme has indicated that it is a key for Europe's manufacturing to enhance technological innovations to make their products and production activities more affordable, accessible and sustainable, so as for the entire European manufacturing value chains to multiply the societal, economic and environmental benefits (EU Manufacturing Visions 2020/2030). In other developed and emerging economies, such as the USA, China, Japan and South Korea, high reliability and environmental friendliness are also highly expected by industries to conform to stricter environmental regulations and keep resource utilization to minimum levels, as well as maintaining the high production quality and productivity.

From 2011 to 2017, sponsored by European Commission and national funding agencies of the UK and China, the contributors of the chapters have collaborated in six major research projects for sustainable product life cycle development. In these projects, sustainability issues have been investigated from the following two perspectives: (1) process planning and scheduling for sustainable manufacturing (sustainable manufacturing) and (2) recycling and disassembly process planning (remanufacturing). In this book, relevant innovative sustainable technologies sponsored by these projects in the two aspects are reported. In particular, the book focuses on the applications of sustainable computer numerical controlled (CNC) machining and waste electronic product remanufacturing management. Theoretical analysis and algorithm design are presented, as well as the state-of-the-art survey, technical implementation details and case studies. The book is featured by including some industrial case studies, industrial experiments and system implementation in manufacturing enterprises in Europe and the world. Meanwhile, multidisciplinary technologies used to support the research from the areas of artificial intelligence, big data analytics, computational intelligence, manufacturing simulation and optimization, etc., are systemically integrated.

We believe that the book is a timely update in the era of sustainable and smart manufacturing. It offers a valuable resource for researchers in sustainable manufacturing, remanufacturing and product life cycle management communities, as well as practicing engineers and decision-makers in industry and all those interested in sustainable product development and Industry 4.0.

Coventry, UK

Weidong Li
Professor in Manufacturing
Sheng Wang
Senior Researcher in Manufacturing

Note

Some chapters are rewritten based on previous publications. Copyright has been transferred for publishing this book.

Chapter 2 is rewritten from the work of Energy-aware Integrated Process Planning and Scheduling for Job Shops based on a Modified Genetic Algorithm, Proceedings of the Institution of Mechanical Engineers, Part B: Journal of Engineering Manufacture, 2015, 229, 13–26 (authors: M. Dai, D. B. Tang, Y. Xu, W. D. Li).

Chapter 3 is rewritten from the work of A Hybrid Honey-bee Mating Optimization and Simulated Annealing Approach for Sustainable Process Planning and Scheduling, Integrated Computer Aided Engineering, 2015, 22, 311–326 (authors: X. X. Li, W. D. Li, X. T. Cai, F. Z. He).

Chapter 4 is rewritten from the work of A Systematic Approach of Process Planning and Scheduling Optimization for Sustainable Machining, Journal of Cleaner Production, 2015, 87, 914–929 (authors: S. Wang, X. Lu, X. X. Li, W. D. Li).

Chapter 5 is rewritten from the work of Experimental Investigation and Multi-objective Optimization Approach for Low-carbon Milling Operation of Aluminium, Proceedings of the Institution of Mechanical Engineers, Part C: Journal of Mechanical Engineering Science, 2017, 231(15), 2753–2772 (C. Y. Zhang, W. D. Li, P. Y. Jiang, P. H. Gu).

Chapter 10 is rewritten from the work of A Semantic Information Services Framework for Sustainable WEEE Management Towards Cloud-based Remanufacturing, ASME Transactions Journal of Mechanical Science and Engineering, 2015, 137(6) (authors: K. Xia, L. Gao, L. H. Wang, W. D. Li, K. M. Chao).

Chapter 11 is rewritten from the work of Selective Disassembly Planning for Waste Electrical and Electronic Equipment with Case Studies on Liquid Crystal Displays, Robotics and Computer-Integrated Manufacturing, 2013, 29 (4), 248–260 (authors: W. D. Li, K. Xia, L. Gao, K. M. Chao).

Chapter 12 is rewritten from the work of A Systematic Selective Disassembly Approach for Waste Electrical and Electronic Equipment with Case Study on

Liquid Crystal Display Televisions (LCD-TVs), Proceedings of the Institution of Mechanical Engineers, Part B: Journal of Engineering Manufacture, 2017, 231(13), 2261–2278 (G. Q. Jin, W. D. Li, S. Wang, S. M. Gao).

Chapter 13 is rewritten from the work of Disassembly Sequence Planning Using a Simplified Teaching-Learning-Based Optimization Algorithm, Journal of Advanced Engineering Informatics, 2014, 28(4), 518–527 (authors: K. Xia, L. Gao, W. D. Li, K. M. Chao).

About the Book

The book reports the latest research and development of some international collaborative projects from 2011 to 2017 sponsored by European Commission and international collaborative research projects among the U.K., Spain and China. In this book, innovative optimization technologies for sustainable manufacturing and remanufacturing are introduced. The book focuses on applications of sustainable machining and e-waste product remanufacturing management. Theoretical analysis and algorithm design are presented, as well as the state-of-the-art survey, implementation details and case studies. The book is featured by including some industrial case studies and system implementation in manufacturing enterprises in Europe and the world.

The book offers a valuable resource for researchers in sustainable manufacturing, remanufacturing and product lifecycle management communities, as well as practicing engineers and decision makers in industry and all those interested in sustainable product development and Industry 4.0.

Contents

About the Editors

Prof. Weidong Li is a full professor in Manufacturing, Coventry University, UK. He has more than twenty years' experience in computer-aided design, manufacturing informatics, cloud manufacturing and sustainable manufacturing. His research has been sponsored by a number of research and development projects from the UK EPSRC, EU and European industries. In the research area, he has published three books and more than 150 research papers.

Dr. Sheng Wang is a senior researcher in Manufacturing, Coventry University, UK. She got her Ph.D. from Queen Mary University of London in Computer Science and Electronic Engineering. In the past 5 years, she has participated in a number of European Commission-sponsored projects in sustainable manufacturing.

Chapter 1
Introduction

W. D. Li and S. Wang

Abstract In this chapter, introduction to sustainable manufacturing and remanufacturing management is given. For sustainable manufacturing, characteristics and technical trends for modern manufacturing processes, sustainable trend for product development lifecycle, and process planning for sustainable manufacturing are briefly discussed. For remanufacturing manufacturing, technologies of recycling and disassembly process planning for Waste Electric and Electronic Equipment (WEEE) are introduced. This chapter is the base for the following chapters of the book for detailed technology development.

Keyword Sustainable manufacturing · remanufacturing · Process planning Recycling · Disassembly

1.1 Characteristics and Technological Trends in Modern Manufacturing Processes

Modern manufacturing industries have shown clear trends in recent years—away from long-standing and well-established products and relevant production stable over many years, away from comprehensive trusts which will cover all the processes of production, and also away from the single economic consideration of production. Instead, companies increasingly focus on their core manufacturing competencies, develop and produce customized products, enter more often alliances especially with small- and medium-sized enterprises (SMEs) for manufacturing and resource optimization, and integrate environmental and social responsibilities into their production operations. For instance, precision, automotive, and aerospace manufacturing has demonstrated the following characteristics (shown in Fig. 1.1 as well):

W. D. Li (✉) · S. Wang
Faculty of Engineering, Environment and Computing,
Coventry University, Coventry, UK
e-mail: weidong.li@coventry.ac.uk

W. Li and S. Wang (eds.), *Sustainable Manufacturing and Remanufacturing Management*, https://doi.org/10.1007/978-3-319-73488-0_1

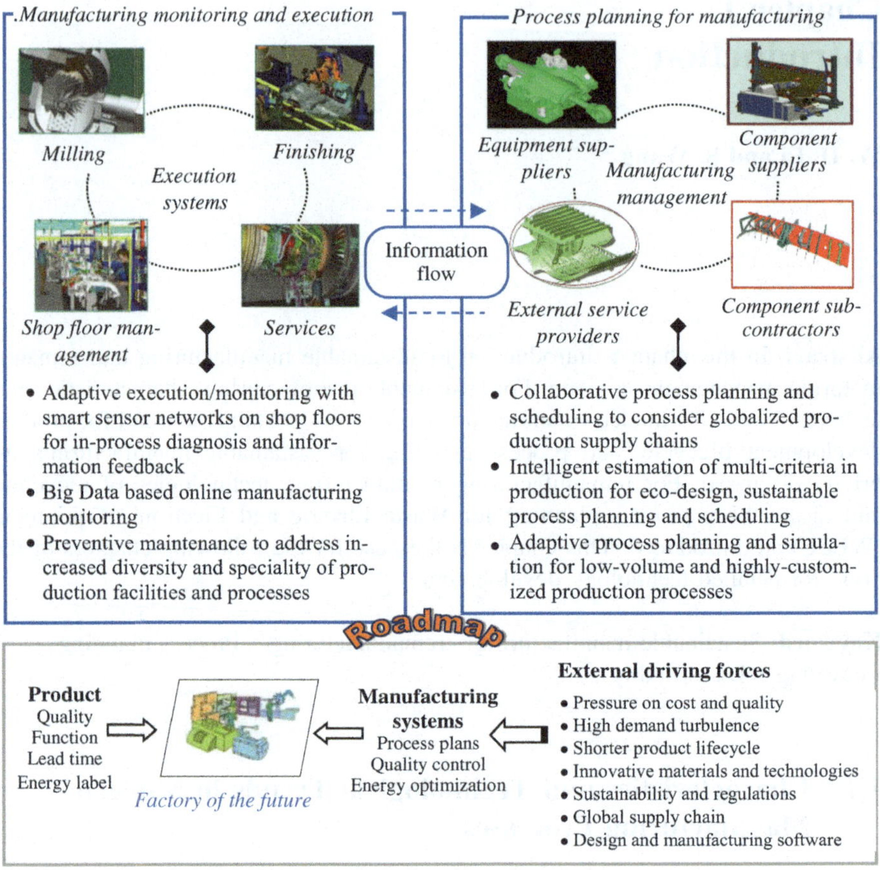

Fig. 1.1 Future factories with innovative information and intelligent technologies

- A faster clock speed of market changes and technological innovations, which leads to coevolution of products, manufacturing processes, and manufacturing systems, demanding more efficient configuration and re-configuration of manufacturing processes and systems to facilitate product varieties and customization [1, 2];
- Keener competitiveness for cost-effectiveness and tighter regulations on energy consumption efficiency, which expects multi-criteria optimization solutions for manufacturing to achieve "Competitive Sustainable Development" [3];
- Industrial Product-Service System and "Extended Producer Responsibilities (EPR)" concepts, which require the high reliability and environmental friendliness of products and relevant manufacturing processes for better service and extended product lifecycle management [4, 5];
- Introduction of sensors and networked communication technologies into shop floors for smart in-process diagnostics and efficient human intervention in order to improve the robustness and adaptability of processes and systems [6];

- Original Equipment Manufacturers (OEMs) work with global partners and suppliers, which are mainly SMEs, more collaboratively to achieve entire manufacturing value chain optimization.

1.2 Sustainable Trend in Product Development Lifecycle

The rapid development of the global economies has been hindered by the unprecedented challenges of the scarcity of fossil fuel resources and ecological damage. To balance the multifaceted dimensions of economic growth, resource utilization, and environmental protection, major economies in the world have developed action plans and schemes for improving the efficiency of energy consumption. For instance, Europe has set out ambitious goals to achieve at least 20% energy saving by 2020.

Product development and manufacturing are the major economical pillars of the world. The increased demand for welfare and new products has incurred more production activities, while the relevant processes are energy intensive making the sector one of the primary energy consumption sources. A series of product regulations (e.g., eco-design of Energy-using Products (EuP) Directive) have been proposed in detail, requiring OEMs and their supply chains to adapt to total energy-efficient management. The roadmap of the European Horizon 2020 Factories of the future program has indicated that it is a key for Europe's manufacturing to enhance technological innovations to make their products and production activities more affordable, accessible, and sustainable, so as for the entire European manufacturing value chains to multiply the societal, economic, and environmental benefits (EU Manufacturing Visions 2020/2030 [7]). In other developed and emerging economies such as USA, China, Japan, and South Korea, high reliability and environmental friendliness are also highly expected by industries to conform to stricter environmental regulations, while keeping resource utilization to minimum levels [8]. On the other hand, due to the complexity of product development, manufacturing, and lifecycle processes, the high cost of establishing energy consumption information and decision-making models are major barriers crippling companies to achieve sustainability. Such a scenario is illustrated in Fig. 1.2.

From 2011 to 2017, sponsored by European Commission and the international cooperative programs in the UK and China, the author and collaborators have participated in six major research projects for sustainable product lifecycle development. In the projects, the sustainability issues have been investigated from the following two perspectives:

- Process planning for sustainable manufacturing;
- Recycling and disassembly process planning.

In the following sections, the above two perspectives will be outlined.

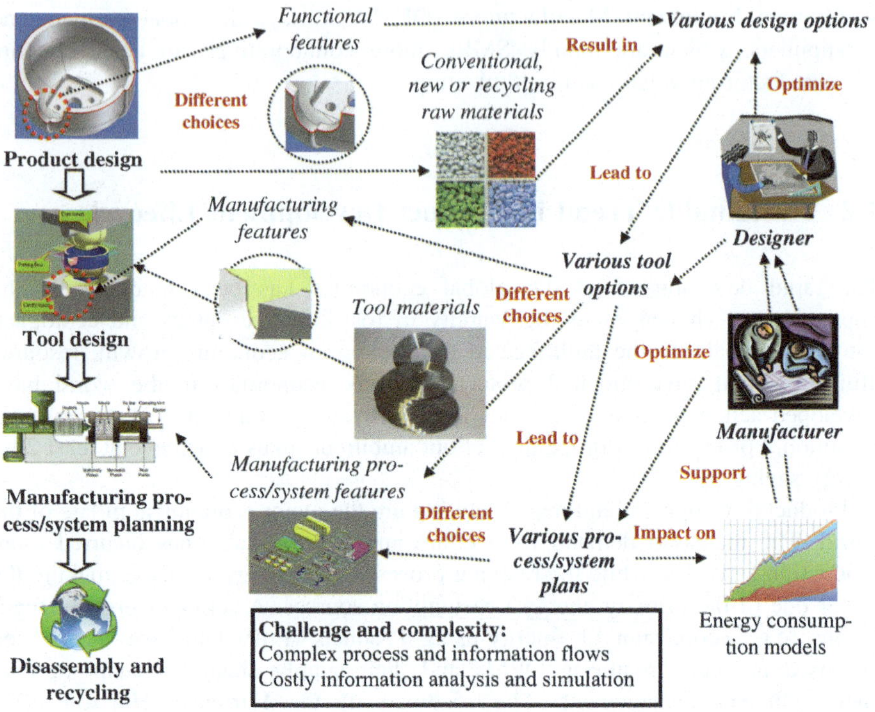

Fig. 1.2 Challenge and complexity to establish effective energy-efficiency models to support product development and manufacturing

1.3 Process Planning for Sustainable Manufacturing

Process planning is the essential decision-making process to determine the manufacturing quality, lead time, and sustainability. Manufacturing processes involve various knowledge-intensive decision-making activities. For the process planning systems used in companies, process plans are based on pre-defined manufacturing conditions/constraints and represented in relatively rigid structures. It is resource-intensive, time-consuming, and error-prone to update process plans flexibly when dynamic changes in production lines take place. Therefore, the current systems limit the flexibility and changeability of production during process planning, scheduling, and quality control, leading to the difficulty to realize sustainable manufacturing strategies effectively. It is expected to develop innovative systems to support process planning to enhance the adaptability of production lines under dynamic conditions. Moreover, intelligent algorithms are embedded to optimize manufacturing parameters to ensure zero defect and overall optimization of productivity, cost, and energy consumption during production.

In this book, the investigated manufacturing process will be focused on the process of Computerized Numerical Controlled (CNC) machining. Various innovative frameworks and algorithms of process planning and scheduling for sustainable CNC machining have been developed. For instance, a proposed framework consists of the following procedures:

- Process planning is organized as an innovative decision-making process, consisting of macro-planning, dynamic scheduling, and micro-planning. The process is shown in Fig. 1.3. The design of the process is aimed to enhance the adaptability, responsibility, and robustness of process planning under dynamic conditions to facilitate deviation intervention and defect prediction/prevention during machining. For instance, when the working machine becomes abnormal and a machine replacement is needed, another available machine from the candidate machine pool will be chosen for replacement. The macro-planning will be kept the same while the micro-planning is needed for the new machine to fine-ture machining process parameters. Based on the design of the above

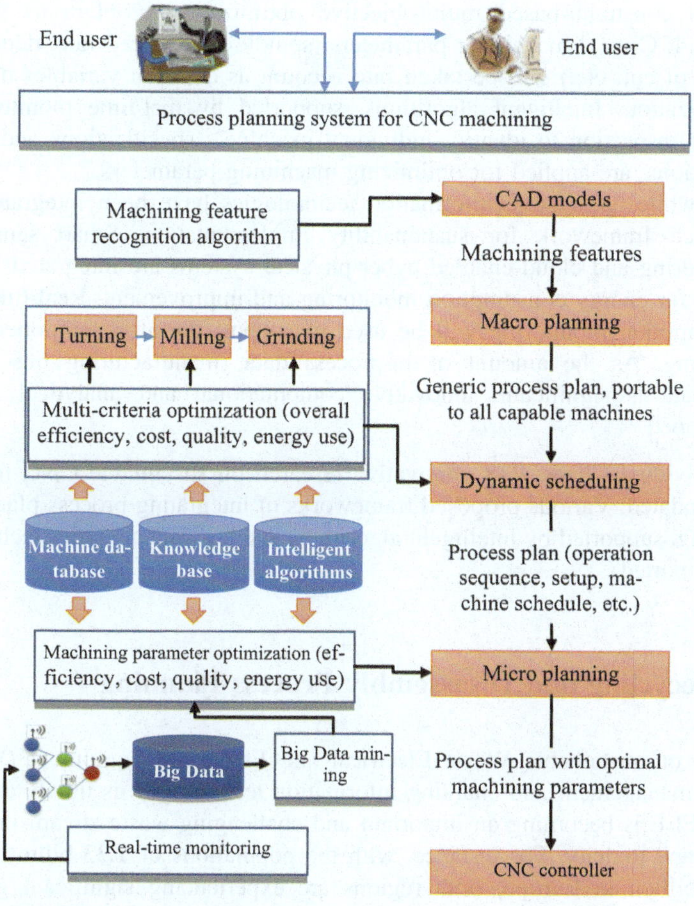

Fig. 1.3 Procedures of process planning for sustainable multi-stage machining

innovative process, the efficiency of process planning will be improved, and the sustainability will be achieved through refined machining process parameters.

- At the macro-planning stage, a product represented by machining features (m-features) is grouped and sequenced against tool approach directions, datum references, and/or machining constraints. Those m-features having no direct constraints remain parallel in sequence. The result of the planning is a generic nonlinear process plan, portable to all capable machines. At the following dynamic scheduling stage, the generic process plans are dispatched to one or more machines according to the availability of machines. The above process is supported by a set of networked databases and knowledge bases for routing, machine assignment, line balancing, and sustainability optimization.
- At the micro-planning stage, refinement of detailed machining parameters in terms of productivity, production quality, and energy efficiency takes place. The operation details of the corresponding m-feature will be specified. At this stage, the performance indicators and constraints in machining are incorporated into a constraint-based multi-objective optimization problem as shown in Fig. 1.4. Critical machining parameters (spindle speed, feed rate, depth of cut, width of cut, etc.) will be taken into account as decision variables during the optimization. Intelligent algorithms, supported by real-time monitoring and in-line inspection to identify individual machine's specifications and working conditions, are applied for optimizing machining parameters.
- Meanwhile, innovative information technologies have been integrated in the research framework for sustainability implementation. Smart sensor-based monitoring and cloud-enabled cyber-physical systems are integrated into shop floors for energy consumption monitoring and improvement. Real-time energy consumption monitoring will be used to support the aforementioned process planning. As the amount of in-process data (manufacturing big data [9]) big data is significant, innovative computational and analytical tools are developed.

In individual Chaps. 2–8, innovative research on sustainable CNC machining will be updated. Various proposed frameworks of integrating process planning and scheduling, supported by intelligent algorithms and big data analytics technologies, will be reported.

1.4 Recycling and Disassembly Process Planning

Research on sustainable Waste Electrical and Electronic Equipment (WEEE or e-waste) management and enabling information technologies is timely and significant. WEEE is becoming an important and challenging waste stream in terms of quantity and toxicity. For instance, with the populations of 1.33 billion in China and 0.5 billion in Europe, both regions are experiencing significant growth of WEEE per year. There are approximately seven million tons of WEEE generated in

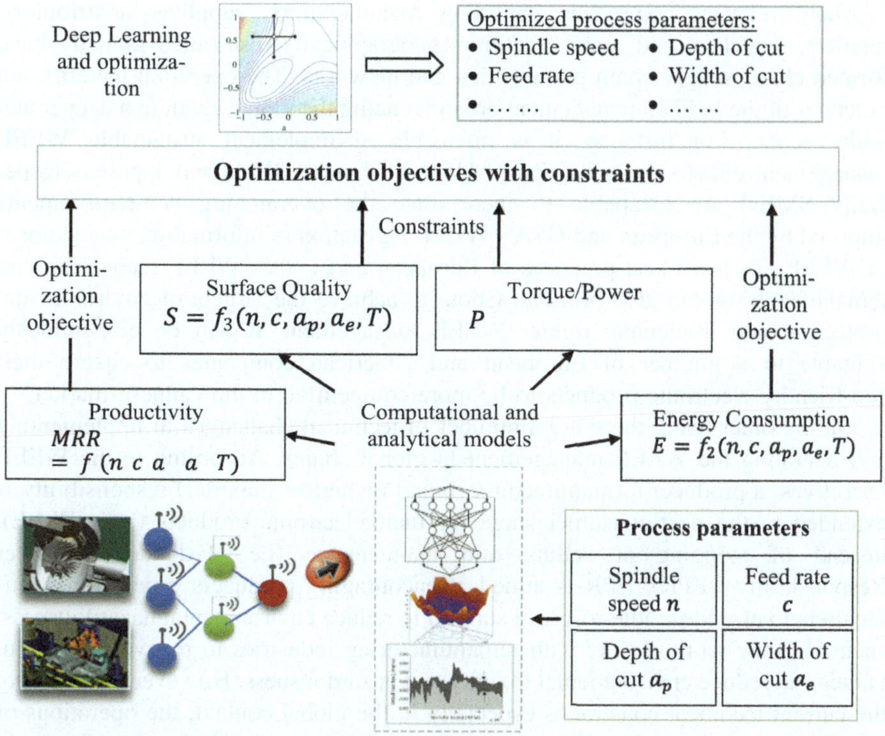

Fig. 1.4 Optimization for the micro-planning stage

Europe per year [10]. In China, 1.1 million tons of WEEE is generated per year, and China is the second in the world in the landfill and incineration of WEEE [11]. Due to the rapid technical innovations and shorter usage lifecycle of electronic products, WEEE is growing much faster than any other municipal waste streams. In order for the Earth to be cleaner, sustainability has become a critical driving force shaping the future of WEEE lifecycle management patterns.

An important research issue is to develop processes and technologies to extend the value and usefulness of WEEE (e.g., recovering and remanufacturing) and to better manage WEEE after service to generate less or even zero environmental impact and CO_2 footprint. It is envisaged that in future all WEEE need to be traceable, manageable, recyclable, recoverable, and remanufacturable. The WEEE Directive was enacted as the European law in 2003, and the EU member states were required to transpose the provisions into national laws by August 2004. As one of the biggest electronic and component production nations in the world, China has realized the serious environmental issues from WEEE and addressed them as a rising priority. The Chinese government issued its WEEE Directive which has been enforced from 2011. Globally, the WEEE Directive has also implemented in USA, Japan, South Korea, etc.

Many relevant companies including manufacturers, suppliers, distributors, retailers, recyclers, and remanufacturers geographically distributed globally have formed closer supply chain partnerships and networks. The operation patterns and practices of the WEEE management are influencing the global chain in a deeper and wider scope. For instance, it is infeasible to implement sustainable WEEE management effectively in Europe or USA if Chinese OEMs and suppliers (especially SMEs) are incapable to incorporate the overarching eco-requirements imposed by the European and USA's WEEE legislation in information management of WEEE. To learn best practices of European and USA's WEEE recovering and remanufacturing can also promote China to achieve the targets of environmental protection and implement tighter WEEE management. It will be tremendously valuable to a number of European and American companies to ensure their eco-friendly electronic products to be more competitive in the Chinese market.

On the other hand, there are a number of technical challenges in implementing and operating the WEEE management in global chains. According to the WEEE Directives, a producer's (manufacturer, brand owner, or importer) responsibility is extended to the post-consumer stage for their electronic products (i.e., WEEE), instead of stopping at selling and maintenance (i.e., Extended Producer Responsibility—EPR). EPR is aimed at encouraging producers especially manufacturers to provide cradle-to-grave support to reduce environmental hazardousness, such that they work closely with remanufacturing industries to recover maximum values and reduce environmental toxicity and hazardousness. However, considering the current technical constraints especially in the global context, the operations of WEEE management (i.e., the producer compliance scheme) do not effectively achieve the aim of the WEEE Directives and EPR. The scenarios are illustrated in Figs. 1.5 and 1.6.

One of the major reasons is there are no distributed information management and sensible remanufacturing strategies. This is expanded below.

According to the WEEE producer compliance scheme, producers are required to report to the WEEE management authorities for product quantity, recycle, and treatment information periodically. Owing to the active e-commerce and more personalized electronic product design, online transactions and mass-customized productions are more frequent. This brings challenges to record the dynamic, low-volume, and varied WEEE information. In the current operation of WEEE management, authorities will be the information bottleneck so that information exchange across the entire WEEE chains is inefficient or even impossible not only internationally and nationally but also regionally. Closer supply chain-spanning information linking between manufacturers, suppliers, distributors, retailers, recyclers, and remanufacturers should take place. In the current situation, there is almost no any distributed information services deployed in the Internet to enable the convenient and secured retrieving of WEEE information. To tackle the issue, it is imperative to develop an effective distributed information architecture to support the dynamic and robust information management in global WEEE chains.

Remanufacturing of WEEE, which has been viewed as a hidden green giant during WEEE management and treatment, is attracting increasing attentions of

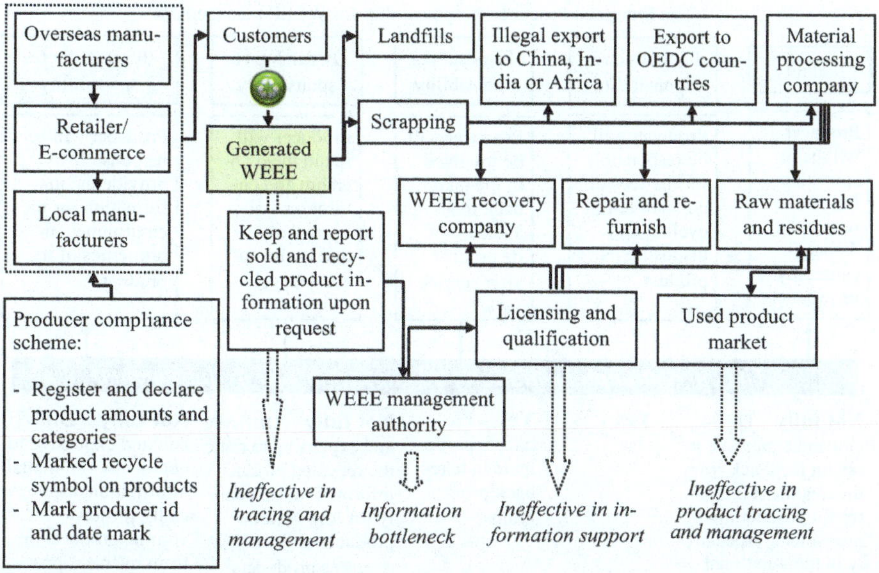

Fig. 1.5 WEEE recycling and recovery mechanisms

researchers and practitioners in recent years [12]. A good remanufacturing strategy will be one of the stronger driving forces for industries to adopt in their practices to balance economic and environmental targets, and close gaps between shorter innovation cycles and longer lives of WEEE.

On the other hand, remanufacturing legislative initiatives are underway in Europe and USA to ensure OEMs and suppliers to provide free access to remanufacturing information facilities in global chains. However, the end-of-life information flows of WEEE between product manufacturers and WEEE remanufacturing enterprises have not been effectively established. Owing to the lack of the information and strategies, remanufacturing enterprises usually have a limited understanding of WEEE due to their diversity. It becomes difficult to recover the larger proportion of components and materials at reasonable cost and to achieve high environmental protection targets. For the same reason, manufacturers have limited feedback from remanufacturing enterprises to further improve their design toward eco-targets. Smart remanufacturing is therefore paramount to share remanufacturing information and use modern artificial intelligent algorithms to incorporate the multi-objectives to balance the economic and environmental aspects so that sensible remanufacturing strategies can be applied to recover diverse WEEE with specific remanufacturing methods.

In this book, from Chaps. 9 to 13, innovative research developed for WEEE remanufacturing management will be discussed.

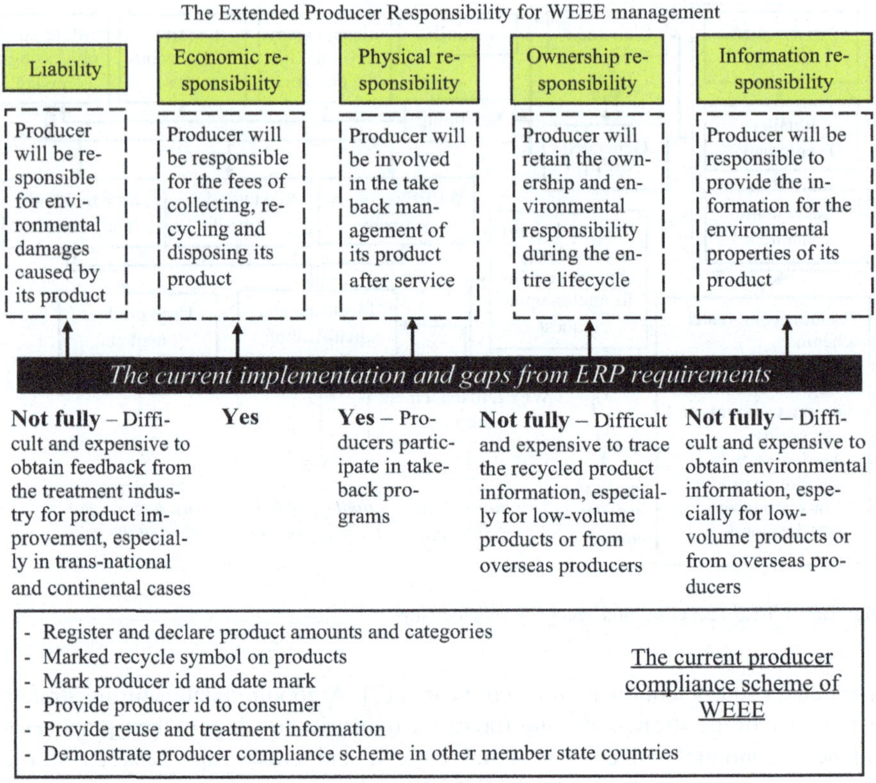

Fig. 1.6 Gaps between EPR and the producer compliance scheme

References

1. Tolio, T., Ceglarek, D., ElMaragphy, H. A., Fischer, A., Hu, S. J., Laperriere, L., et al. (2011). SPECIES—Co-evolution of products, processes and production systems. *CIRP Annals—Manufacturing Technology, 59,* 672–693.
2. Hu, S. J., Ko, J., Weyand, L., ElMaragphy, H. A., Lien, T. K., Koren, Y., et al. (2011). Assembly system design and operations for product variety. *CIRP Annals—Manufacturing Technology, 60,* 715–733.
3. Jovane, F., Yoshikawa, H., Alting, L., Boer, C. R., Westkamper, E., Williams, D., et al. (2008). The incoming global technological and industrial revolution towards competitive sustainable manufacturing. *CIRP Annals—Manufacturing Technology, 57,* 641–659.
4. Meier, H., Roy, R., & Seliger, G. (2010). Industrial product-service systems—IPS2. *CIRP Annals—Manufacturing Technology, 59,* 607–627.
5. Mayers, C. K. (2007). Strategic, financial, and design implications of extended producer responsibility in Europe: A producer case study. *Journal of Industrial Ecology, 11,* 113–131.
6. Bengtsson, N., Michaloski, J., Proctor, F., Shao, G., & Venkatesh, S. (2010). Towards data-driven sustainable machining—Combining MTConnect production data and discrete event simulation. In *Proceedings of ASME MSEC 2010*, Pennsylvania, MSE010-34178, October 12–15.

7. EFFRA. (2013). Factories of the future—Multi-annual roadmap for the contractual PPP under Horizon 2020, http://www.effra.eu/attachments/article/129/Factories%20of%20the%20Future%202020%20Roadmap.pdf. Last accessed on 12 Mar 2018.
8. Bilgea, P., Badurdeenb, F., Seligera, G., & Jawahirb, I. S. (2016). A novel manufacturing architecture for sustainable value creation. *CIRP Annals—Manufacturing Technology, 65,* 455–458.
9. Dubey, R., Gunasekaran, A., Childe, S. J., Wamba, S. F., & Papadopoulos, T. (2016). The impact of big data on world-class sustainable manufacturing. *The International Journal of Advanced Manufacturing Technology, 84,* 631–645.
10. Walther, G., Steinborn, J., Spengler, T. S., Luger, T., & Herrmann, C. (2010). Implementation of the WEEE-directive—Economic effects and improvement potentials for reuse and recycling in Germany. *International Journal of Advanced Manufacturing Technology, 47,* 461–474.
11. Hicks, C., Dietmar, R., & Eugster, M. (2005). The recycling and disposal of electrical and electronic waste in China—legislative and market responses. *Environmental Impact Assessment Review, 25,* 459–471.
12. Zlampareta, G. I., Ijomahb, W., Miao, Y., Awasthi, A. K., Zeng, X. L., & Li, J. H. (2017). Remanufacturing strategies: A solution for WEEE problem. *Journal of Cleaner Production, 149,* 126–136.

Chapter 2
Energy-aware Integrated Process Planning and Scheduling for Job Shops

M. Dai, D. B. Tang, Y. C. Xu and W. D. Li

Abstract Process planning that is based on environmental consciousness and energy-efficient scheduling currently plays a critical role in sustainable manufacturing processes. Despite their interrelationship, these two topics have often been considered to be independent of each other. It, therefore, would be beneficial to integrate process planning and scheduling for an integrated energy-efficient optimization of product design and manufacturing in a sustainable manufacturing system. This chapter proposes an energy-aware mathematical model for job shops that integrates process planning and scheduling. First, a mixed integrated programming model with performance indicators such as energy consumption and scheduling makespan is established to describe a multi-objective optimization problem. Because the problem is strongly non-deterministic polynomial-time hard (NP-hard), a modified genetic algorithm is adopted to explore the optimal solution (Pareto solution) between energy consumption and makespan. Finally, case studies of energy-aware integrated process planning and scheduling are performed, and the proposed algorithm is compared with other methods. The approach is shown to generate interesting results and can be used to improve the energy efficiency of manufacturing processes at the process planning and scheduling levels.

Keywords Sustainable manufacturing · Process planning and scheduling
Energy consumption · Makespan · Genetic algorithm

M. Dai
College of Mechanical and Electrical Engineering, Yangzhou University, Yangzhou, China

D. B. Tang (✉)
College of Mechanical and Electrical Engineering, Nanjing University of Aeronautics and Astronautics, Nanjing, China
e-mail: d.tang@nuaa.edu.cn

Y. C. Xu
School of Engineering, Aston University, Birmingham, UK

W. D. Li
Faculty of Engineering, Environment and Computing, Coventry University, Coventry, UK

W. Li and S. Wang (eds.), *Sustainable Manufacturing and Remanufacturing Management*, https://doi.org/10.1007/978-3-319-73488-0_2

2.1 Introduction

Due to increasing environmental awareness, energy efficiency has become indispensable as a productivity criterion in manufacturing processes. Since the beginning of the industrial revolution, the industrial sector has consumed large amounts of energy for production. Manufacturing enterprises are responsible for approximately 33% of global total energy consumption and 38% of greenhouse gas emissions [1]. Moreover, the need for sustainable development to attain economic, ecological and social goals is presenting new challenges to manufacturing companies [2]. It is, therefore, important that the manufacturing community has access to systems that can improve energy efficiency in manufacturing processes by reducing energy consumption using new technologies and techniques [3, 4].

Process planning and scheduling are two of the most significant elements of manufacturing processes. Process planning and scheduling are usually implemented sequentially. The former is used to plan manufacturing resources (e.g. machines and tools) and operations of jobs based on cost-effective criteria, such as manufacturing cost; the latter is used to determine how and when to assign the operations of all jobs to the manufacturing resources in terms of constraints, such as time feasibility and resource availability. Due to the increasing importance of energy savings, environmental factors like energy consumption are increasingly taken into account along with traditional performance criteria (time, quality and cost) in the process planning and scheduling levels. Recently, many interesting studies have explored energy-efficient manufacturing with respect to environment-friendly process planning strategies and energy-aware scheduling approaches. Unfortunately, these two factors have been considered independently. As a result, energy consumption in process planning systems has not obtained optimal energy savings due to resource constraints in the scheduling systems. In addition, energy-efficient scheduling with regard to energy-saving information, which is generated after process planning strategies are implemented, cannot be fed back to the process planning system to improve it. It is, therefore, necessary to integrate these factors to enhance energy-efficient manufacturing.

The following sections of this article are organized as follows. Related studies are introduced in section 'Related Work'. An energy-aware integrated process planning and scheduling (EIPPS) problem is addressed in section 'EIPPS'. A modified heuristic algorithm optimization approach for the EIPPS problem is described in section 'A Modified Genetic Algorithm Optimization Approach'. Case studies are presented in section 'Case Studies', and the conclusions are summarized in section 'Conclusions and Future Research'.

2.2 Related Work

2.2.1 Energy-Efficient Process Planning

Process planning describes the transformation of raw materials into products through planning the operations of a product based on machining features, the identification of manufacturing resources that are available to the operations and the determination of the machining sequence in terms of cost-effective indicators. It has a direct influence on the design and manufacturing of products, which are closely related to environmental impacts such as energy consumption. Many studies of energy-efficient process planning have been performed. Sheng and Srinivasan presented an environmentally conscious multi-objective process planning approach that is based on manufacturing features and explored local and global optimum process planning with respect to process energy consumption, process time, waste mass and surface quality factors at the micro- and macro-planning levels [5–7]. Singh et al. developed a framework and models of environmental process planning to assess configurations of a product and its related environmental impact during the advanced product quality planning (APQP) process [8]. Kai et al. [9] developed a multi-objective mathematical model for environmentally supportive process planning that simultaneously considers cost, time and environmental impact. Newman et al. introduced energy consumption as an objective of process planning for computer numerical control (CNC) machining and confirmed that energy consumption can be used as a performance indicator in a multi-criteria process planning system [10]. Yin et al. [11] proposed a new process planning approach that considers environmental factors, such as carbon emissions, and obtained a comparatively green and economical process plan. In summary, a process plan provides a framework to make energy-efficient decisions that minimize energy consumption while maintaining expected production objectives such as time, cost and quality.

2.2.2 Energy-Efficient Scheduling

Scheduling generally involves planning how and when to assign the operations of all jobs to manufacturing resources based on the priority of the jobs, the availability of machines and tools, and time constraints. When scheduling takes into account environmental impacts like energy consumption as an optimization objective, it is energy-efficient. Recently, research on minimizing energy consumption in manufacturing processes through scheduling has been gradually increasing. One of the best-known studies of the impact of scheduling on energy efficiency was Mouzon et al. [12], who proposed a multi-objective mathematical programming model and several algorithms to investigate the scheduling of jobs on a single CNC machine with the goals of reducing energy consumption and total completion time. Mouzon and Yildirim [13] outlined a multi-objective optimization schedule with the objective of minimizing the total energy consumption and the total tardiness on a machine using a greedy randomized adaptive search algorithm. Fang et al. [14] provided a new mixed integer linear programming model to schedule a classical flow shop that incorporated the peak total power consumption, the carbon footprint and the makespan. Bruzzone et al. [15] reported an energy-aware scheduling algorithm that is based on a mixed integer programming formulation to realize energy savings for a flexible flow shop that was required to maintain the original job assignment and sequencing. Zhang et al. [16] developed a mathematical model to minimize energy consumption and improve scheduling efficiency for the dynamic scheduling problem in a flexible manufacturing system. Liu et al. [17] considered reducing the total wasted energy consumption using a branch and bound algorithm in a permutation flow shop scheduling problem. Dai et al. [18] also explored the multi-objective energy-efficient scheduling problem with the makespan and energy consumption of manufacturing processes as objectives. The energy-aware schedules in these studies were developed after process planning generating the operations of all jobs. Because the assumption that all of the manufacturing resources will be available at the process planning level might not be valid at the scheduling level, the scheduling results might not be optimized.

2.2.3 Integrated Process Planning and Scheduling

Process planning and scheduling play critical roles in linking product design and manufacturing and must be integrated on the basis of their complementary relationship in manufacturing processes. The preliminary concept of integrated process planning and scheduling (IPPS) was first proposed by Chryssolouris et al. [19, 20]. Numerous studies by several researchers have subsequently explored IPPS, and three main models for IPPS have been developed: nonlinear process planning, closed-loop process planning and distributed process planning. The research objectives for IPPS mostly focus on traditional production performance indicators, such as the makespan, the job tardiness, the balanced level of machine utilization and the manufacturing cost [21–26]. However, most of these process planning and scheduling methods have paid little attention to the energy efficiency of manufacturing processes. This study proposes an energy-aware mathematical model for IPPS to achieve energy-saving design and manufacturing of a product based on a nonlinear process planning method. The proposed model simultaneously generates the energy-efficient process planning strategy of all jobs and the energy-efficient scheduling strategy. The EIPPS strategies are deployed by the multi-objective optimization function with two objectives: the energy consumption and the makespan on the job shop floor. In addition, a modified genetic algorithm (GA) is adopted to determine the optimal solution of the model.

2.3 EIPPS

2.3.1 Problem Description

According to the definition of the IPPS problem [22], the EIPPS formulation can be described as follows. There is a set of n jobs $J = \{1, 2,\ldots, n\}$ which are to be processed on a set of m machines $M = \{1, 2,\ldots, m\}$. Each job $j \in J$ is characterized by a set of $o_{lj}(l \in G_j, j \in J)$ operations $O_{lj} = \{1, 2, \ldots, o_{lj}\}$ including a set of g_j alternative process plans $G_j = \{1, 2, \ldots, g_j\}$ which have to be executed by selecting a desirable process plan. The operations of all jobs are to be assigned to available manufacturing resources like machines and tools so as to determine a schedule based on some constraints among the operations. Furthermore, one operation of a job in different process plans could be processed on different machines with different power (energy) consumptions or on the same machine with different processing parameters like cutting speeds. Thus, the operations of each job $j \in J$ on machine $i \in M$ have processing time and corresponding energy consumption. The objective of the EIPPS is to assign jobs to machines and to determine the processing sequence of operations on each machine in order to search for optimal or near-optimal results (Pareto solution) between the energy consumption and maximum completion time, i.e. makespan. The connection of both process planning and scheduling functions is maintained through the operations of the jobs, which is illustrated in Fig. 2.1

The EIPPS should satisfy the following constraints:

1. All jobs and machines are available at time zero.
2. Each machine should not process more than one job at a time.
3. The different operations of each job have to be processed sequentially on machines.
4. Pre-emption is not allowed for executing each operation on machines; i.e. once one operation is started, it must be finished without interruption.
5. There are no precedence relationships between operations of different jobs, but there are precedence relationships between different operations of one job.
6. The power (energy) consumption can be calculated for each job processed on each machine.

2.3.2 Mathematical Model

The energy-aware modelling of the IPPS problem is defined in this section. This article considers a scheduling problem that arises on a job shop floor. The mixed integer programming model is constructed to minimize the energy consumption in the development of process planning and scheduling while satisfying the possible optimization of the makespan. The notations used to describe the model are as follows:

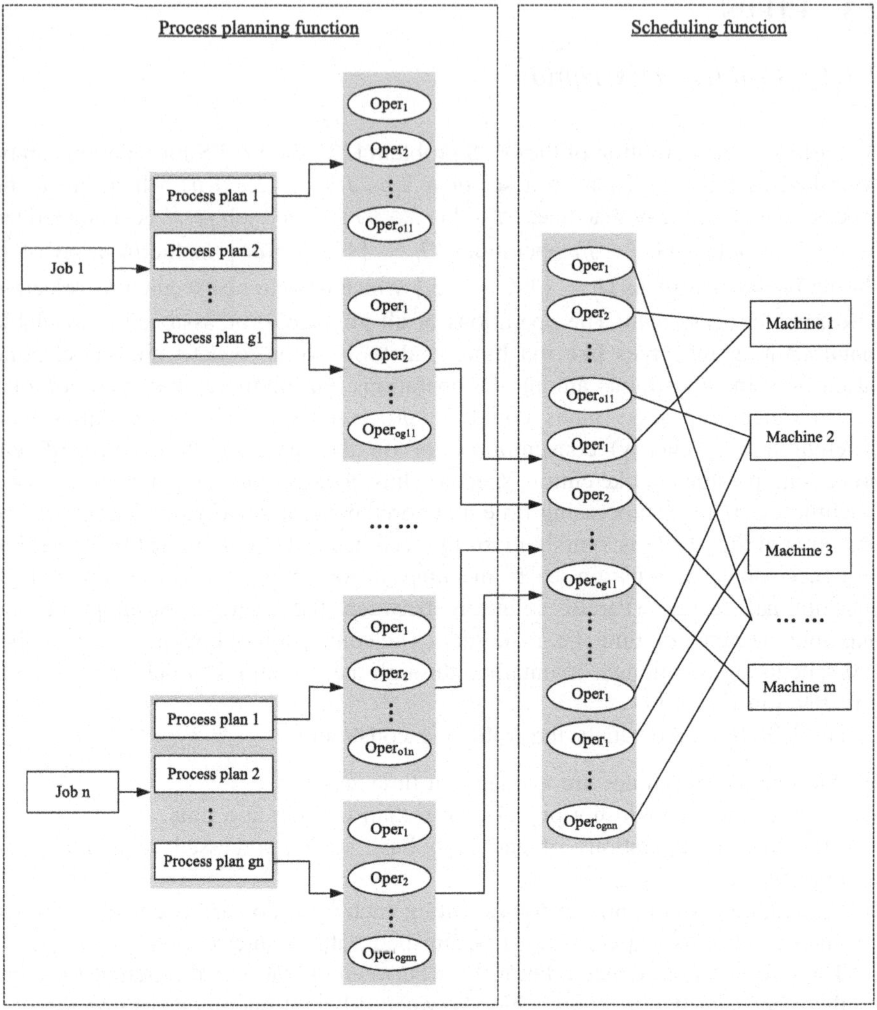

Fig. 2.1 Interactions of the process planning and scheduling

- *i, i'* is the index for the position processed on the machine;
- *l, r* is the index for the process plan;
- *j, p* is the index for the job;
- *k, k', q* is the index for the operation;
- *m, m'* is the index for the machine;
- J is the set of jobs, $J = \{1, 2, \ldots, n\}$;
- M is the set of machines, $M = \{1, 2, \ldots, m\}$;

- O_{lj} is the set of the operations of the alternative process plan l of job j, $O_{lj} = \{1, 2,\ldots, o_{lj}\}$, where o_{lj} is the maximum number of the operations of the alternative process plan l of job j;
- G_j is the set of alternative process plans of job j, $G_j = \{1, 2,\ldots, g_j\}$, where g_j is the maximum number of alternative process plans of job j;
- P_m is the set of operations processed on machine m, $P_m = \{1, 2,\ldots, p_m\}$, where p_m is the maximum number of operations processed on machine m;
- $T_{klj}^{m}\,(k \in O_{lj}, l \in G_j, j \in J, m \in M)$ is the processing time of operation k in the lth alternative process plan of job j on machine m;
- $S_{klj}^{im}(i \in P_m, k \in O_{lj}, l \in G_j, j \in J, m \in M)$ is the start time of operation k, which is the ith position processed on machine m, in the lth alternative process plan of job j;
- $C_{klj}^{im}(i \in P_m, k \in O_{lj}, l \in G_j, j \in J, m \in M)$ is the completion time of operation k, which is the ith position processed on machine m, in the lth alternative process plan of job j;
- $C_{\max}$ is the completion time of the last operation in the schedule, i.e. the makespan of the schedule;
- $\mathrm{Pc}_{klj}^{m}(k \in O_{lj}, l \in G_j, j \in J, m \in M)$ is the cutting power consumption of the kth operation, which belongs to the lth alternative process plan of job j, processed on machine m;
- $\mathrm{Pu}_{klj}^{m}(k \in O_{lj}, l \in G_j, j \in J, m \in M)$ is the unload power consumption of the kth operation, which belongs to the lth alternative process plan of job j, processed on machine m;
- L is a very large positive number;
- $X_{lj}(l \in G_j,\, j \in J)$ is equal to 1 if the lth alternative process plan is picked for job j, and 0 otherwise;
- $Y_{klj}^{im}(i \in P_m, k \in O_{lj}, l \in G_j, j \in J, m \in M)$ is equal to 1 if operation k in the lth alternative process plan of job j is the ith position processed on machine m, and 0 otherwise;
- $Z_{kljqrp}^{m}(k, q \in O_{lj,rp}, l, r \in G_{j,p}, j, p \in J, m \in M)$ is equal to 1 if operation k in the lth alternative process plan of job j precedes operation q in the rth alternative process plan of job p on machine m, and 0 otherwise;

The following is a multi-objective mixed integer programming mathematical model that considers (1) minimizing the total energy consumption (f_1) and (2) minimizing the makespan (f_2).

1. Minimizing the total energy consumption

The energy consumption model is based on existing research work on energy-efficient manufacturing processes [18]. The total power consumption of manufacturing processes can be divided into three types: basic power consumption, unload power consumption and cutting power consumption; the basic power is used for maintaining the normal operation of machine components, including the power consumption of motor drive components, main spindle drive components, servo feed drive components and auxiliary components, such as hydraulic components, cooling and lubrication components, control components, and periphery components; the unload power corresponds to activities such as workpiece loading, unloading, positioning and clamping, and cutting tools changing, and the cutting power corresponds to the actual cutting operation. For the energy-efficient manufacturing processes considered in this paper, the primary contributors to the total energy consumption are unload power consumption and cutting power consumption. Under this assumption, the objective of minimizing the total energy consumption which consists of direct energy consumed by removing material volume in productive modes and indirect energy usage in non-productive modes like standby stage can be expressed as:

$$
\begin{aligned}
f_1 = \min \sum_{k \in O_{lj}} \sum_{l \in G_j} \sum_{j \in J} \sum_{i \in P_m} \sum_{m \in M} \left(\left(\alpha \cdot (\mathrm{Pc}_{klj}^{m})^2 + \beta \cdot \mathrm{Pc}_{klj}^{m} + \mathrm{Pu}_{klj}^{m} \right) \cdot \mathrm{T}_{klj}^{m} \cdot X_{lj} \cdot Y_{klj}^{im} \right) \\
+ \sum_{k,q \in O_{lj,rp}} \sum_{l,r \in G_{j,p}} \sum_{j,p \in J} \sum_{i \in P_m} \sum_{m \in M} \Big(\mathrm{Pu}_{klj}^{m} \cdot \Big((C_{klj}^{im} - \mathrm{T}_{klj}^{m}) \cdot X_{lj} \cdot Y_{klj}^{im} \\
- C_{qrp}^{(i-1)m} \cdot X_{rp} \cdot Y_{qrp}^{(i-1)m} \Big) \cdot X_{lj} \cdot X_{rp} \cdot Z_{kljqrp}^{m} \Big)
\end{aligned}
\quad (2.1)
$$

where the first part on the right side of the equation is direct energy consumption for removing material volume in productive stage, α, β are the coefficients of the load power consumption, and they can be obtained by using the equations of linear regression based on the idle power consumption within the different spindle speeds [27]; the second part on the right side of the equation is indirect energy consumption such as energy for standby.

2. Minimizing the makespan

The objective of minimizing the makespan which is the maximum completion time of all jobs can be defined as:

$$
f_2 = \min \max_{j \in J} \left(C_{klj}^{im} \cdot X_{lj} \cdot Y_{klj}^{im} \right) \quad (2.2)
$$

The objectives are subjected to the constraints as follows:

(1) For the last operation in the lth alternative process plan of job j on machine m:

$$
C_{\max} \geq \left(C_{klj}^{P_m m} \cdot X_{lj} \cdot Y_{klj}^{P_m m} \right) - L(1 - X_{lj}), \ k \in O_{lj}, \ l \in G_j, \ j \in J, \ m \in M \quad (2.3)
$$

(2) The earliest completion time of one operation:

$$
\begin{aligned}
C_{klj}^{im} \cdot X_{lj} \cdot Y_{klj}^{im} = S_{klj}^{im} \cdot X_{lj} \cdot Y_{klj}^{im} + \mathrm{T}_{klj}^{m} \cdot X_{lj} \cdot Y_{klj}^{im}, \\
i \in P_m, \ k \in O_{lj}, \ l \in G_j, \ j \in J, \ m \in M
\end{aligned}
\quad (2.4)
$$

(3) One machine can process only one job at a time:

$$\begin{aligned} & S_{qrp}^{im} \cdot X_{rp} \cdot Y_{qrp}^{im} + L \cdot \left(1 - X_{lj} \cdot X_{rp} \cdot Y_{qrp}^{im} \cdot Y_{klj}^{(i-1)m} \cdot Z_{kljqrp}^{m}\right) \\ & \geq \mathrm{T}_{klj}^{m} \cdot X_{lj} \cdot Y_{klj}^{(i-1)m} + S_{klj}^{im} \cdot X_{lj} \cdot Y_{klj}^{(i-1)m}, \\ & i \in P_m : i > 2,\ k, q \in O_{lj,rp},\ l, r \in G_{j,p},\ j, p \in J,\ m \in M \end{aligned} \tag{2.5}$$

(4) The different operations of the job cannot be performed at the same time:

$$\begin{aligned} & S_{klj}^{im} \cdot X_{lj} \cdot Y_{klj}^{im} - S_{k'lj}^{i'm'} \cdot X_{lj} \cdot Y_{k'lj}^{i'm'} + L \cdot (1 - X_{lj}) \geq \mathrm{T}_{k'lj}^{m'} \cdot X_{lj} \cdot Y_{k'lj}^{i'm'}, \\ & i, i' \in P_m,\ k, k' \in O_{lj} : k > k',\ l \in G_j,\ j \in J,\ m, m' \in M \end{aligned} \tag{2.6}$$

(5) Each job can select only one alternative process plan:

$$\sum_{l \in G_j} X_{lj} = 1, \quad j \in J \tag{2.7}$$

(6) Each operation can select only one machine:

$$\sum_{m \in M} Y_{klj}^{im} = 1, \quad i \in P_m,\ k \in O_{lj},\ l \in G_j,\ j \in J \tag{2.8}$$

(7) The time should be nonnegative:

$$C_{klj}^{im} \cdot X_{lj} \cdot Y_{klj}^{im} \geq 0, \quad i \in P_m,\ k \in O_{lj},\ l \in G_j,\ j \in J,\ m \in M \tag{2.9}$$

Constraint (2.3) ensures that the completion time of the operation, which is the last position processed on one machine, is not greater than the makespan of the schedule. Constraint (2.4) notes that an operation cannot pre-empt another on a machine. Constraint (2.5) is the machine constraint; it allows each machine to process a maximum of one operation at once. Constraint (2.5) represents the operational constraints and prevents different operations for a job from being executed simultaneously. Constraint (2.7) states that only one alternative process plan can be selected for each job. Constraint (2.8) ensures that each operation will be processed by only one machine tool at a time; in other words, it does not allow an operation to be executed on more than one machine at any time. Constraint (2.9) ensures that the completion time of each operation will be nonnegative.

2.4 A Modified Genetic Algorithm Optimization Approach

For a job shop floor with various resources available to the operations of all given jobs, there is an enormous search space for determining the energy-aware modelling of the IPPS problem. The mixed integer programming formula for the EIPPS problem described above is a multi-objective function with constraints. Because process planning and scheduling individually are NP-hard, their integration is also NP-hard [28]. Hence, it is imperative to explore optimal or near-optimal solutions based on intelligent algorithms to facilitate the search and optimization process according to the IPPS criteria. In this study, a modified GA that combines a GA with a simulated annealing algorithm (SAA) is adopted to search for the optimal solutions of the objective functions discussed above. A GA is a searching technique that is based on the process of natural evolution [29]. GAs have been widely applied to combinational and other optimization problems, such as scheduling problems. One of the remarkable advantages of GAs is being able to quickly and efficiently obtain a good solution for an objective function in a complex solution space, but a major disadvantage is potentially being trapped in a local optimum, which is called premature convergence. An SAA is a local searching technique that is analogous to annealing in solids. One of the most famous studies of SAAs is Kirkpatrick et al., who successfully applied an SAA to combinational optimization problems like travelling salesman problems [30]. Two of its prominent advantages are avoiding convergence to a local optimum and efficiently determining the global optimum of an objective function in a complex solution space. Therefore, this article proposes to incorporate the strengths of an SAA into a GA. The modified GA for the EIPPS problem is illustrated in Fig. 2.2.

2.4.1 Representation

Each chromosome that is based on the multi-layer encoding scheme in process planning and scheduling is composed of two-layer strings as shown in Fig. 2.3. One represents an alternative process plan gene-string, and the other represents a scheduling plan gene-string. In the first layer encoding, the hth position denotes job h, and the element of each position corresponds to a selected process plan of the job. For example, the fourth position of the process plan gene-string represents job 4, and job 4 picks the first alternative process plan (Fig. 2.3). The gene element of the fourth position is equal to 1. Therefore, the length of the alternative process plan gene-string is dependent on the number of jobs. In the second layer encoding, the encoding scheme is based on an operation representation, which means that different appearances of the same job number denote different operations of the job, and the pth appearance represents the pth operation of the job. Specifically, there are five different appearances for job 4 in the scheduling plan gene-string, which

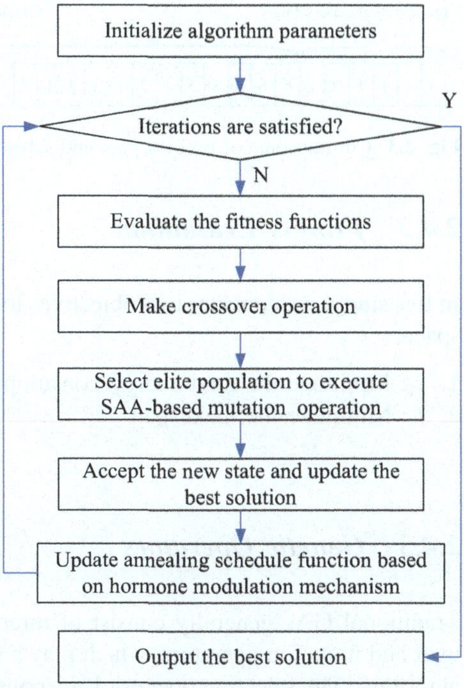

Fig. 2.2 Flow chart of the modified genetic algorithm

means that job 4 has five operations. The first appearance denotes the first operation of job 4; the second denotes the second operation of job 4 and so on. In addition, the length of each alternative process plan for a job could be different. To generate a feasible initial population, the length of each chromosome is set as follows. First, the length of an alternative process plan gene-string is set as the total number of jobs. Second, given that the number of operations $\left(o_{lj}(l \in G_j, j \in J)\right)$ of job j which has the maximum operations among g_j $(j \in J)$ alternative process plans is expressed as $\max_l\{o_{lj}\}$. Then, the total length of the scheduling plan gene-string is equal to the sum of the maximum length of each job, i.e. $\sum_{j=1}^{n} \max_l\{o_{lj}\}$. Hence, the total length of the chromosome is equal to $\sum_{j=1}^{n} \max_l\{o_{lj}\} + n$. If the length of the selected process plan gene-string of job j is not corresponding with the maximum number $\left(\max_l\{o_{lj}\}\right)$ in the decoding process, the elements of operations of job j are removed from the last operation position to the first one until the length is satisfied with the selected one. In Fig. 2.3, there are six jobs, and the maximum operations of each job are all five. Thus, the total length of the chromosome is equal to 36. Assume that the first process plan of job 4 has only four operations, which means the last operation of job 4 in the chromosome is to be removed in the decoding process.

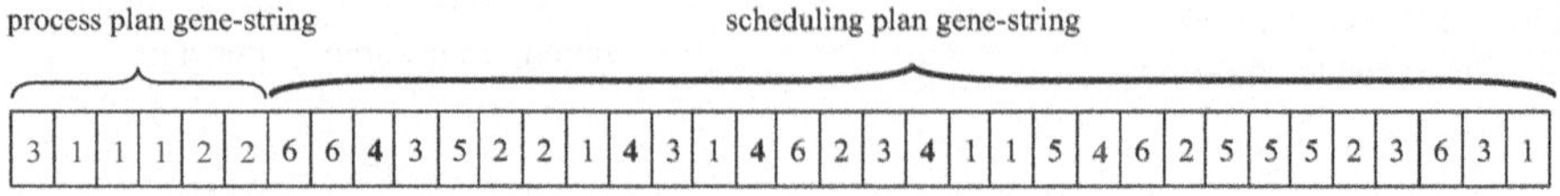

Fig. 2.3 Chromosome of process plan and scheduling

2.4.2 *Fitness Evaluation*

In this study, we consider two objectives to find a set of efficient results in a solution space:

1. f_1: Minimize the total energy consumption
2. f_2: Minimize the makespan

2.4.3 *Genetic Operators*

Traditional GAs generally consist of three genetic operators, the selection, crossover and mutation operators, which play a significant role in the performance of the algorithm. The selection operator is responsible for picking elite chromosomes from the current population to generate the next population, which is used for the crossover and mutation operations. The crossover operator is the main way to produce new chromosomes with the parents

1. Selection operator
   In this study, the rank-based selection approach is employed to produce excellent individuals for the next generation. In the rank-based selection mechanism, the best individuals are chosen with the desired probability from the parents and offspring chromosomes. In this way, an elite population can be obtained because the mechanism will only accept improvements.
2. Crossover operator
   The procedure of the crossover operation is designed as follows, and a crossover instance is shown in Fig. 2.4.

   Step 1 Choose two parent individuals P1 and P2 at random, and construct two empty offspring individuals O1 and O2.

   Step 2 Generate the alternative process plan gene-strings of O1 and O2.

   Step 2.1 Select a crossover point for a pair of process plan gene-strings at random, and each process plan gene-string can be separated as right and left parts from the crossover point.

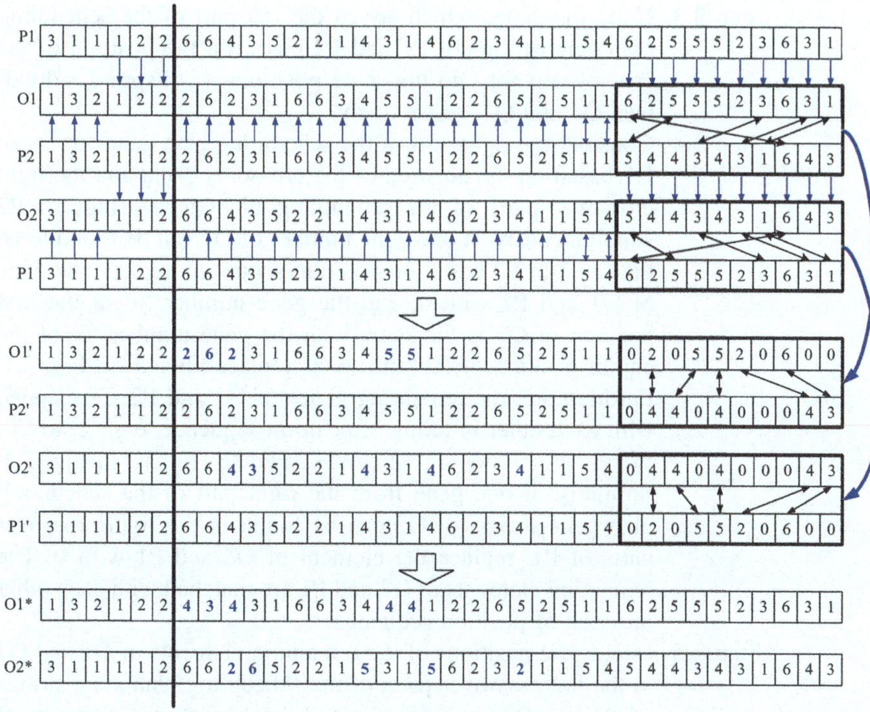

Fig. 2.4 Crossover for a pair of chromosomes of process plan and scheduling

Step 2.2 Copy the genes which are in the right part of the alternative process plan gene-strings of P1 and P2 based on the position of the crossover point into the same positions of offspring individuals O1 and O2, respectively.

Step 2.3 Copy the genes which are in the left part of the alternative process plan gene-strings of P1 and P2 based on the position of the crossover point into the same positions of offspring individuals O2 and O1, respectively.

Step 3 Generate the scheduling plan gene-strings of O1 and O2

Step 3.1 Select a crossover point for a pair of scheduling plan gene-strings at random.

Step 3.2 Copy the genes which are in the right part of the scheduling plan gene-strings of P1 and P2 based on the position of the crossover point into the same positions of offspring individuals O1 and O2, respectively.

Step 3.3 Copy the genes which are in the left part of the scheduling plan gene-strings of P1 and P2 based on the position of the crossover point into the same positions of offspring individuals O2 and O1, respectively.

Step 3.4 Consider the right part of the scheduling plan gene-strings of O1 based on the position of the crossover point and the right crossover parts of the scheduling plan gene-strings of P2 simultaneously. If one gene number of O1 can be found to be the same in P2 based on position sequence, replace the gene of O1 and P2 with 0; e.g. the gene number '6' of the first position of O1 is the same with the gene number '6' of the eighth position of P2, and '6' is replaced by '0' in Fig. 2.4. Furthermore, the remaining genes of O1 and P2 are mapped with each other in terms of position sequence, e.g. '2' to '4', '5' to '4', '5' to '4', '2' to '4', and '6' to '3', in Fig. 2.4. Similarly, if one gene from the right part of the scheduling plan gene-strings of O2 can be found in the right crossover parts of P1, replace the element of O2 and P1 with 0. The remaining elements of O2 and P1 are matched with each other in terms of position sequence.

Step 3.5 Decide the positions of the remaining elements of O1 and O2 in the left crossover parts of the scheduling plan gene-strings of O1 and O2, respectively, and then update the elements of the positions with the matched elements of O1 and O2, respectively (see Fig. 2.4).

3. SAA-based mutation operator

A mutation operator is required to generate schedules with new information after the crossover operator. Several approaches for the mutation operator, such as uniform and non-uniform mutation and immunity-based mutation operators, have been used to solve complex global optimization problems. In this study, a new SAA-based mutation operator is designed for the algorithm. In the SAA, a temperature parameter plays a significant role in guiding and controlling the performance of the algorithm. When the temperature reaches a high level, the SAA has a high probability of accepting a new state, while the search space becomes very small at low temperatures, and the SAA has a low probability of accepting a new state. Hence, the new mutation operator can improve the search ability and search efficiency of the algorithm by developing a novel nonlinear annealing control function based on up-regulatory Hill functions in the solution space. The major process of the SAA-based mutation operator can be outlined as follows:

Step 1 Initialize SAA parameters.

Step 1.1 Get the initial temperature T_0, the final temperature T_1, the current temperature T and set $T = T_0$, where T_0 is given as: $T0 = -100 \times (\max(\text{Objv}) - \min(\text{Objv}))/\log \text{Pa}$, where max(Objv) is the maximum value of the objective function, min(Objv) is the minimum value of the objective function, and Pa is the initial acceptance probability.

Step 1.2 Select an initial schedule S_0 and set the current schedule $S = S_0$, the best schedule $S^* = S_0$.

Step 1.3 Calculate the function value of the current schedule $f(S) = f(S_0)$.

Step 1.4 Set $k = 0$.

Step 2 While the stop criterion is not satisfied, do:

Step 2.1 Generate a feasible schedule S' in the neighbourhood of the current schedule S by using adjacent swapping strategy.

Step 2.2 Calculate $\Delta = (f(S') - f(S))/f(S) \times 100$

Step 2.3 If $\Delta \leq 0$, set $S = S'$, $f(S) = f(S')$ and update the best schedule $S^* = S'$.

Step 2.4 Else, generate a random number $r \in (0, 1)$, if $r < \exp(-\Delta/T)$, set $S = S'$, $f(S) = f(S')$ and update the best schedule $S^* = S'$.

Step 3 Generate an annealing control function based on an up-regulatory Hill function, which can keep hormone regulation adaptive and stable at a good accuracy and fast speed [31]. The expression is described in the following:

$$T(k+1) = \alpha * F_{\text{down}}(k) - k * \Delta T / \exp(k)$$

where, $F_{\text{down}}(k) = 1/(1 + k^n)$ and $\Delta T = T(k+1) - T(k)$, $k = k + 1$

Step 4 Return the best schedule and the corresponding function value.

Step 5 Stop

2.5 Case Studies

The proposed modified genetic algorithm is carried out by utilizing MATLAB programming language. The experimental tests are carried out on a personal computer with Intel Pentium (R) with 1 GB memory and 3.20 GHz processor, and its operating system is Windows XP. Two experiments are conducted: one is to reveal the effectiveness and performance of the proposed approach and the other is to discuss the EIPPS model.

2.5.1 Performance Evaluation of the Algorithm

To demonstrate the performance of the proposed modified GA, two problems that were given by Dong and Sun [32] and Moon et al. [33] are presented in the first experiment. The objective of this section is to minimize the total completion time (i.e. makespan). The first example includes 10 jobs and 10 machines in process planning and scheduling. Each job has three alternative process plans, and each operation of jobs can be executed on different machines. The alternative process plans and processing time are shown in Table 2.1. The proposed algorithm is compared with several different algorithms, including the genetic multi-objective SAA of Mohammadi et al. [26], the immune GA of Dong and Sun [32], and the particle optimization algorithm of Zhu et al. [34]. The experimental results are shown in Table 2.2, and one of the best schedules is illustrated with the Gantt chart, as shown in Fig. 2.5.

As shown in Table 2.2, which compares the experimental results of the four algorithms, the best makespan for problem 1 is obtained by the proposed algorithm, which gives a makespan value of 27. It is clear that the proposed algorithm outperforms those of Dong and Sun [32], Zhu et al. [34] and Mohammadi et al. [26]. The optimal process plan is 2-2-2-2-3-2-3-1-2-2, which means that job 1 selects the first process plan, job 2 selects the second process plan and so on, and the corresponding processing times are shown in bold in Table 2.1. The optimal scheduling results can be described with the Gantt chart in Fig. 2.5. The proposed modified GA is more effective than the other algorithms; its convergence iteration time is 170, while the convergence iteration times for the other methods are 245, 225 and 185, respectively.

The works of Moon et al. [33] and Mohammadi et al. [26] are given as the second example to demonstrate the performance of the proposed modified GA. The problem consists of five jobs and five machines, and the makespan is set as the objective function; each job has multiple alternative process plans, and each operation of jobs can be processed on different machines. The relevant data are shown in Table 2.3. Table 2.4 shows the experimental results obtained from the different algorithms, and the scheduling results are shown in the Gantt chart in Fig. 2.6. The experimental results show that the best makespan is 14 and is obtained with the proposed approach, which is superior to those of Moon et al. and Mohammadi et al. The optimal process plan is 2-2-9-1-3, and the corresponding processing time is shown in bold in Table 2.3.

Table 2.1 Job-related information of problem instance 1

Job number	Process plan number	Machine number of operation sequences	Processing time
J1	*1*	*M9-M4-M3-M5-M10-M7-M1-M6*	*4-2-2-3-2-3-4-3*
	2	**M10-M1-M2-M4-M8-M7-M3-M9-M5-M6**	**3-2-2-2-3-2-3-2-2-2**
	3	M3-M1-M2-M9-M7-M8-M4-M6-M5	3-3-2-2-2-2-3-3-3
J2	1	M8-M6-M9-M5-M4	4-5-5-4-2
	2	***M2-M1-M4-M3-M7***	***5-5-3-5-2***
	3	M10-M3-M5-M2	6-5-5-4
J3	1	M3-M4-M7-M8	5-5-4-4
	2	***M9-M6-M5-M4***	***6-4-4-4***
	3	M1-M2-M10-M9	3-5-5-5
J4	1	M8-M7-M4-M3-M9	4-5-3-4-4
	2	**M6-M5-M2-M8-M7**	**3-5-4-4-4**
	3	*M10-M6-M4-M9*	*5-6-4-5*
J5	1	M3-M2-M7-M8-M6-M9-M10	3-3-2-3-2-3-3
	2	M2-M10-M9-M7-M5-M4-M6	4-2-3-3-3-2-2
	3	***M1-M9-M7-M4-M5-M6***	***4-3-3-3-2-4***
J6	1	M7-M2-M5	6-6-5
	2	**M6-M9-M10**	**6-5-6**
	3	*M4-M7-M8*	*5-6-6*
J7	*1*	*M1-M2-M9*	*4-7-6*
	2	M3-M4-M5	6-6-5
	3	**M5-M7-M10**	**5-6-6**
J8	***1***	***M8-M10-M2***	***5-6-6***
	2	M5-M7-M9	4-7-6
	3	M10-M6-M3	5-6-6
J9	1	M7-M4-M9-M10	5-6-5-5
	2	***M3-M5-M8-M2***	***6-6-5-4***
	3	M1-M3-M5-M7	5-5-6-5
J10	*1*	*M5-M3-M7*	*6-6-5*
	2	**M4-M1-M9**	**6-5-6**
	3	M10-M5-M8	5-6-6

Table 2.2 Experimental results of problem instance 1

	Dong and Sun	Zhu et al.	Mohammadi et al.	Proposed algorithm
Makespan	30	29	28	27
CPU time (s)	245	225	185	170

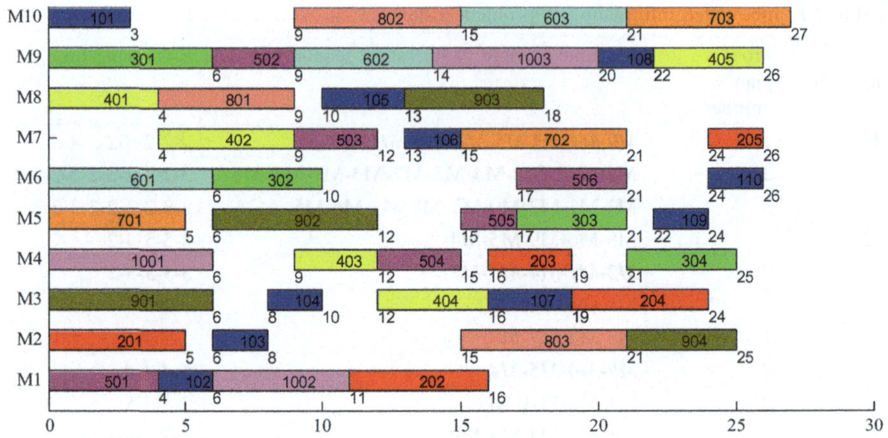

Fig. 2.5 Gantt chart of problem instance 1

2.5.2 Performance Evaluation of the Energy-Aware IPPS

To evaluate the EIPPS model, the extended job shop instances that involve power consumption for the machines are developed based on the problem given by Dong and Sun [32] that was presented previously. Assume that all of the machines in the study are automated, the cutting power on each machine is set to the same value, and the data for the unload power consumption of each machine can be obtained by the method developed by Liu et al. [35], which is shown in Table 2.5. For the problem given by Dong and Sun [32], the makespan value was obtained using the optimization approach in the experiment described above. According to the optimized plan of process planning and scheduling, the total energy consumption is calculated as 806.01. Because of the requirements for factory production, the decision-maker mainly considers the production efficiency with the objective of production time; here, the energy consumption could be reduced as much as possible, while the completion time could be delayed without affecting the delivery time, and the value of the makespan factor can be set from 0.9 to 1. The proposed algorithm was run 10 times for the EIPPS problem. Figure 2.7 shows the points on the Pareto efficient frontier for the energy consumption and makespan of the bi-objective mixed integer programming model, and each point presents the average of these 10 runs.

Figure 2.7 describes the Pareto efficient frontier for the problem given by Dong and Sun [32]. The experimental results show that a conflicting relationship exists between the energy consumption and makespan. Because the makespan is considered to be the dominant solution, average values of the makespan as short as approximately 27.5 can be obtained at the expense of high energy consumption. The corresponding average energy consumption is 813.15. On the other hand, if a longer production time can be accepted without affecting the delivery time, the

Table 2.3 Job-related data of problem instance 2

Job number	Process plan number	Machine number of operation sequences	Processing time
J1	1	M1-M2	5-5
	2	**M2-M2**	**3-5**
J2	1	M3-M5	6-4
	2	**M4-M5**	**5-4**
J3	1	M2-M3-M2	4-2-5
	2	M2-M4-M2	4-3-5
	3	M1-M3-M2	5-2-5
	4	M1-M4-M2	5-3-5
	5	M2-M2-M3	4-5-2
	6	M2-M2-M4	4-5-3
	7	M1-M2-M3	5-5-2
	8	M1-M2-M4	5-5-3
	9	**M2-M1-M3**	**5-5-2**
	10	M2-M1-M4	5-5-3
	11	M2-M2-M3	5-4-2
	12	M2-M2-M4	5-4-3
J4	**1**	**M3-M4**	**4-5**
J5	1	M1-M1-M3-M4	4-2-5-4
	2	M1-M2-M3-M3	4-4-5-4
	3	**M1-M1-M3-M5**	**4-2-5-3**
	4	M1-M2-M3-M5	4-4-5-3
	5	M2-M1-M3-M3	3-2-5-4
	6	M2-M1-M3-M5	3-2-5-3
	7	M2-M2-M3-M3	3-4-5-4
	8	M2-M2-M3-M5	3-4-5-3
	9	M1-M3-M1-M3	4-5-2-4
	10	M1-M3-M1-M5	4-5-2-3
	11	M1-M3-M2-M3	4-5-4-4
	12	M1-M3-M2-M5	4-5-4-3
	13	M2-M3-M1-M3	3-5-2-5
	14	M2-M3-M1-M5	3-5-2-3
	15	M2-M3-M2-M3	3-5-4-4
	16	M2-M3-M2-M5	3-5-4-3
	17	M3-M1-M1-M3	5-4-2-4
	18	M3-M1-M1-M5	5-4-2-3
	19	M3-M1-M2-M3	5-4-4-4
	20	M3-M1-M2-M5	5-4-4-3
	21	M3-M2-M2-M3	5-3-4-4
	22	M3-M2-M2-M5	5-3-4-3
	23	M3-M2-M1-M3	5-3-2-4
	24	M3-M2-M1-M5	5-3-2-3

Table 2.4 Experimental results of problem instance 2

	Moon et al.	Mohammadi et al.	Proposed algorithm
Makespan	16	15	14

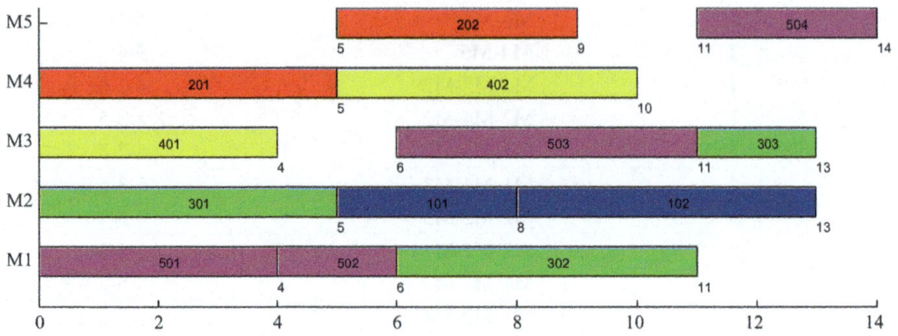

Fig. 2.6 Gantt chart of problem instance 2

Table 2.5 Data for the unload power consumption of each machine

	M1	M2	M3	M4	M5	M6	M7	M8	M9	M10
Unload power	2.40	3.36	2.00	1.77	2.20	7.50	2.00	1.77	2.20	7.50

energy consumption can be reduced to 654.36 with a makespan that is increased by approximately 20%. The total energy consumption can be decreased by 19.53%. The Pareto efficient frontier between the energy and makespan of one run for the problem is shown in Fig. 2.8. Note that the same makespan will correspond to different levels of energy consumption due to alternative process planning and scheduling plans, and that the minimum energy consumption can be selected while maintaining the same production time. For example, when the makespan value reaches 28, the associated energy consumption has minimum and maximum values of 717.74 and 784.13, respectively. Thus, the optimal objective with the minimum value of 717.74 is employed as a production plan, which indicates an 8.47% improvement in the total energy consumption. The results of the bi-objective optimization problem and the results of the single-objective optimization problem are compared in Fig. 2.8. In the bi-objective optimization problem, the optimal makespan value is 27, while the corresponding minimum energy consumption value is 785.96. In the single-objective optimization problem, the optimal makespan value is 27, while the associated energy consumption value is 806.01. The reduction of the total energy consumption can reach approximately 2.5% compared to the single-objective optimization result. Furthermore, the optimal process plan is 1-2-2-3-3-3-1-1-2-1. The corresponding processing time is shown in italics in Table 1, and the Gantt chart of the scheduling results is shown in Fig. 2.9.

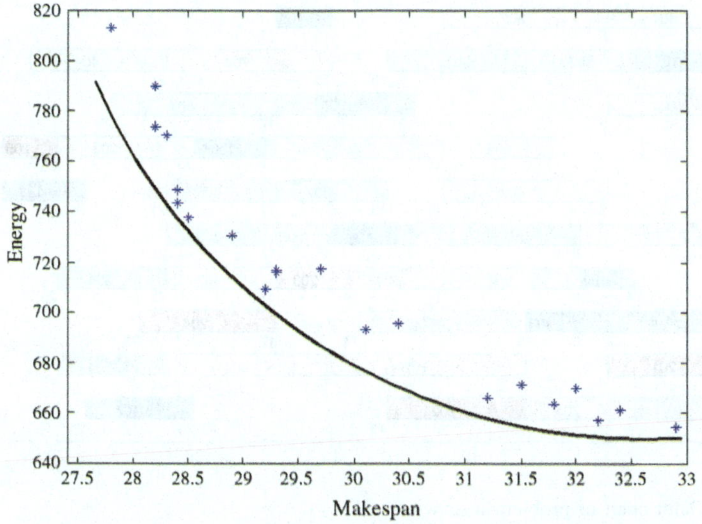

Fig. 2.7 Plots of energy and makespan for the problem instance

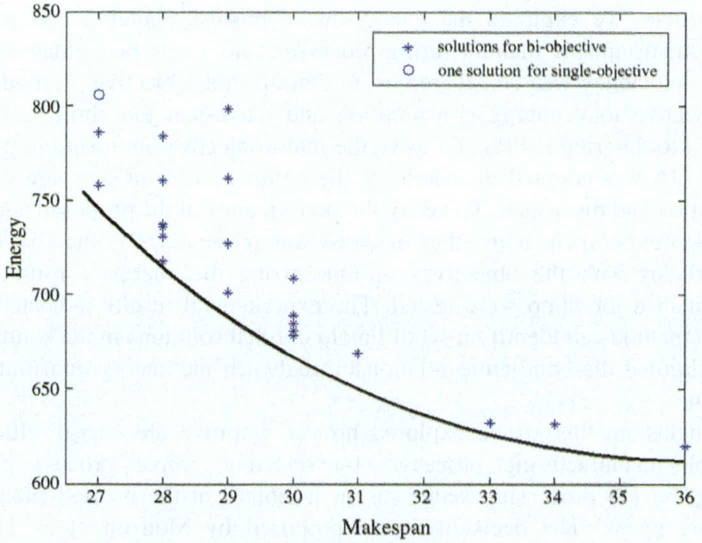

Fig. 2.8 Pareto efficient frontier between the energy and makespan

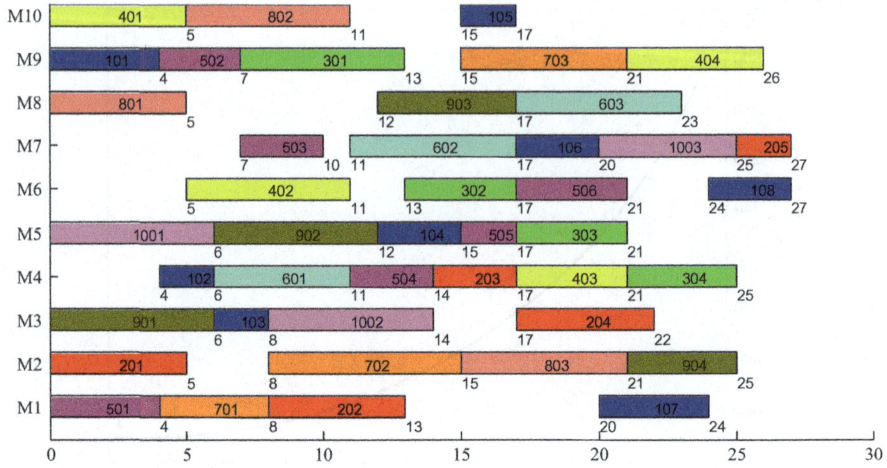

Fig. 2.9 Gantt chart of problem instance 1

2.6 Conclusions and Future Research

In this article, we explored the energy-aware process planning and scheduling problem in sustainable manufacturing processes, and a new performance indicator of energy efficiency was considered as an optimization objective. A model for the multi-objective total energy consumption and makespan job shop problem was developed to describe EIPPS. To solve the multi-objective optimization problem, a modified GA was adopted to search for the optimal solutions between the energy consumption and makespan. To verify the performance of the proposed algorithm, a comparison experiment with other methods was presented. Bi-objective optimization problems with the objectives of minimizing the energy consumption and makespan in a job shop were tested. The experimental results indicated that the proposed method can identify a set of Pareto optimal solutions in the solution space and highlighted the conflicting relationship between the energy consumption and makespan.

In conclusion, this article explored how to improve the energy efficiency of sustainable manufacturing processes by selecting proper process plans and changing the job processing sequence on machines at the process planning and scheduling levels. The decision model proposed by Mouzon et al. [12] is an effective and efficient method for reducing energy consumption by turning off and then turning on idle machines. It is, therefore, worth studying methods of obtaining better energy-efficient optimization results in process planning and scheduling. In addition, unexpected events that may occur in real manufacturing processes, such as machine failures, rush orders and job cancellations, should be considered in energy-aware process planning and scheduling problems. Minimizing the energy consumption in dynamic scheduling problems should be investigated further in the future.

References

1. IEA. (2008). *Worldwide trends in energy use and efficiency*. Paris, French.
2. Herrmann, C., Bergmann, L., & Thiede, S. (2009). Methodology for the design of sustainable production systems. *International Journal of Sustainable Manufacturing, 1*(4), 376–395.
3. Apostolos, F., Alexios, P., Georgios, P., Panagiotis, S., & George, C. (2013). Energy efficiency of manufacturing processes: A critical review in 46th CIRP Conference on Manufacturing Systems, CIRP CMS 2013, May 29–30, 2013. Setubal, Portugal: Elsevier.
4. Duflou, J. R., Sutherland, J. W., Dornfeld, D., Herrmann, C., Jeswiet, J., Kara, S., et al. (2012). Towards energy and resource efficient manufacturing: A processes and systems approach. *CIRP Annals—Manufacturing Technology, 61*(2), 587–609.
5. Sheng, P., & Srinivasan, M. (1995). Multi-objective process planning in environmentally conscious manufacturing: A feature-based approach. *CIRP Annals—Manufacturing Technology, 44*(1), 433–437.
6. Srinivasan, M., & Sheng, P. (1999). Feature-based process planning for environmentally conscious machining—Part 1: Microplanning. *Robotics and Computer-Integrated Manufacturing, 15*(1), 257–270.
7. Srinivasan, M., & Sheng, P. (1999). Feature-based process planning in environmentally conscious machining—Part 2: Macroplanning. *Robotics and Computer-Integrated Manufacturing, 15*(1), 271–281.
8. Singh, S., Goodyer, J., & Popplewell, K. (2007). Integrated environmental process planning for the design and manufacture of automotive components. *International Journal of Production Research, 45*(18–19), 4189–4205.
9. Kai, J., Zhang, H. C., Balasubramaniam, P., & Nage, S. (2009). A multiple objective optimization model for environmental benign process planning. In *16th International Conference on Industrial Engineering and Engineering Management, 2009. IE&EM'09.*
10. Newman, S. T., Nassehi, A., Imani-Asrai, R., & Dhokia, V. (2012). Energy efficient process planning for CNC machining. *CIRP Journal of Manufacturing Science and Technology, 5*(2), 127–136.
11. Yin, R., Cao, H., Li, H., & Sutherland, J. W. (2014). A process planning method for reduced carbon emissions. *International Journal of Computer Integrated Manufacturing*, 1–12.
12. Mouzon, G., Yildirim, M. B., & Twomey, J. (2007). Operational methods for minimization of energy consumption of manufacturing equipment. *International Journal of Production Research, 45*(18–19), 4247–4271.
13. Mouzon, G., & Yildirim, M. B. (2008). A framework to minimise total energy consumption and total tardiness on a single machine. *International Journal of Sustainable Engineering, 1*(2), 105–116.
14. Fang, K., Uhan, N., Zhao, F., & Sutherland, J. W. (2011). A new approach to scheduling in manufacturing for power consumption and carbon footprint reduction. Amsterdam: Elsevier.
15. Bruzzone, A. A. G., Anghinolfi, D., Paolucci, M., & Tonelli, F. (2012). Energy-aware scheduling for improving manufacturing process sustainability: A mathematical model for flexible flow shops. *CIRP Annals—Manufacturing Technology, 61*(1), 459–462.
16. Zhang, L., Li, X., Gao, L., Zhang, G., & Wen, X. (2012). Dynamic scheduling model in FMS by considering energy consumption and schedule efficiency. In 2012 IEEE 16th International Conference on Computer Supported Cooperative Work in Design, CSCWD 2012, May 23–25, 2012. Wuhan, China: IEEE Computer Society.

17. Liu, G.-S., Zhang, B.-X., Yang, H.-D., Chen X., & Huang G. Q. (2013). A branch-and-bound algorithm for minimizing the energy consumption in the PFS problem. *Mathematical Problems in Engineering*, 2013.
18. Dai, M., Tang, D., Giret, A., Salido, M. A., & Li, W. D. (2013). Energy-efficient scheduling for a flexible flow shop using an improved genetic-simulated annealing algorithm. *Robotics and Computer-Integrated Manufacturing, 29*(5), 418–429.
19. Chryssolouris, G., Chan, S., & Cobb, W. (1984). Decision making on the factory floor: An integrated approach to process planning and scheduling. *Robotics and Computer-Integrated Manufacturing, 1*(3–4), 315–319.
20. Chryssolouris, G., Chan, S., & Suh, N. P. (1985). An Integrated Approach to Process Planning and Scheduling. *CIRP Annals—Manufacturing Technology, 34*(1), 413–417.
21. Amin-Naseri, M. R., & Afshari, A. J. (2012). A hybrid genetic algorithm for integrated process planning and scheduling problem with precedence constraints. *International Journal of Advanced Manufacturing Technology, 59*(1–4), 273–287.
22. Guo, Y. W., Li, W. D., Mileham, A. R., & Owen, G. W. (2009). Applications of particle swarm optimization in integrated process planning and scheduling. *Robotics and Computer-Integrated Manufacturing, 25*(2), 280–288.
23. Li, W. D., & McMahon, C. A. (2007). A simulated annealing-based optimization approach for integrated process planning and scheduling. *International Journal of Computer Integrated Manufacturing, 20*(1), 80–95.
24. Li, X., Gao, L., Shao, X., Zhang, C., & Wang, C. (2010). Mathematical modeling and evolutionary algorithm-based approach for integrated process planning and scheduling. *Computers & Operations Research, 37*(4), 656–667.
25. Mohammadi, G., Karampourhaghghi, A., & Samaei, F. (2012). A multi-objective optimization model to integrating flexible process planning and scheduling based on hybrid multi-objective simulated annealing. *International Journal of Production Research, 50*(18), 5063–5076.
26. Morad, N., & Zalzala, A. (1999). Genetic algorithms in integrated process planning and scheduling. *Journal of Intelligent Manufacturing, 10*(2), 169–179.
27. Liu, F., Xu, Z.-J., & Dan, B. (1995). *Energy performance of machining systems with its application*. Beijing: China Machine Press.
28. Khoshnevis, B., & Chen, Q. (1991). Integration of process planning and scheduling functions. *Journal of Intelligent Manufacturing, 2*(3), 165–175.
29. Holland, J. H. (1992). *Adaptation in natural and artificial systems: An introductory analysis with applications to biology, control, and artificial intelligence*. Cambridge, MA: MIT Press.
30. Kirkpatrick, S., Gelatt, C. D., & Vecchi, M. P. (1983). Optimization by simulated annealing. *Science, 220*(4598), 671–680.
31. Rene Thomas R. D. A. (1990). Biological Feedback. CRC Press: Boca Raton, FL.
32. Dong, C. Y., & Sun, S. D. (2007). Immune genetic algorithm job scheduling process and collaborative optimization. *Mechanical Science and Technology, 26*(6), 761–766.
33. Moon, C., Lee, Y. H., Jeong, C. S., & Yun, Y. (2008). Integrated process planning and scheduling in a supply chain. *Computers & Industrial Engineering, 54*(4), 1048–1061.
34. Zhu, H., Ye, W., & Bei, G. (2009). A particle swarm optimization for integrated process planning and scheduling. In 2009 IEEE 10th International Conference on Computer-Aided Industrial Design and Conceptual Design: E-Business, Creative Design, Manufacturing—CAID and CD'2009, November 26–29, 2009. Wenzhou, China: IEEE Computer Society.
35. Liu, Y., Dong, H., Lohse, N., Petrovic, S., & Gindy, N. (2014). An investigation into minimising total energy consumption and total weighted tardiness in job shops. *Journal of Cleaner Production, 65*, 87–96.

Chapter 3
A Hybrid Optimization Approach for Sustainable Process Planning and Scheduling

X. X. Li, W. D. Li, X. T. Cai and F. Z. He

Abstract Process planning and scheduling are important stages in manufacturing, and good strategies can significantly improve the energy performance of manufacturing to achieve sustainability. In this paper, an innovative optimization approach has been developed to facilitate sustainable process planning and scheduling. In the approach, honeybee mating and annealing processes are simulated to optimize multi-objectives including energy consumption, makespan, and the balanced machine utilization. Experiments on practical cases show that the optimization results from this approach are promising in comparison with those from a genetic algorithm, a honeybee mating optimization algorithm, ant colony optimization, and a simulated annealing algorithm, respectively.

Keywords Honeybee mating optimization · Simulated annealing
Sustainable manufacturing · Process planning · Scheduling

3.1 Introduction

With the rapidly growing production demands, manufacturing has become one of the largest energy-consuming sectors [1]. Statistics have shown that the greenhouse gas emitted from the usage of energy sources such as electricity, coal, oil, and gas during manufacturing accounts for more than 37% even 50% of the world's total

X. X. Li
College of Informatics, Huazhong Agricultural University, Wuhan
People's Republic of China

W. D. Li (✉)
Faculty of Engineering, Environment and Computing, Coventry University, Coventry, UK
e-mail: weidong.li@coventry.ac.uk

X. T. Cai · F. Z. He
School of Computer Science and Technology, Wuhan University, Wuhan, People's Republic of China

W. Li and S. Wang (eds.), *Sustainable Manufacturing and Remanufacturing Management*, https://doi.org/10.1007/978-3-319-73488-0_3

greenhouse gas emissions [2]. In order to balance the multi-faceted dimensions of economic growth and environmental protection, a series of regulations and guidelines, such as the European standard EN 16001:2010 [3], life-cycle carbon labeling outlined by the life-cycle assessment frameworks of the ISO 14040: 2006 and ISO 14044: 2006 [4, 5], and the Publicly Available Specification 2050: 2008 (PAS 2050) [6], have been developing. For manufacturing companies, based on the regulations and guidelines, energy efficiency can be improved and the greenhouse gas emission level can be minimized so as to embrace "Competitive Sustainable Development" and shoulder "Extended Producer Responsibilities (EPR)" [7] effectively. Meanwhile, energy saving is also increasingly vital for customers in choosing products. For instance, a survey indicated that 67% of UK consumers are more likely to buy a product with a lower life-cycle energy consumption and 44% would switch to such a greener product even if the brand/model is not their first choice [8]. Therefore, these economic, environmental, and competitive factors are motivating manufacturing companies to take measures to minimize their energy consumption and achieve sustainable manufacturing.

A large number of studies [9] have indicated that careful process planning and scheduling for manufacturing systems will realize great energy savings. That is, process planning and scheduling are critical functions not only to minimize cost, improve adaptability, responsiveness, and robustness but also to enhance the sustainability of manufacturing processes. Therefore, effective process planning and scheduling are imperative to achieve sustainable manufacturing.

In order to realize sustainable manufacturing by optimizing process planning and scheduling, a multi-objective optimization problem of minimizing overall energy consumption, makespan, and machine utilization in a job shop is considered in this paper. An innovative optimization approach for sustainable process planning and scheduling is presented. In this approach, a five-phase energy consumption model and an effective hybrid honeybee mating optimization and simulated annealing (HBMO-SA) algorithm are developed to optimize the overall energy consumption in process planning and scheduling.

This research is a comprehensively extended version from a previous work of authors [10], and significant improvements are reflected from the following aspects. First, a detailed literature survey is given. Second, a model of integrated process planning and scheduling for dynamic manufacturing is presented in detail. Finally, the developed HBMO-SA approach is benchmarked with other classic heuristic algorithms to demonstrate its merits and effectiveness.

The remainder of this paper is organized as follows. In Sect. 3.2, related work is reviewed. In Sect. 3.3, process planning and scheduling are first represented and then its energy consumption model is presented. Then, the HBMO-SA approach is presented in Sect. 3.4. In Sect. 3.5, case studies and comparisons with other heuristic algorithms are given. Finally, a conclusion is drawn in Sect. 3.6.

3.2 Related Work

In recent years, both sustainable manufacturing and process planning and scheduling have attracted many researchers' attention. This section presents a review of the related work from the viewpoint of sustainable manufacturing and heuristic algorithms applied to sustainable manufacturing applications.

3.2.1 Sustainable Manufacturing

The research works focusing on sustainable manufacturing can be viewed from four different perspectives, i.e., machining process, machine design, eco-product development, and manufacturing system.

(1) Machining process

Most of the research works were dedicated to quantifying the energy consumed in machining. Some research works focus on developing machine-specific models of unit process energy consumption. In the existing models, the most representative one was developed by Gutowski et al. [11]. The specific energy consumption (SEC) of machining processes was modeled as a function of the process rate by analyzing the unit process energy for the processes in an exergy framework [11]. However, in the model, the specifications for the fixed power P0 and the constant k were not given. Taking up the missing specification in the model of Gutowski et al., researchers developed some improved energy consumption models such as the unit process energy consumption model created by Kara and Li [12], the SEC model provided by Diaz et al. [13], and the energy consumption model for milling processes developed by Li et al. [14].

Different from the research works on modeling unit process energy consumption, some other research works concentrated on modeling the total energy consumption in machining. For instance, Diaz et al. modeled the total energy consumption in machining as a function of average power and the processing time [13]. However, the impact of machine tools, workpiece materials, and cutting variables were neglected. Mori et al. [15] presented an energy consumption model which involved the energy consumption of several processes: positioning and acceleration of the spindle, returning the spindle to the tool change position after machining and stopping the spindle. He et al. divided the total energy consumption of Numerical Control (NC) machining into five parts, i.e., the energy consumption of spindle, axis feed, tool change system, coolant pump, and the fixed energy consumption. Each part can be estimated by the corresponding power characteristics and the parameters extracted from the NC codes [16]. However, this model can only be used to help NC code designers make decisions regarding energy-efficient NC programs because it is based on the NC programming but ignores some other energy factors consumed by chillers, swarf conveyors, and lubrication of machines, etc. Aiming at predicting direct electrical energy requirements in machining tool-paths, Balogun and Mativenga [17] divided the energy states of machine tools into three categories: basic, ready, and cutting states, where the energy is demanded to activate required machine components, making the axis and tool to be ready for action and remove workpiece material, respectively. However, this model is limited because the energy consumed by the machine axis has not been taken into account. Duflou et al. [18] studied the energy efficiency on the machining factory level. However, the energy consumption of preparation phase in scheduling is neglected.

(2) Machine design

The research works focus on developing and designing more energy-efficient machines and equipment. The most representative work is the standard on energy saving for machine tools drafted by the International Standard Organization (i.e., ISO/WD14955-1) [19]. Two informative annexes related to environmentally relevant improvements and well-tried mechanical and electrical components are listed in the standard.

(3) Eco-product development

The related works concentrate on modeling and reducing the embodied product energy (EPE) during manufacturing to support energy-efficient manufacturing. A modeling framework was introduced by Seow et al. [20] to represent the total energy required to manufacture a unit product. Another model was built by Kara et al. [21] to assess the impact of global manufacturing on the EPE.

(4) Manufacturing system

The energy efficiency of a manufacturing system is improved mainly by two methods. One is simulation-based method which improves the energy efficiency by simulating electricity consumption and the process chain [22]. The other is optimization-based method which improves the energy efficiency by optimizing the single machine scheduling [23] and the flow shop scheduling [24].

From the literature survey, it can be observed that energy consumption and sustainability are related to the different stages of product development and manufacturing life cycle. Hence, as an important sector of manufacturing, process planning and scheduling are of great importance for energy saving.

3.2.2 Algorithms for Process Planning and Scheduling

In the past decade, a number of research works appeared to develop heuristic algorithms [25]. Among these algorithms, GA is the most popular one [26–31]. It was used to generate the feasible sequences of operations and identify the optimal tool sequence in process planning for machining (e.g., milling). It is proved that GA has a good global search capability [32–38]. However, the algorithm is liable to be trapped in a local optimum. SA is another algorithm widely used in process planning and scheduling. It was applied to search the optimal solution in process planning and scheduling [39]. SA can identify a good solution quickly but may fluctuate around the local optima due to the lack of the memory mechanism. Additionally, several swarm intelligence algorithms such as ACO [40], particle swarm optimization (PSO) [41–46], and HBMO [47] were applied to process planning and scheduling. Compared with the other most-known heuristic algorithms such as GA, SA, and ACO, HBMO has a better performance in computational effectiveness and stability.

HBMO is a swarm intelligence algorithm which was developed by modeling the mating behavior of honeybee swarm. It is known that the intelligent behaviors of bees include mating, foraging, dance, nest site selection, and so on. Hence, different intelligence algorithms can be obtained by modeling different behaviors of honeybee swarm. For instance, the artificial bee colony (ABC) algorithm [48] was developed by simulating foraging behavior of honeybees. Different from these algorithms, HBMO is inspired by the mating behavior of honeybees.

After HBMO was first presented by Abbass [49] to address the propositional satisfiability problems, the algorithm was applied to various domains. Koudil et al. [50] applied HBMO to solve partitioning and scheduling problems in code design. Curkovic and Jerbic [51] used HBMO to address the nonlinear diophantine equation benchmark problem. Haddad et al. applied HBMO to optimize reservoir operation and distribution systems. Amiri and Fathian [52] improved HBMO to solve a real-world problem of an Internet bookstore market segmentation and then applied the algorithm in clustering. Marinakis et al. [53] used a hybrid algorithm combining HBMO and greedy randomized adaptive search procedure to solve the vehicle routing problem. Niknam et al. [54] presented an improved HBMO for multi-objective placement of renewable energy resources. Wen et al. [47] first applied HBMO to address process planning problem and obtained a good process plan with minimal global machining cost in reasonable time.

Although HBMO can be used to solve optimization problems and good solutions can be obtained quickly, it often converges to local optima. Therefore, an improved HBMO method is required to overcome this shortcoming.

3.3 Sustainable Process Planning and Scheduling Approach

3.3.1 Representations for Process Planning and Scheduling

Process planning and scheduling, which bridge Computer-Aided Design (CAD) and Manufacturing Execution Systems (MES), are critical functions to minimize cost, improve adaptability, responsiveness, robustness, and sustainability of manufacturing processes.

The major considerations in process planning include: (1) generating machining operations based on the features of a part to meet desired functional specifications and achieve good manufacturability, (2) identifying machining resources applicable to the operations, and (3) determining the setup plan and operation sequence according to some criteria such as makespan, energy efficiency, and so on. Therefore, a process plan for a part can be represented by a series of machining operations, applicable resources for the operations, setup plans, operation sequence, etc. A setup can be generally defined as a group of operations that are manufactured on a single machine with the same fixture.

Based on the generated process plans of parts, the scheduling task is to assign the parts and their machining operations to specific machines to be executed in different time slots, targeting at a good shop floor performance, such as the shortest makespan, the most balanced machine utilization, the least total tardiness. A Gantt chart has been popularly used to represent a schedule of a group of parts, illustrated in Fig. 3.1. In the Gantt chart, the order in which the parts and their operations are carried out is laid out and the dependencies of the tasks are managed. The *X*-axis of the Gantt chart represents time. Each row in the *Y*-axis represents a machine and the specific arrangement for the operations of the parts on the machine. A machine is comprised of a number of time slots, which can be further classified into idle time slots, preparation time slots for machining operations (further including the setup time, the machine change time, or the tool change time), and machining time slots of operations.

As a consequence of the above, a part can be manufactured by different process plans. A group of alternative process plans can be generated using two strategies: processing flexibility and operation sequencing flexibility. Processing flexibility

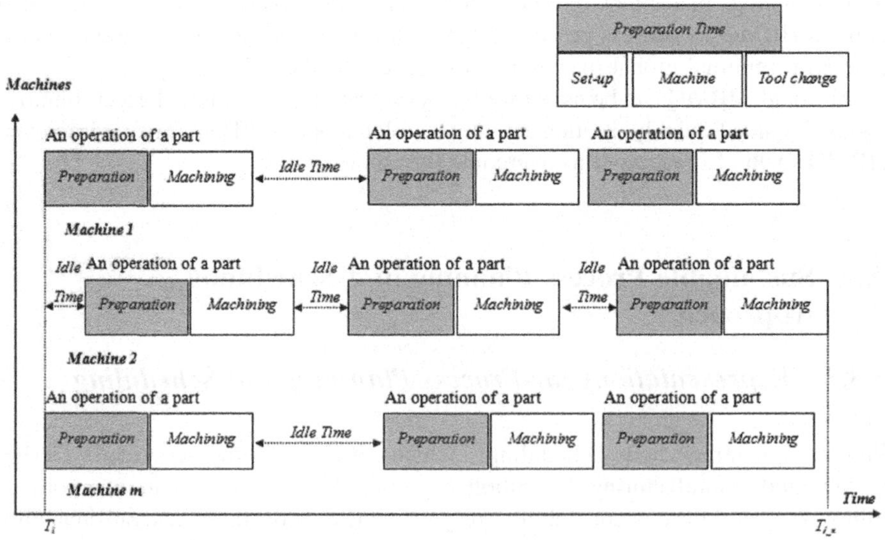

Fig. 3.1 Gantt chart for scheduling parts and machining operations

refers to the possibility of performing an operation on alternative machines with alternative setups or tools. Operation sequencing flexibility corresponds to the possibility of interchanging the sequence in which the operations are performed. For a group of parts, alternative schedules can be created based on scheduling flexibility, which relates to the possibility of arranging different schedules to manufacture the parts and the operations.

3.3.2 Energy Consumption Modeling for Process Planning and Scheduling

It has been known that there are a group of machines in the machining resources. Following the process planning and scheduling solution, one or more operations will be executed on the same machine. For a machine, its power profile is illustrated in Fig. 3.2, which consists of startup phases, idle phases, preparatory phases, working phases, and shutdown phases.

Hence, the energy consumption of a machine can be separated into the corresponding five segments.

The energy consumption during the startup phase can be computed as:

$$\mathrm{E}(M_i).\mathrm{setup} = \int_{T_i}^{T_{i\_1}} P_{\mathrm{startup}}(M_i)\mathrm{d}t \tag{3.1}$$

where $E(M_i)$.startup represents the energy consumed during the startup phase, $P_{\mathrm{startup}}(M_i)$ represents the power demand of the ith machine during the startup phase, T_i and T_{i_1} stand for the start and end time of the startup phase.

The energy consumption during the idle phase can be formulated as:

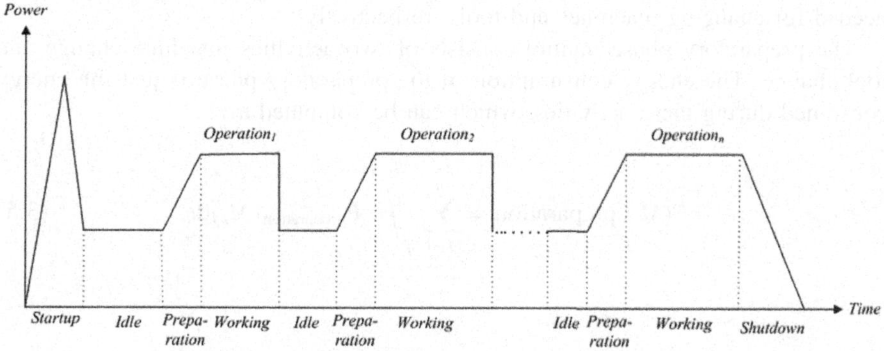

Fig. 3.2 Different phases of energy consumption for a machine

$$E(M_i).\text{idle} = P_{\text{idle}}(M_i) \times T_{\text{idle}}(M_i) \tag{3.2}$$

where $E(M_i)$.idle stands for the energy consumption of the idle phase, $P_{\text{idle}}(M_i)$ is the power demand of the ith machine during the idle phase, which is the sum of the power demand of all the components in the machine and usually a constant value. $T_{\text{idle}}(M_i)$ stands for the total idle time of the ith machine and can be represented as:

$$T_{\text{idle}}(M_i) = \sum_{j=1}^{n} T_{j,j+1} \tag{3.3}$$

where n stands for the number of the operations to be executed on the ith machine, and $T_{j,j+1}$ stands for the idle time between the jth and $(j + 1)$th operations to be executed on the ith machine. $T_{j,j+1}$ can be obtained by the following pseudo codes:

```
If  Tj.end< Tj+1.start  then
// The j-th operation ends before the (j+1)-th
// operation and the machine will be idle.
 Tj,j+1=Tj+1.start-Tj.end
Else                // There is no waiting.
 Tj,j+1 =0
Endif
```

Here, T_j.end and T_{j+1}.start stand for the jth operation's end time and $(j + 1)$th operation's start time, respectively. Let O be the operation in the process plan, which shares the same part with the $(j + 1)$th operation and is followed by the $(j + 1)$th operation. T_{j+1}.start does not equal to O's end time because O and the $(j + 1)$th operation may be executed on different machines or tools. That is, the time consumed by changing machines should be considered. Thus, T_{j+1}.start can be represented as:

$$T_{j+1}.\text{start} = T_o.\text{end} + T_{\text{machinechange}} + T_{\text{toolchange}} \tag{3.4}$$

where T_o.end and $T_{\text{machine change}}$ and $T_{\text{tool change}}$ stand for O's end time and the time needed for changing machines and tools, respectively.

The preparatory phase mainly consists of two activities: machine change and tool change. The energy consumption of the preparatory phase is just the energy consumed during these activities, which can be computed as:

$$E(M_i).\text{preparation} = \sum_{j=1}^{n} \int_{T_{i\_2}}^{T_{i\_3}} P_{\text{preparation}}(M_I)\text{d}t \tag{3.5}$$

where $E(M_i)$.preparation represents the energy consumed during the preparatory phase, n stands for the number of the operations to be executed on the ith machine, T_{i_2} and T_{i_3} stand for the start and end time to prepare the jth operation, $P_{preparation}$ is the power needed during the preparation for the jth operation, which can be the sum of P_{idle} and the machine change power or the tool change power.

The energy consumption during the working phase can be represented as:

$$E(M_i).\text{working} = \sum_{j=1}^{n} E_j(M_i).\text{working} \tag{3.6}$$

where $E(M_i)$.working stands for the energy consumption of the working phase, n stands for the number of the operations to be executed on the ith machine, and $E_j(M_i)$.working stands for the energy consumption of the jth operation to be executed on the ith machine. $E_j(M_i)$.working can be computed as:

$$E_j(M_i).\text{working} = V_J \times \text{SEC} \tag{3.7}$$

where V_j is the volume of the material removed by the jth operation, and SEC is the energy consumed by removing 1 cm^3 material. SEC can be computed by using the following formulation [12].

$$\text{SEC} = C_0 + \frac{C_1}{\text{MRR}} \tag{3.8}$$

where the coefficient C_0 is related to the workpiece material, tool geometry, and spindle drive characteristics; C_1 depends on how the machine tool is designed including its motor and transmission system; MRR stands for material removal rate for the working phase, which is influenced mainly by four types of factors, i.e., tool conditions, workpiece material, cutting parameters, and cutting environment. For C_0 and C_1, multiple machines' corresponding coefficients were provided by Kara and Li [12]. MRR can be obtained using the cutting volume and time.

The energy consumed during the shutdown phase can be computed as:

$$E(M_I).\text{shutdown} = \int_{T_{I\_1}}^{T_{I\_2}} P_{\text{shutdown}}(M_i)d \tag{3.9}$$

where $E(M_i)$.shutdown represents the energy consumption of the shutdown phase, T_{i_4} and T_{i_5} stand for the start and end time for switching off the ith machine, respectively, and $P_{\text{shutdown}}(M_i)$ stands for the power consumption of the ith machine during the shutdown phase.

Based on the energy consumption of the above phases, the total energy consumption of a machine can be represented below:

$$E(M_I) = E(M_I).\text{startup} + E(M_I).\text{idel} \\ + E(M_I).\text{preparation} + E(M_I).\text{working} + E(M_I).\text{shutdown} \quad (3.10)$$

where $E(M_i)$ stands for the total energy consumption of the ith machine.

Therefore, if there are m machines to be used in the process planning and scheduling, the overall energy consumed by all the machines to machine all the parts is:

$$E_{\text{Group}} = \sum_{i=1}^{m} E(M_i) \quad (3.11)$$

3.3.3 *Process Planning and Scheduling Criteria*

The energy consumption for a process planning and scheduling, as an essential criterion, has been defined above. Some other criteria based on time to evaluate the performances of process plans and schedules are defined in the following, which consists of the makespan and the balanced level of the machine utilization. In order to present the criteria, two assumptions are made. One is the number of machine (m). The other is the number of operations (n) to be executed on the machine.

Makespan means the maximum interval time spent to machine all the parts. It can be defined in the following:

$$\text{Makespan} = \text{Max}_{i=1}^{m}(T(M_i)) \quad (3.12)$$

where $T(M_i)$ is the total utilization time of the ith machine, which is composed of multiple startup, idle, preparation, and working phases. That is, $T(Mi)$ can be represented as:

$$T(M_i) = T(M_i).\text{startup} + T(M_i).\text{preparation} \\ + T(M_i).\text{idle} + T(M_i).\text{working} \quad (3.13)$$

where $T(M_i)$.startup, $T(M_i)$.preparation, $T(M_i)$.idle, and $T(M_i)$.working stand for the time of the corresponding phases. Let O_{ij} be the jth operation to be executed on the ith machine.

$$T(M_i).\text{startup} = \sum_{j=1}^{m} O_{IJ}.\text{startup} \times O_{IJ}.\text{startup\_index} \quad (3.14)$$

$$O_{ij}.\text{startup} = \begin{cases} 0 & \text{if the } i\text{th machine has been started up} \\ 1 & \text{if not} \end{cases} \quad (3.15)$$

where *Oij*.startup represents whether the *i*th machine is started up, and *Oij*.startup_index is the time index for each start up.

$$T(M_i).\text{preparation} = \sum_{j=1}^{n} (O_{ij}.\text{MC\_}T + O_{ij}.\text{TC\_}T) \quad (3.16)$$

where O_{ij}.MC_*T* and O_{ij}.TC_*T* represent the time for machine change and tool change, respectively.

$T(M_i)$.idle can be computed by using Eq. 3.3. $T(M_i)$.working is just the time used to execute all the operations on the *i*th machine. Thus, it can be computed as:

$$T(M_i).\text{working} = \sum_{j=1}^{n} (O_{IJ}.\text{Working\_}T) \quad (3.17)$$

where O_{ij}.Working_*T* represents the time used by the *i*th machine to execute the *j*th operation.

The standard deviation concept is introduced here to evaluate the balanced machine utilization.

$$X = \frac{\sum_{i=1}^{m} T(M_i)}{m} \quad (3.18)$$

$$\text{Utilization\_level} = \sqrt{\sum_{i=1}^{m} (T(M_i) - X)^2} \quad (3.19)$$

Based on the above functions, the weighted additive utility function is used to solve multi-objective optimization problem. The total weighted performance criteria (TWPC) can be described as:

$$\text{TWPC} = w_1 E_{\text{Group}} + w_2 \text{Makespan} + w_3 \text{Utilization\_level} \quad (3.20)$$

where w_1, w_2 and w_3 are the weights. The value of each weight is between 0 and 1, and the sum of them equals to one.

3.4 Hybrid HBMO-SA Algorithm

3.4.1 Overview

HBMO is a recently developed evolutionary algorithm. It is inspired by the process of real honeybee mating and has been applied in some combinatorial optimization problems such as the traveling salesman problem, vehicle routing problem, and the process planning problem. Furthermore, all the results have shown that better local optimum solutions can be found quickly using HBMO. However, it has also been observed that HBMO is liable to converge to local optima. Fortunately, SA can be used to compensate for this shortcoming because it can accept some probability. Consequently, in this paper, the strengths of HBMO and SA are combined to achieve the global optimization effectively.

The proposed HBMO-SA consists of two phases: HBMO phase and SA phase. In the HBMO phase, the honeybee mating process is simulated to generate a population. Good chromosomes in the population are then selected as the initial current process plans and schedules for the SA to search the optimal or near-optimal process plans and schedule. The flowchart of the proposed hybrid HBMO-SA is shown in Fig. 3.3.

In the hybrid HBMO-SA, the encoding scheme for each individual consists of four parts: operation sequencing, machine sequencing, tool sequencing, and tool approaching direction (TAD) sequencing. In the operation sequencing, each position corresponds to an operation number. That is, the ith element of the operation sequence represents the operation that will be executed. The second part is machine sequencing where the ith position corresponds to the machine number used by the ith operation. The third part is tool sequencing that is just the tool number used by the corresponding operation. The fourth part is TAD sequencing used by the operations.

In addition, the reciprocal of the objective function is chosen as the fitness function (cf. Eq. 3.21) to follow the rule that the individual with a greater fitness has a higher chance to be chosen. The fitness function for a solution is as follows:

$$\text{Fitness} = \frac{1}{\text{TWPC}} \tag{3.21}$$

3.4.2 *Honeybee Mating Algorithm: Phase 1*

The HBMO algorithm has been developed by simulating the honeybee mating process. In the mating process, the queen flies far from the nest. During the mating flight, the strong drones catch up with the queen and mate with her. After the queen's spermatheca is full of sperms, she will fly back to the nest and lay eggs. Each time the queen lays eggs, she randomly retrieves a sperm from her spermatheca to fertilize the eggs and a set of broods are generated. Then, the workers will take care of and improve the broods (e.g., feeding them with royal jelly). If a brood is better than the queen, it will be the new queen and starts its mating flight. The final queen obtained after all the mating flights is just the optimization result.

As a consequence of the above, the HBMO algorithm should contain a number of different procedures which correspond to the different phases of the honeybee mating process. The HBMO algorithm can be described as follows:

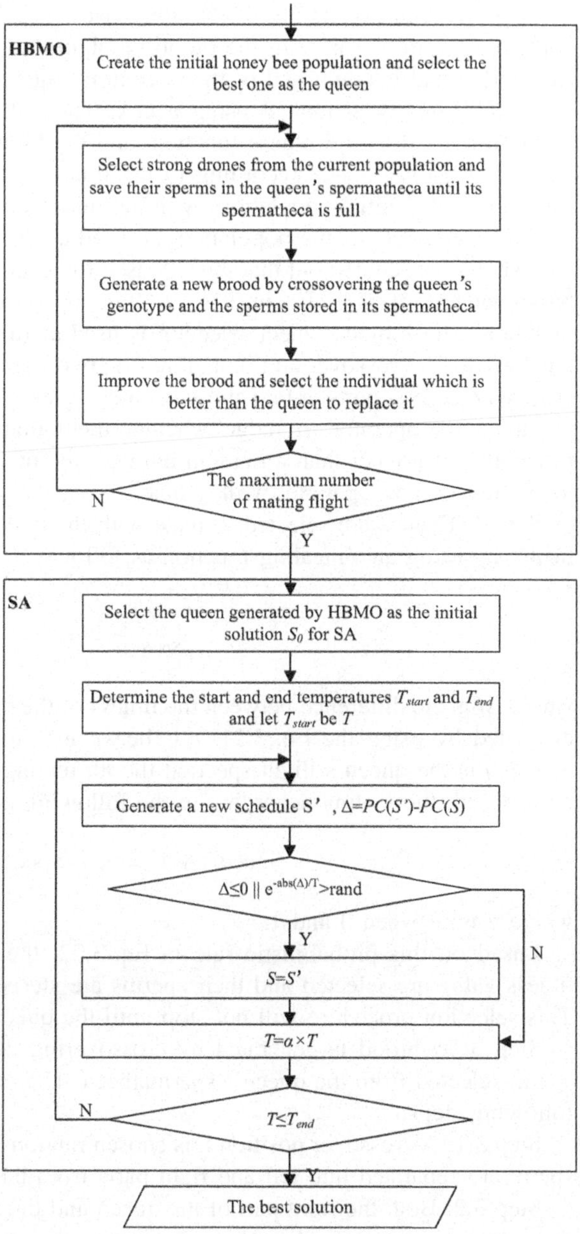

Fig. 3.3 Flowchart of the hybrid HBMO-SA algorithm

Step 1: Initialization consists of the following steps:

Step 1.1: The parameters of HBMO, including size of queen's spermatheca, number of drones, workers, number of mating flights, the queen's flight speed at the tth mating Speed(t), and its drop coefficient α, are initialized, respectively.

Step 1.2: The population of the honeybees (i.e., initial solutions) is created to configure the initial hive. In the proposed algorithm, the initial population, where each individual is composed of the operation sequence, and the selected machine, tool, and TAD is generated using a GA. Then, the reciprocal of the objective function is used as the fitness function, and the fitness values of all members are calculated by Eq. 3.21. According to the fitness values, the queen (i.e., the best schedule in the initial population) with the maximum fitness value is selected. All the other members of the population are used as drones.

It should be pointed out that the GA used to generate the initial population is the basic version of GA. It consists of three operators: selection, crossover, and mutation. The roulette wheel selection is used as the selection operator to choose individuals for crossover and mutation. The two-point crossover is implemented to crossover the first parts of two chromosomes. A specified probability is then used in the mutation operator to judge whether the mutation will be executed. If the probability is greater than a random number, the machine, tool, and TAD used by the corresponding operation will mutate.

Step 2: Drones are selected to mate with the queen. A drone mates probabilistically by using an annealing function as follows:

$$\frac{-|\Delta f|}{e^{\mathrm{Speed}(t)}} > r \tag{3.22}$$

where Δf is the difference between the fitness of the drone and queen, which can be evaluated by using the Eq. 3.21, r is between 0 and 1 and randomly generated, Speed(t) is the queen's flight speed at the tth mating, and the queen's flight speed decays with the mating according to the following equation:

$$\mathrm{Speed}(t+1) = \alpha \times \mathrm{Speed}(t) \tag{3.23}$$

where α is between 0 and 1.

Based on this probabilistic rule in Eq. 3.22, the strong drones with the great fitness value are selected and their sperms are stored in the queen's spermatheca. This selection procedure will not stop until the queen's spermatheca is full.

Step 3: A brood is generated by crossovering the queen's genotype with the sperm selected from the queen's spermatheca. The crossover operator includes the following steps:

Step 3.1: A crossover position P is chosen randomly. The queen and the selected sperm are separated into left and right parts from the crossover point.

Step 3.2: Both the right part of the queen and the left part of the selected sperm are copied to generate a brood.

Step 3.3: The brood is checked to find out those genes that are redundant or lost.

Step 3.4: The redundant genes in the brood are replaced with the lost ones. The machines, tools, and TADs are adjusted according to the operations. The new brood is obtained.

Step 4: For each brood, a worker (i.e., a local search heuristics) is chosen randomly to improve it. If the improved brood (i.e., the new solution) has a greater fitness than that of the current queen, it will replace the queen. All the other broods will take part in the next mating flight as drones. The improvement is achieved by three operators: mutation, adjacent swapping, and shift.

- The mutation operator is achieved mainly by two steps. First, an operation in the current solution is selected randomly. Then, the corresponding machine and tool are renewed from the candidate lists.
- The adjacent swapping is realized by exchanging two adjacent operations in the current solution. Meanwhile, the corresponding machines, tools, and TADs are exchanged.
- The shift operator is done by removing an operation from its present position to insert it at another position. The machine, tool, and TAD used by the operation will also be removed to the corresponding positions.

Based on the above operators, four workers are generated. The first three workers correspond to mutation, adjacent swapping, and shift, respectively. The fourth worker simultaneously employs two operators: mutation and adjacent swapping.

Step 5: If the number of mating flight is still not more than the maximum which has been defined in Step 1.1, a new mating flight will begin. That is, Steps 2.4 will be repeated until the number of mating flight reaches the largest number.

3.4.3 Simulated Annealing: Phase 2

In the second phase of the hybrid HBMO-SA approach, the SA algorithm is used to jump out of local optima and achieve a better-localized search. Its processes are described as follows.

Step 1: The individual with the maximum fitness value is first selected from the populations generated by HBMO. The selected individual is used as the initial current schedule S_0 for the SA.

Step 2: The start and end temperatures T_{start} and T_{end} are determined. T_{start} is taken as the current temperature T.

Step 3: A new temporary schedule S' is generated by using two types of mutation strategies which are listed as follows:

- Two operations in the current schedule S are chosen randomly and exchanged. The machines, tools, and TADs used by the operations are exchanged too.
- An operation in S is selected. Then, the machine, tool, and TADs used by the selected operation are changed by referring the candidate list.

Step 4: One of the performance criterion functions defined above is used to compute the difference between the performance criteria of S' and S. Let PC represent a performance criterion.

$$\Delta = PC(S') - PC(S)$$

If $\Delta \leq 0 \,||\, e^{-abs(\Delta)/T} > rand \qquad //0 < rand < 1$

$$S = S'$$

End

$T = \alpha \times T \qquad //0 < \alpha < 1$

Step 5: Steps 3.4 will be repeated until $T \leq T_{end}$.

3.5 Case Studies and Discussions

In this section, three groups of parts from practical manufacturing are presented to validate the effectiveness and feasibility of the proposed approach. In the case studies, the machines tested by Kara and Li were used as the machining resources [12]. The experiments were performed on the Windows 7 operating system with Intel Dual-core CPU at 2.10 GHz and 3.00 GB of main memory. The simulation was carried out by using the MATLAB programming language.

In order to select the appropriate parameters, a number of different alternative values of the proposed algorithm's parameters were tested. The selected parameters of HBMO are: size of queen's spermatheca equals to 60, number of drones equals to 100, number of mating flights equals to 400, and the initial speed(t) and α in Eq. (3.23) are 1000 and 0.85, respectively. For the SA, the selected parameters are: the start and end temperatures T_{start} and T_{end} equal to 1000 and 3.5116e−025, respectively, and α equals to 0.9.

It should be pointed out that some assumptions [39, 55] that are commonly used in most of the previous research works regarding scheduling were still taken for this study. The assumptions include:

- Parts are independent, and part preemption is not allowed;
- The sequence of the operations of each part complies with manufacturing constraints;
- All parts, machines, and tools are available at time zero simultaneously;
- Each operation is performed on a single machine, and each machine can only execute an operation at a time;
- Machines are continuously available for production;

- If a machine or a tool is broken down, or a new part is inserted, the algorithm can restart and generate new process plans and a schedule due to the efficient optimization performance of the algorithm;
- The time for a setup is identical and independent of specific operations. The time for a machine change or a tool change follows the same assumption.

3.5.1 Case Study 1

The first case consists of three parts which are shown in Fig. 3.4. This group of parts has 38 machining features.

For the parts, various experiments were conducted for three different conditions.

- $PC_1 = \text{Makespan}$
  where PC_1 represents the first performance criterion (to achieve the minimum makespan).
- $PC_2 = \text{Energy\_Consumption}$
  where PC_2 represents the second performance criterion (to achieve the minimum energy consumption).
- The third performance criterion is used to optimize energy consumption and the balanced utilization simultaneously, and thus consists of both of them. It can be represented as:

$$PC_3 = w_1 \times \text{Energy\_Consumption} + w_3 \times \text{Utilization\_Level}$$

where PC_3 stands for the third performance criterion, and w1 and w3 stand for the weights of the two criteria. Both w_1 and w_3 are set 0.5 in the experiments to take the energy consumption and utilization into account simultaneously.

The proposed HBMO-SA was conducted 15 times under different conditions, respectively, to verify its performance. The average convergence curves of HBMO-SA under different conditions are illustrated in Figs. 3.5, 3.6, and 3.7,

Fig. 3.4 First group of parts

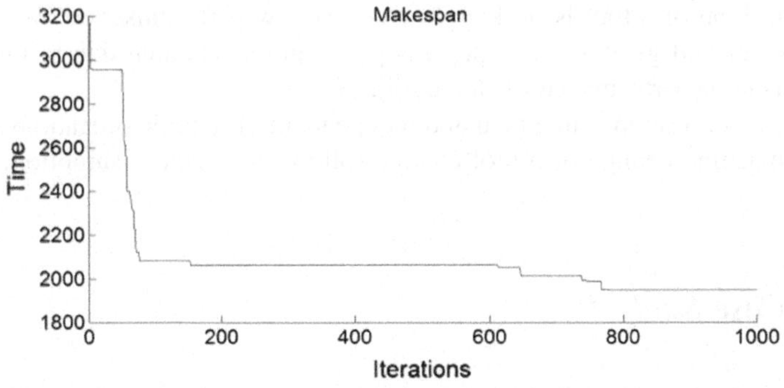

Fig. 3.5 Convergence curve for case 1 under makespan

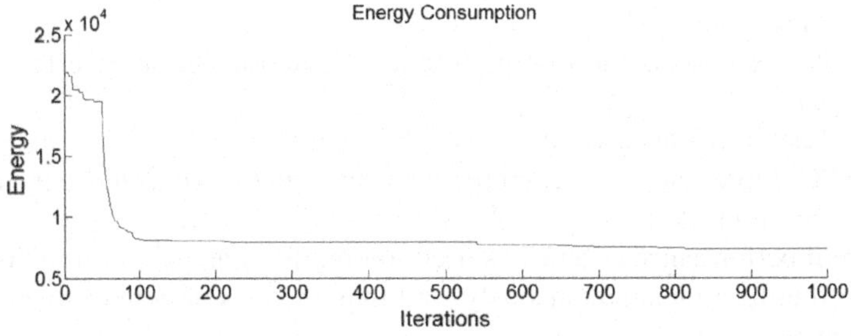

Fig. 3.6 Convergence curve for case 1 under energy consumption

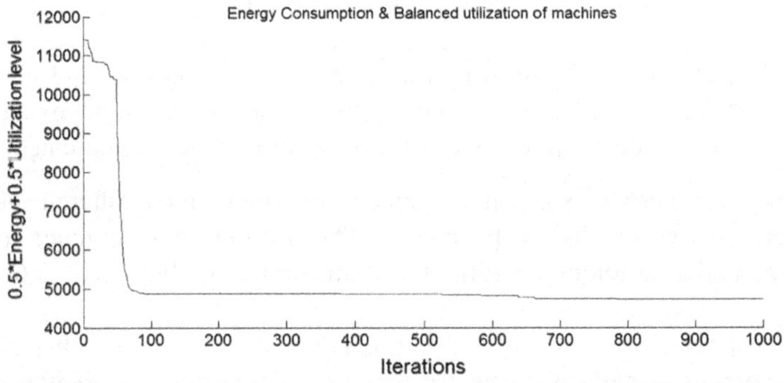

Fig. 3.7 Convergence curve for case 1 for energy consumption/utilization level

respectively. It can be observed that the HBMO drops fast in the first phase of the proposed algorithm and converges to an intermediate solution. Based on this intermediate solution, a further optimization is obtained by the SA in the second phase.

To explore the relation between energy consumption and the other performance criteria, energy consumption versus makespan and energy consumption vs balanced

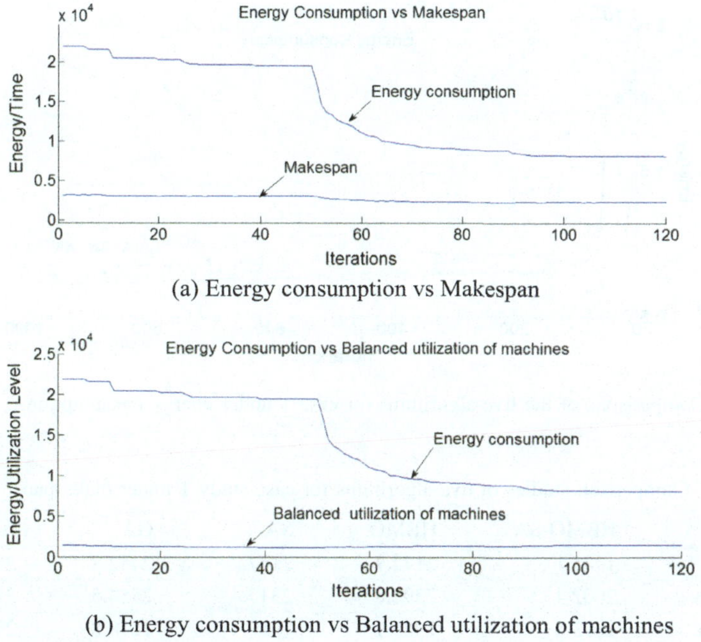

Fig. 3.8 Comparison of two criteria for case 1 under the same condition

utilization of machines were obtained under Condition 2. As shown in Fig. 3.8a, they follow the similar trends since the numbers of machine changes and tool changes and the reduced idle time under Condition 2 contribute to both energy consumption and makespan. However, it is not the linear relation since the energy consumption is also affected by other factors such as power, material removal rate. In Fig. 3.8b, the energy consumption and the balanced utilization of machines were obtained under Condition 2. Energy saving can be achieved through the intensive utilization of energy-efficient machines. It is conflicted with the criterion for the balanced utilization of machines. Therefore, the targets for the lower energy consumption and the more balanced utilization of machines are in contradiction, which is clearly reflected in Fig. 3.8b.

The algorithm was further compared with four popular evolutionary algorithms: HBMO, SA, ACO, and GA. In GA, the operators including selection, crossover, and mutation are used to improve the populations gradually [30]. The number of generations equals to 1000, and the crossover and mutation rates equal to 0.8 and 0.6, respectively. SA's parameters mainly include the start and end temperatures and factor α [39]. T_{start} and T_{end} equal to 1000 and 1.7479e−043, respectively, and α equals to 0.9. An ACO algorithm was modeled based on the behavior of ant seeking a path between their colony and a source of food [40]. Its parameters include the number of ants n, evaporation factor ρ, and weight parameters α and β. n equals to 100. ρ equals to 0.1. α and β are 1 and 2, respectively. For HBMO, the number of

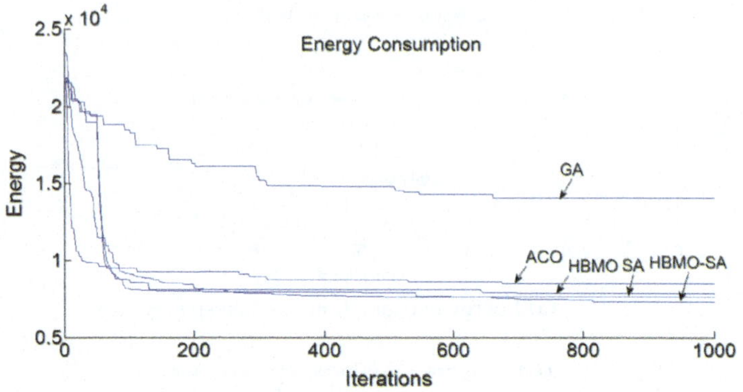

Fig. 3.9 Comparisons of the five algorithms for case 1 under energy consumption

Table 3.1 Comparison studies of five algorithms for case study 1 under makespan

	HBMO-SA	HBMO	SA	GA	ACO
Mean	1945.1	2172.7	2147.7	2442.8	2928.5
Maximum	2002.4	2392.8	2317.4	2554.5	3128.2
Minimum	1893.6	1985.1	1980.1	2278.8	2743.7

mating flights is 1000, and other parameters are the same as the corresponding ones used by the hybrid HBMO-SA. Figure 3.9 shows the comparisons among the proposed algorithm: HBMO, SA, ACO, and GA under Condition 2. All the algorithms were executed for 15 times. The results obtained by the proposed algorithm were better than those obtained by other algorithms. In addition, the CPU time consumed by the proposed algorithm and HBMO were 126 and 287 s, respectively. The proposed algorithm showed a better performance in efficiency since the inner loop in the SA adopted in its second phase iterates only 1 time to obtain a better efficiency than the HBMO under the aforementioned parameters. In addition, the comparisons among HBMO-SA, HBMO, SA, GA, and ACO under Condition 1 are displayed in Table 3.1. From the results of Table 3.1, HBMO-SA got smaller makespan in mean, maximum, and minimum. The comparison studies in this case showed that HBMO-SA exceeds HBMO, SA, GA, and ACO in effectiveness and stability.

3.5.2 *Case Study 2*

The second case study employed another three parts (c.f. Fig. 3.10) to further validate the proposed approach. The parts have 48 machining features.

The above three conditions in Condition 1 were taken into account again. All the algorithms were conducted for 15 times under these conditions. The optimization

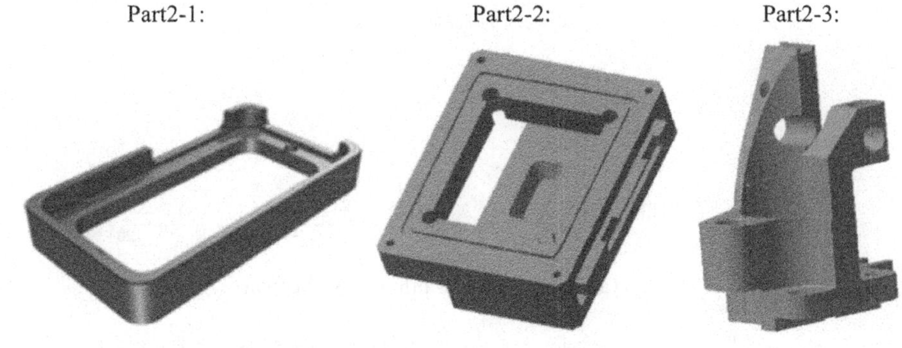

Fig. 3.10 Second group of parts

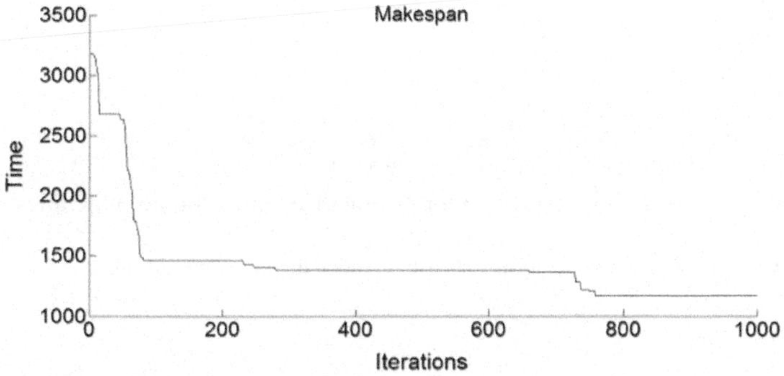

Fig. 3.11 Convergence curve for case 2 under makespan

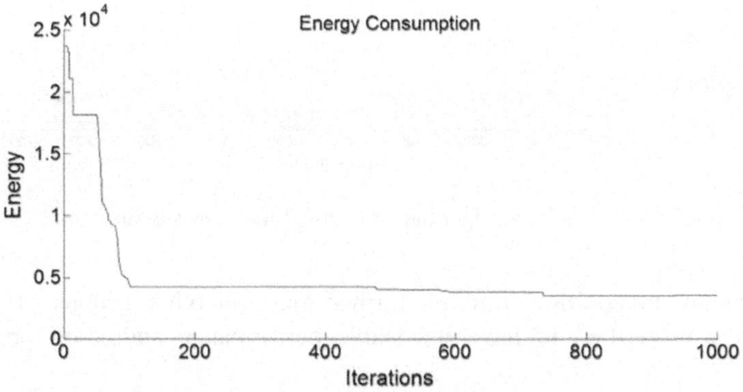

Fig. 3.12 Convergence curve for case 2 under energy consumption

results are shown in Figs. 3.11 and 3.12. The trend was similar to that in case study 1. As illustrated in Fig. 3.13, the similar trend between energy consumption and makespan, energy consumption and the balanced utilization of machines were

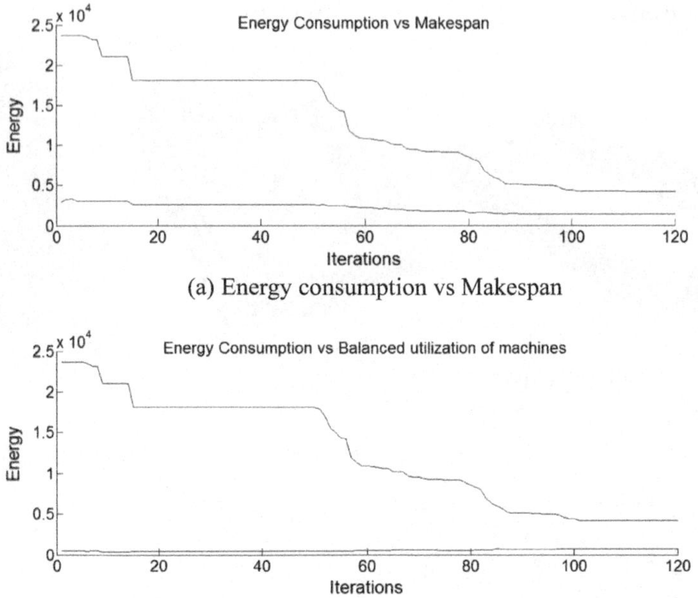

(a) Energy consumption vs Makespan

(b) Energy consumption vs Balanced utilization of machines

Fig. 3.13 Comparison of two criteria for case 2 under the same condition

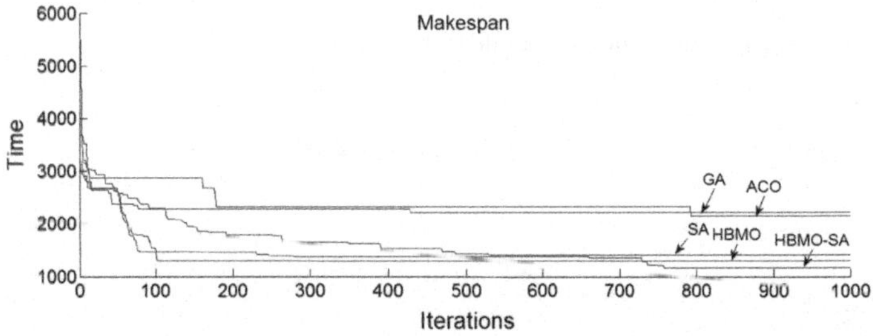

Fig. 3.14 Comparisons of the five algorithms for case 2 under makespan

verified again. In addition, through further trials on other groups of parts, the algorithm was verified to have the stable performance and good optimization results.

The comparisons among the proposed algorithm, HBMO, SA, ACO, and GA under condition makespan are shown in Fig. 3.14. From the results, it can be observed that HBMO converged to a good solution more quickly than other algorithms. The local optima can be avoided by SA. The hybrid HBMO and SA adopted in the approach of this paper combined the advantages of HBMO and SA and thus can achieve better-optimized results.

3.5.3 Case Study 3

The third case study utilized the parts shown in Fig. 3.15. This group of parts consists of 48 machining features.

The proposed algorithm was executed for 15 times under the above three conditions defined in case study 1 to verify its performance of the parts. The optimization results are shown in Figs. 3.16, 3.17, 3.18, and 3.19. It can be observed that the results for this group of parts followed the similar trends for the parts in the above two groups. It was verified again that the presented algorithm had the stable performance and good optimization results. The comparison results shown in Fig. 3.20 also demonstrated that HBMO-SA had a good performance in solving process planning and scheduling problem.

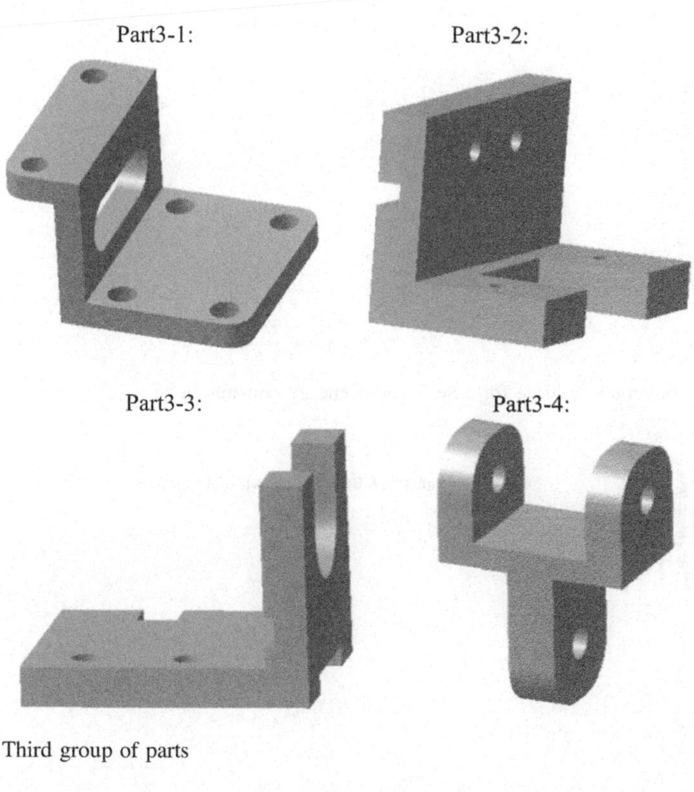

Fig. 3.15 Third group of parts

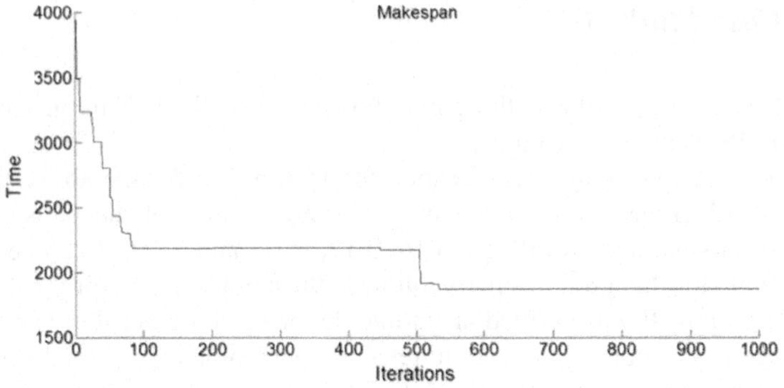

Fig. 3.16 Convergence curve for case 3 under makespan

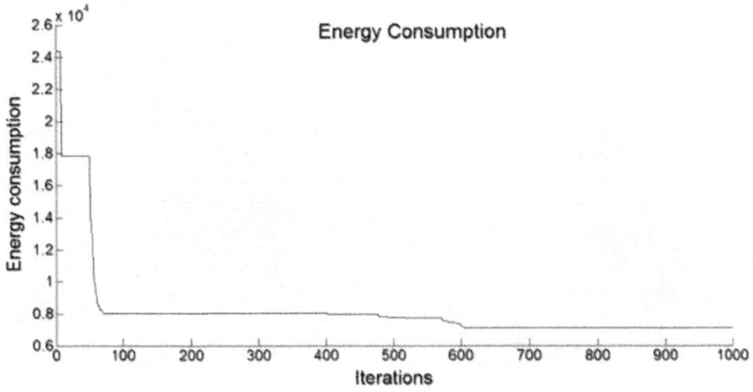

Fig. 3.17 Convergence curve for case 3 under energy consumption

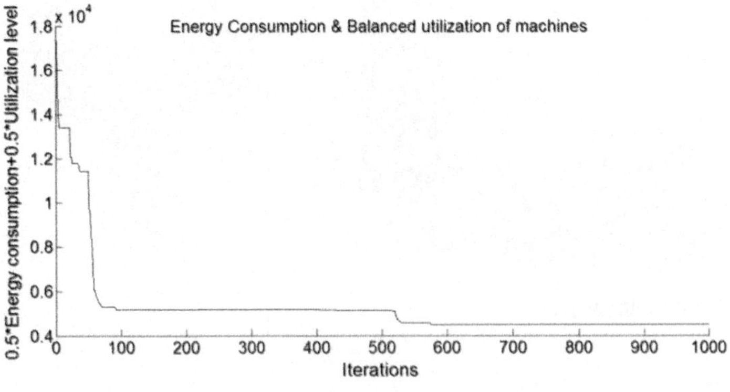

Fig. 3.18 Comparisons of the five algorithms for energy consumption

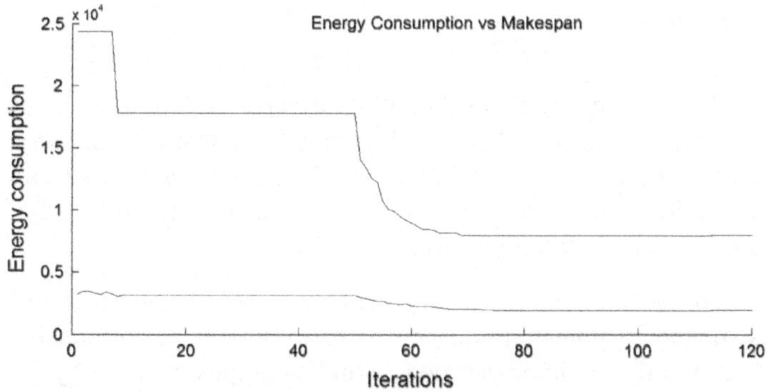

(a) Energy consumption vs Makespan

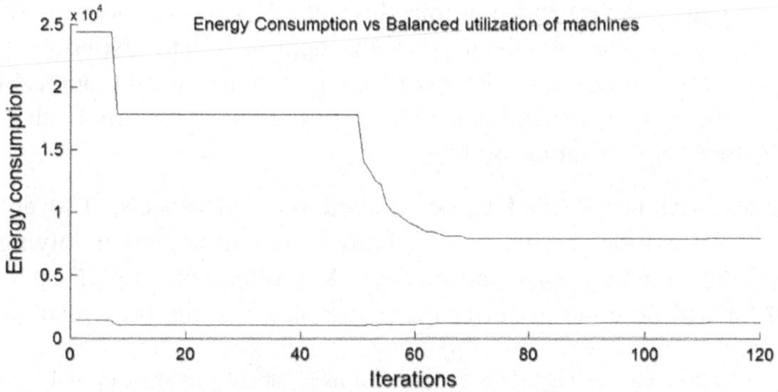

(b) Energy consumption vs Balanced utilization of machines

Fig. 3.19 Comparison of two criteria for case 3 under the same condition

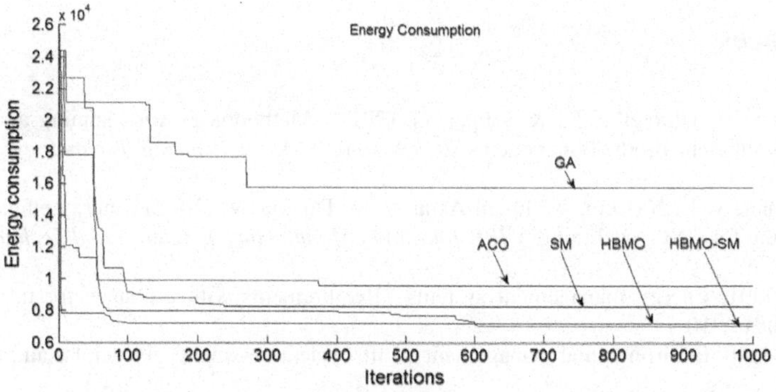

Fig. 3.20 Comparisons of the five algorithms for case 3 under energy consumption

3.6 Conclusions

Sustainable process planning and scheduling are critical to reduce its energy consumption and achieve sustainable development of machining processes. In this paper, sustainable process planning and scheduling are explored and presented in a hybrid optimization approach. In summary, the contributions of the presented approach are from the following aspects:

- Systematic models and strategies for dynamic process planning and scheduling in terms of energy consumption and other important performance criteria are established. On the basis of the models and strategies, the energy consumption of machining processes is effectively considered to achieve sustainability.
- Compared with some other optimization algorithms, the developed HBMO-SA approach can achieve better-optimized results. Various cases are used to verify and demonstrate the effectiveness of the approach. The experimental benchmarking demonstrates that the approach is promising and outperforms GA, HBMO, and SA. The approach can be further used to solve other complex combinatorial optimization problems.

Future research is intended to be focused on two aspects. The energy consumption model will be applied in more complex environments involving machine breakdown and existing jobs cancelation. A further test and discussion about hybridization will be done to verify the performance of the presented algorithm.

Acknowledgements This research was carried out as a part of the Smarter and CAPP-4-SMEs projects which are supported by the 7th European Community Framework Programme under the grant agreement No 610675 (PEOPLE-2013-IAPP-610675) and No 314024 (FP7-2012-NMP-ICT-FoF). The paper reflects only the authors' views, and the Union is not liable for any use that may be made of the information contained therein.

References

1. Weinert, N., Chiotellis, S., & Seliger, G. (2011). Methodology for planning and operating energy-efficient production systems. *CIRP Annals—Manufacturing Technology, 60*(1), 41–44.
2. Newman, S. T., Nassehi, A., Imani-Asrai, R., & Dhokia, V. (2012). Energy efficient process planning for CNC machining. *CIRP Journal of Manufacturing Science and Technology, 5*(2), 127–136.
3. EN 16001: Energy management systems—Requirements with guidance for use, European Standard 2010.
4. ISO 14040: Environmental management—Life cycle assessment—Principles and framework, www.iso.org.
5. ISO 14044: Environmental management—Life cycle assessment—Requirements and guidelines, www.iso.org.
6. Sinden, G. (2009). The contribution of pas 2050 to the evolution of international greenhouse gas emission standards. *International Journal of Life Cycle Assessment, 14*(3), 195–203.

7. Jovane, F., Yoshikawa, H., Alting, L., Boër, C. R., Westkamper, E., & Williams, D. (2008). The incoming global technological and industrial revolution towards competitive sustainable manufacturing. *CIRP Annals—Manufacturing Technology, 57*(2), 641–659.
8. Zhang, X. F., Zhang, S. Y., Hu, Z. Y., Yu, G., Pei, C. H., & Sa, R. N. (2012). Identification of connection units with high GHG emissions for low-carbon product structure design. *Journal of Cleaner Production, 27*(6), 118–125.
9. Behrendt, T., Zein, A., & Min, S. (2012). Development of an energy consumption monitoring procedure for machine tools. *CIRP Annals—Manufacturing Technology, 61*(1), 43–46.
10. Li, X. X., Li, W. D., Cai, X. T., & He, F. Z. (2013). A honey-bee mating optimization approach for sustainable manufacturing process planning and scheduling. In *Proceedings of the International Conference on Computer Supported Cooperative Work in Design. Whistler, BC, Canada* (pp. 465–470), June 27–29, 2013.
11. Gutowski, T., Dahmus, J., & Thiriez, A. (2006). Electrical energy requirements for manufacturing processes. In *13th CIRP International Conference on Life Cycle Engineering* (Vol. 31(1), pp. 623–638).
12. Kara, S., & Li, W. (2011). Unit process energy consumption models for material removal processes. *CIRP Annals—Manufacturing Technology, 60*(1), 37–40.
13. Diaz, N., Redelsheimer, E., & Dornfeld D. (2011). Energy consumption characterization and reduction strategies for milling machine tool use. *Glocalized Solutions for Sustainability in Manufacturing* 263–267.
14. Li, L., Yan, J., & Xing, Z. (2013). Energy requirements evaluation of milling machines based on thermal equilibrium and empirical modelling. *Journal of Cleaner Production, 52,* 113–121.
15. Mori, M., Fujishima, M., Inamasu, Y., & Oda, Y. (2011). A study on energy efficiency improvement for machine tools. *CIRP Annals—Manufacturing Technology, 60*(1), 145–148.
16. He, Y., Liu, F., Wu, T., Zhong, F. P., & Peng, B. (2011). Analysis and estimation of energy consumption for numerical control machining. *Proceedings of the Institution of Mechanical Engineers Part B Journal of Engineering Manufacture, 226*(2), 255–266.
17. Balogun, V. A., & Mativenga, P. T. (2013). Modelling of direct energy requirements in mechanical machining processes. *Journal of Cleaner Production, 41,* 179–186.
18. Duflou, J. R., Kellens, K., Renaldi, Guo Y., & Dewulf, W. (2012). Critical comparison of methods to determine the energy input for discrete manufacturing processes. *CIRP Annals—Manufacturing Technology, 61*(1), 63–66.
19. ISO/WD 14955-1 Environmental Evaluation of Machine Tools Part 1: Energy-saving Design Methodology for Machine Tools, International Organization for Standardization, Geneva, Switzerland.
20. Seow, Y., & Rahimifard, S. (2011). A framework for modelling energy consumption within manufacturing systems. *CIRP Journal of Manufacturing Science and Technology, 4*(3), 258–264.
21. Kara, S., Manmek, S., & Herrmann, C. (2010). Global manufacturing and the embodied energy of products. *CIRP Annals—Manufacturing Technology., 59*(1), 29–32.
22. Herrmann, C., & Thiede, S. (2009). Process chain simulation to foster energy efficiency in manufacturing. *CIRP Journal of Manufacturing Science and Technology, 1*(4), 221–229.
23. GillesMouzon, Yildirim M. (2008). A framework to minimise total energy consumption and total tardiness on a single machine. *International Journal of Sustainable Engineering, 1*(2), 105–116.
24. Bruzzone, A. A. G., Anghinolfi, D., Paolucci, M., & Tonelli, F. (2012). Energy-aware scheduling for improving manufacturing process sustainability: A mathematical model for flexible flow shops. *CIRP Annals—Manufacturing Technology, 61*(1), 459–462.
25. Siddique, N., & Adeli, H. (2013). *Computational intelligence: Synergies offuzzy logic.* Neural Networks and Evolutionary Computing: Wiley.
26. Adeli, H., & Hung, S. L. (1995). *Machine Llearning—Neural networks, genetic algorithms, and fuzzy sets.* New York: Wiley.

27. Adeli, H., & Sarma, K. C. (2006). *Cost optimization of structures: Fuzzy logic, genetic algorithms, and parallel computing*. Wiley, New York.
28. Hejazi, F., Toloue, I., Jaafar, M. S., & Noorzaei, J. (2013). Optimization of earthquake energy dissipation system by genetic algorithm. *Computer-Aided Civil and Infrastructure Engineering, 28*(10), 796–810.
29. Jia, L., Wang, Y., & Fan, L. (2014). Multiobjective bilevel optimization for production-distribution planning problems using hybrid genetic algorithm. *Integrated Computer Aided Engineering, 21*(1), 77–90.
30. Lin, D. Y., & Ku, Y. H. (2014). Using genetic algorithms to optimize stopping patterns for passenger rail transportation. *Computer-Aided Civil & Infrastructure Engineering, 29*(4), 264–278.
31. Zhu, W., Hu, H., & Huang, Z. (2014). Calibrating rail transit assignment models with genetic algorithm and automated fare collection data. *Computer-Aided Civil and Infrastructure Engineering, 29*(7), 518–530.
32. Campomanes-Álvarez, B. R., Cordón, O., & Damas, S. (2013). Evolutionary multi-objective optimization for mesh simplification of 3D open models. *Integrated Computer-Aided Engineering, 20*(4), 375–390.
33. Joly, M. M., Verstraete, T., & Paniagua, G. (2014). Integrated multifidelity, multidisciplinary evolutionary design optimization of counterrotating compressors. *Integrated Computer Aided Engineering, 21*(3), 249–261.
34. Luna, J., Mar, Romero J., Romero, C., & Ventura, S. (2014). Reducing gaps in quantitative association rules: A genetic programming free-parameter algorithm. *Integrated Computer-Aided Engineering, 21*(4), 321–337.
35. Molina-García, M., Calle-Sánchez, J., González-Merino, C., Fernández-Durán, A., & Alonso, J. I. (2014). Design of in-building wireless networks deployments using evolutionary algorithms. *Integrated Computer-Aided Engineering, 21*(4), 367–385.
36. Pedrino, E. C., Roda, V. O., Kato, E. R. R., Tsunaki, R. H., Tsunaki, R. H., & Tsunaki, R. H. (2013). A genetic programming based system for the automatic construction of image filters. *Integrated Computer-Aided Engineering, 20*(3), 275–287.
37. Alonso, P., & Montes, S. (2014). On the use of fuzzy partitions to protect data. *Integrated Computer-Aided Engineering, 21*(4), 355–366.
38. Reyes, O., Morell, C., & Ventura, S. (2014). Evolutionary feature weighting to improve the performance of multi-label lazy algorithms. *Integrated Computer-Aided Engineering, 21*(4), 339–354.
39. Li, W. D., & Mcmahon, C. A. (2007). A simulated annealing-based optimization approach for integrated process planning and scheduling. *International Journal of Computer Integrated Manufacturing, 20*(1), 80–95.
40. Forcael, E., González, V., Orozco, F., Vargas, S., Pantoja, A., & Moscoso, P. (2014). Ant colony optimization model for tsunamis evacuation routes. *Computer-Aided Civil & Infrastructure Engineering, 29*(10), 723–737.
41. Hsu, W. Y. (2013). Application of quantum-behaved particle swarm optimization to motor imagery EEG classification. *International Journal of Neural Systems, 23*(06), 63–331.
42. Iacca, G., Caraffini, F., & Neri, F. (2014). Multi-strategy coevolving aging particle optimization. *International Journal of Neural Systems, 24*(01), 1450008.
43. Shafahi, Y., & Bagherian, M. (2013). A customized particle swarm method to solve highway alignment optimization problem. *Computer-Aided Civil & Infrastructure Engineering, 28*(1), 52–67.
44. Tao, H., Zain, J. M., Ahmed, M. M., Abdalla, A. N., & Jing, W. (2012). A wavelet-based particle swarm optimization algorithm for digital image watermarking. *Integrated Computer Aided Engineering, 19*(1), 81–91.
45. Wu, J. W., Tseng, J. C. R., & Tsai, W. N. (2014). A hybrid linear text segmentation algorithm using hierarchical agglomerative clustering and discrete particle swarm optimization. *Integrated Computer Aided Engineering, 21*(1), 35–46.

46. Zeng, Z., Xu, J., Wu, S., & Shen, M. (2014). Antithetic method-based particle swarm optimization for a queuing network problem with fuzzy data in concrete transportation systems. *Computer-aided Civil & Infrastructure Engineering, 29*(10), 771–800.
47. Wen, X. Y., Li, X. Y., Gao, L., Sang, H. Y. (2014). Honey bees mating optimization algorithm for process planning problem. *Journal of Intelligent Manufacturing, 25*(3), 459–472.
48. Pham, D. T., Castellani, M., & Ghanbarzadeh, A. (2007). Preliminary design using the bees algorithm. In *Proceedings of Eighth International Conference on Laser Metrology, CMM and Machine Tool Performance. LAMDAMAP, Euspen, Cardiff, UK* (pp. 420–429).
49. Abbass, H. A. (2001). Marriage in honey bees optimisation: A haplometrosis polygynous swarming approach. In *Proceedings of the Congress on Evolutionary Computation* (pp. 207–214).
50. Koudil, M., Benatchba, K., Tarabet, A., & Sahraoui, E. B. (2007). Using artificial bees to solve partitioning and scheduling problems in codesign. *Applied Mathematics and Computation, 186*(2), 1710–1722.
51. Curkovic, P., & Jerbic, B. (2007). Honey-bees optimization algorithm applied to path planning problem. *International Journal of Simulation Modelling, 6*(3), 154–164.
52. Fathian, M., & Amiri, B. (2008). A honeybee-mating approach for cluster analysis. *International Journal of Advanced Manufacturing Technology, 38*(7–8), 809–821.
53. Marinakis, Y., Marinaki, M., & Matsatsinis, N. (2008). A hybrid clustering algorithm based on honey bees mating optimization and greedy randomized adaptive search procedure. In *International Conference on Learning and Intelligent Optimization, Berlin, Heidelberg* (pp. 138–152).
54. Niknam, T., Taheri, S. I., Aghaei, J., Tabatabaei, S., & Nayeripour, M. (2011). A modified honey bee mating optimization algorithm for multiobjective placement of renewable energy resources. *Applied Energy, 88*(12), 4817–4830.
55. Li, W. D., Ong, S. K., & Nee, A. Y. C. (2004). Optimization of process plans using a constraint-based tabu search approach. *International Journal of Production Research, 42*(10), 1955–1985.

Chapter 4
A Systematic Approach of Process Planning and Scheduling Optimization for Sustainable Machining

S. Wang, X. Lu, X. X. Li and W. D. Li

Abstract The implementation of sustainability in manufacturing companies, whose activities are usually characterized by high variety and low volume, has been crippled by the lack of effective process planning and scheduling solutions for sustainable management of manufacturing shop floors. To address the challenge, an innovative and systematic approach for machining process planning and scheduling optimization has been developed. This approach consists of a process stage and a system stage, augmented with intelligent mechanisms for enhancing the adaptability and responsiveness to job dynamics in machining shop floors. In the process stage, key operational parameters for machining a part are optimized adaptively to meet multiple objectives and constraints, i.e., energy efficiency of the machining process and productivity as objectives and surface quality as a constraint. In the consecutive system stage, to achieve higher energy efficiency and shorter makespan in the entire shop floor, sequencing/set-up planning of machining features, operations and scheduling for producing multiple parts on different machines are optimized. Artificial neural networks are used for establishing the complex nonlinear relationships between the key process parameters and measured data sets of energy consumption and surface quality. Intelligent algorithms, including pattern search, genetic algorithm, and simulated annealing, are applied and benchmarked to identify optimal solutions. Experimental tests indicate that the approach is effective and configurable to meet multiple objectives and technical constraints for sustainable process planning and scheduling. The approach, validated through industrial case studies provided by a European machining company, demonstrates significant potentials of research applicability in practice.

Keywords Sustainable manufacturing · Computer numerical control machining
Process planning · Process scheduling · Intelligent algorithm · Machining feature

---

S. Wang · X. Lu · X. X. Li · W. D. Li (✉)
Faculty of Engineering, Environment and Computing,
Coventry University, Coventry, UK
e-mail: weidong.li@coventry.ac.uk

W. Li and S. Wang (eds.), *Sustainable Manufacturing and Remanufacturing Management*, https://doi.org/10.1007/978-3-319-73488-0_4

4.1 Introduction

Paramount demands for new products have increasingly incurred more manufacturing activities. In order to balance the multifaceted dimensions of economic growth and environmental protection, a series of regulations and guidelines on lifecycle energy/carbon-related management have been developed in recent years for product design and manufacturing enterprises to embrace "Competitive Sustainable Development" [1] and shoulder "Extended Producer Responsibilities (EPR)" [2]. For instance, the lifecycle carbon labeling scheme, outlined by the ISO 14040: 2006, ISO 14044: 2006, and Publicly Available Specification 2050 (PAS 2050), has been introduced with a bid to stimulate energy efficiency improvement and carbon emission reduction during product lifecycle. Among the various stages of product lifecycle, manufacturing processes are energy intensive making the stage one of the primary energy consumptions and carbon footprint generation sources. Manufacturing processes in factories, in which motors, compressors, and machine systems need to be powered, and adequate heating, ventilation, and air-conditioning equipment need to be maintained, contribute to over 24% of total European energy consumption [3]. Therefore, the effective implementation of manufacturing sustainability is prevalent. The research road map of intelligent manufacturing toward 2020, conducted by an international consortium consisting of researchers from Europe, Japan, Korea, and the USA, has summarized that the energy efficiency indicators of manufacturing on a national or sectional level have been defined, but sustainable process management solutions for companies have not been effectively implemented, and the research is highly imperative (EU FP7 project IMS2020 [4]). Machining such as milling is one of the important manufacturing processes. Cooperations between machining companies and their customers are more project-specific, customer-centric and flexible; the jobs and orders are likely to be diversified and many of them are urgent. As thus, there are many uncertainties and adjustment requirements in shop floors as part of the day-to-day operation planning in companies [5]. However, effective process planning and scheduling solutions, which are adaptive to dynamics in both the machining process and the machine system levels, and multiple criteria such as sustainability, product quality, and productivity are systematically incorporated in the solutions, are lacking.

To address the above issue, an innovative approach of sustainable process planning and scheduling for machining multiple parts using multiple computer numerical control (CNC) machines has been developed. The approach focuses on the milling process and addresses dynamics in the process from the following two aspects: (1) It optimizes the key milling parameters of individual machines for producing individual parts to meet constraint-based multiple objectives, in terms of energy efficiency, surface quality, and productivity; and (2) based on the optimized milling process parameters, an optimized solution of process sequencing, setting-up, and scheduling for machining multiple parts using multiple candidate machines in a shop floor is achieved by considering the criteria of energy consumption and makespan of the machine system.

The innovations of the approach are summarized below:

- The approach provides a systematic, adaptive, and efficient means to optimize machining companies' multi-objectives, such as sustainability, productivity, and makespan, and to meet technical constraints such as the required surface quality and precedence constraints among machining features and operations;
- Machining feature-based sustainable process planning and scheduling is highly desirable as machining features have been used as essential building blocks in modern Computer Aided Manufacturing (CAM) software. This approach supports intelligent decision-making processes for feature-based sustainable process planning and scheduling, and based on that a practical way is paved for the approach to be integrated into modern feature-based CAM systems.

The rest of the chapter is organized as follows. In Sect. 4.2, a literature survey on sustainable machining processes especially milling processes is given. In Sect. 4.3, the system framework of the research is presented. In Sect. 4.4, the constraint-based multi-objective optimization of key milling process parameters is presented. Based on the optimized parameters of individual machines for individual parts, the multi-objective optimization process of a machine system in a dynamic shop floor is described in Sect. 4.5. In Sect. 4.6, case studies and experimental tests are described. Finally, the research is concluded in Sect. 4.7.

4.2 Related Work

In the past decades, research on manufacturing process planning and scheduling has been extensively conducted, and comprehensive surveys can be found from [6]. This chapter focuses on energy efficient process planning and scheduling, and the related state-of-the-art research is summarized below.

4.2.1 Energy Consumption Modeling Based on Key Machining Parameters

The European Machine Tool Builder Association indicates that the machine tool industry has shown strong interests on developing energy efficient manufacturing systems. To support the industry to achieve sustainability, a self-regulatory initiative for identification of measurements for energy performance and resource efficiency of machine tool systems has been proposed by the Association [7]. Aiming at implementing the initiative effectively, researchers have been actively investigating the energy consumption profile of machine tool systems during execution and identifying the key process parameters that affect the consumption profile.

Based on that, optimization strategies are applied for process and system improvement in terms of energy saving.

Abele et al. summarized the total energy demand of a machine tool system during production as: $E_{total} = E_{th} + E_{additional} + E_{periphery}$, where E_{th} is the active energy theoretically needed to obtain the physical process effect, and $E_{additional}$ and $E_{periphery}$ stand for the additional energy demands of the machine tool (e.g., energy to cover efficiency losses or energy for machine functions such as central control) and peripherals (e.g., cutting fluid pump), respectively [8]. Among the energy consumption of a machine tool system, the unit energy consumption demand of a machining process is remaining a challenging research issue. Gutowski et al. [9] classified related energy consumption of manufacturing into the following categories:

- Fixed energy: energy demand of all activated machine components ensuring the operational readiness of the machine;
- Operational energy: energy demand to distinctively operate components enabling the cutting as performed in air-cuts;
- Tool tip energy: energy demand at tool tip to remove the workpiece material;
- Unproductive energy: energy converted to heat mainly due to friction during the material removal.

A series of research work were carried out to detail the energy profile for the aforementioned categories. A summary of the work is given in Table 4.1. Mori et al. [10] developed an empirical model, in which several processes are considered such as positioning and acceleration of the spindle, tool changes, machining, and stop of the spindle. Newman et al. [11] developed empirical models to establish the relationship between cutting parameters, such as depth of cut, feedrate and number of cuts, and power consumption. Two case studies of finish cutting and semifinish cutting of aluminum were used to verify the models. In [12], a torque sensor was mounted onto the cutter and active power consumed by a machining process was calculated, while the total input power to the machine tool system was measured by a power sensor. Based on experimental data, an empirical model was established to estimate the total power and active power for machining, which are used to support the online monitoring system. The Taguchi method was introduced to analyze the relationship among cutting parameters, energy consumption, and surface roughness in order to determine the suitable cutting parameters leading to the minimum energy consumption and the best surface roughness [13]. A Grey Relationship Analysis method was developed for establishing relationships among material removal rate (MRR), machining power, and surface roughness minimization; the response surface methodology (RSM) and the Taguchi method were used for factor effect analysis [14]. Winter et al. investigated the energy performance of a grinding process [15]. The Sensitivity Analysis method was applied to illustrate how cutting parameters, including cutting depth, cutting speed, and dressing speed, affect the energy consumption in order to achieve multi-objective optimization.

Table 4.1 Energy consumption models for machine tool systems

References	Input variables						Optimization objectives				Research methods
	Depth of cut	Spindle speed	Cutting speed	Width of cut	Chip load	M-features	Cutting power	Roughness	Processing time	Other	
[10]		×	×	×			×				Empirical models for case studies of cutting condition changes and deep hole drilling
[16]	×	×	×	×		×	×				Empirical models for usage stages of machining
[17]	×	×	×	×		×	×		×		Empirical models for start-up, idle and usage stages of machining
[11]		×	×		×		×		×		Empirical models and two case studies on semifinish and finishing machining
[12]	×	×	×				×				Least square method (LSM) for machining
[18]	×	×	×	×			×				Empirical models for start-up, idle and usage stages of *machining*
[13]	×	×	×				×	×			Orthogonal array, signal-to-noise (S/N) ratio and analysis of variance (ANOVA)
[14]	×	×	×	×			×	×			Grey Relationship Analysis, response surface methodology (RSM) and the Taguchi method

(continued)

Table 4.1 (continued)

References	Input variables						Optimization objectives				Research methods
	Depth of cut	Spindle speed	Cutting speed	Width of cut	Chip load	M-features	Cutting power	Roughness	Processing time	Other	
[19]	×	×	×	×			×				Empirical models for machining shop floor
[20]	×	×	×	×				×			A hybrid genetic algorithm for sustainable machining optimization
[15]	×		×				×	×		×	Sensitivity analysis method
[21]						×	×		×	×	Assessment of alternative tool paths, identified major opportunities for energy reduction

4.2.2 Energy Consumption Modeling Based on Specific Energy Consumption

The method of the above research is to design and conduct experimental tests to reveal the underlying relationship between the energy performance of a machine tool system and key cutting parameters, qualitatively and quantitatively. Another group of research focuses on developing empirical models based on MRR and specific energy consumption (SEC) to model and estimate the unit process energy consumption of a machining process. The related work is summarized in Table 4.2. The most representative model was developed in [9]. The specific energy requirements for manufacturing processes, i.e., SEC, were modeled as a function of MRR in an energy framework. SEC is defined as the energy consumption in cutting 1 cm^3 material. However, in the model, the specifications for the fixed power P_0 and the constant k were not given. To improve this model, researchers developed enhanced energy consumption models. For instance, Li and Kara used an empirical modeling approach to develop a unit process energy consumption model to characterize the relationship between SEC and machining parameters, and the coefficients in the model were decided through experimental tests [22]. Furthermore, the empirical approach was applied to turning, milling, and grinding processes on different machine tools, [23] focused on turning process while [24] focused on grinding process. Similarly, a case study of a micromachining center was developed in [25] to model the SEC in cutting. This model further confirms the relationship between the energy consumption and MRR. In addition, Li et al. [26] adopted a hybrid modeling method based on thermal equilibrium and empirical modeling to characterize the relationship between process variables and energy consumption for milling processes and experimental tests were conducted to identify the energy-related coefficients for a specific machine. Yan and Li developed the grey relational analysis method to model the relationship between multi-objectives (including energy consumption, production rate, and cutting quality) and key machining parameters (including spindle speed, feedrate, depth of cut, and width of cut) [14]. Meanwhile, the Taguchi method was applied to analyze the influence of machining parameters on the multi-objectives in a qualitative way in order to identify a trade-off among the energy consumption, production rate, and cutting quality based on different combinations of machining parameters.

4.2.3 Energy Consumption Modeling Based on CNC Codes

The set of CNC codes can describe an entire machining process and the working process of the related accessory equipment, and therefore, the entire energy profile can be modeled and estimated. Based on the tool paths from CNC codes (e.g., G-code), the energy consumption assessment for the spindle axis, feed axis, and load/unload cycle was formulated [16]. Based on an established energy consumption

Table 4.2 SEC-based energy models for machining processes

Works	Models or methods
Gutowski et al. [9]	$SEC = P_0/v + k$, where P_0 is the fixed power and k is a constant with units of kJ/cm^3, v is the rate of material processing in cm^3/sec
Kara and Li [23]	$SEC = C_0 + C_1/MRR$, where the coefficients C_0 and C_1 are different among different machine tools and needed to be experimentally determined; MRR is the material removal rate
Diaz et al. [25]	$SEC = k/MRR + b$, where the constant k is related to the unit of power and b represents the steady-state-specific energy
Li et al. [26]	$SEC = k_0 + k_1 \cdot n/MRR + k_2/MRR$, where k_0 is the specific energy requirement in cutting operations, k_1 is the specific coefficient of spindle motor, k_2 is the constant coefficient of machine tools and equals the sum of standby power and the spindle motor's-specific coefficient; n is the spindle speed in rounds/second

model from tool paths, a Web-based system was developed for environmental sustainability monitoring and evaluation [17]. In the system, tool path generation strategies for different machining features were modeled in terms of energy consumption. Based on CNC codes, an energy assessment framework for machining workshop was built up [27]. The energy consumption framework consists of four layers, i.e., workshop layer, task layer, manufacturing unit layer, and machine tool layer. In each layer, major elements affecting energy consumption were considered. In the machine tool layer, machining tasks are executed and the primary energy consumption comes from a machining workshop. In the manufacturing unit layer, the layout and sequence of a set of machine tool systems are designed. The task layer and workshop layer are mainly for particular task's planning and scheduling, and the venue for the manufacturing tasks to take place, in which electricity, heating, ventilation, and air-conditioning equipment are the major energy consumption units. In the machine tool system, cutting force, cutting velocity, cutting depth, and feedrate were used as inputs to establish an empirical model for energy consumption estimation during machining. Some calculation processes in these works are summarized in Table 4.3.

Although a large amount of research works have been reported as above, the following research issues are highly expected for further investigation:

- Modern machining software packages have adopted machining features as essential building blocks, and the energy consumption assessment associated with machining features and the process parameters will be more effective to support decision making in sustainable process planning and scheduling. Research on machining feature-based sustainable process planning was still preliminary, further research is imperative to understand the characteristics of energy consumption influenced by machining features/operations and key process parameters;

Table 4.3 CNC-based energy models for machining processes

Works	Model and methods
Avram and Xirouchakis [16]	$E = E_{as} + E_{\text{run}} + E_{\text{cut}} + E_{ds} = \int_{t0}^{t1} P_{as}\text{d}t + \int_{t1}^{t3} P_{\text{run}}\text{d}t + \int_{t2}^{t3} P_{\text{cut}}\text{d}t + \int_{t3}^{t4} P_{ds}\text{d}t$, where E_{as} and P_{as} are the energy and power requirements for spindle, respectively, E_{run} and P_{run} are the energy and power requirements for the motors before engaging the material cutting, E_{cut} and P_{cut} are the energy and power requirements for the material cutting, and E_{ds} and P_{ds} are the energy and power requirements on spindle unloading; t_0, t_1, t_2, t_3, t_4 are the time spent on the above stages, respectively
Diaz et al. [25]	$E = P_{\text{avg}} * \Delta t = (P_{\text{cut}} + P_{\text{air}}) \cdot \Delta t$, where P_{avg} is the average power demand and composed of a cutting power P_{cut} and air-cutting power P_{air}; Δt is the processing time
Kong et al. [17]	$E_{\text{machining}} = E_{\text{const}} + E_{\text{run-time}} + E_{\text{cut}}$, and $E_{\text{cut}} = K_{\text{cut}} \cdot w \cdot b \cdot z^p \cdot v_f^{1-p} \cdot n^p$, where $E_{\text{machining}}$, $E_{\text{run-time}}$, and E_{cut} represent the total energy of machining process, constant energy consumed by the functions that are not directly related to the machining, run-time energy consumed by a spindle, machine axes and tool changer, and energy consumed by the material removal action of a machine tool, respectively. v_f is the feedrate, n is the spindle speed, w is the width of cut, z is the number of flutes of a cutter, p and K_{cut} are empirically determined fitting constants
Mori et al. [10]	$E = P_1 * (T_1 + T_2) + P_2 \cdot T_2 + P_3 \cdot T_3$, where P_1, P_2 and P_3 are constant, corresponding to the power demand of cutting, positioning the work and accelerating/decelerating the spindle to a specified speed, T_1, T_2, and T_3 are the corresponding times
He et al. [27]	$E = E_{\text{spindle}} + E_{\text{feed}} + E_{\text{tool}} + E_{\text{cool}} + E_{\text{fix}}$, where E_{spindle}, E_{feed}, E_{tool}, E_{cool}, and E_{fix} represent the energy consumed by spindle, feed, tool, cool, and fix
Newman et al. [11]	$E = P/fhD$, where E and P are the energy and power requirements for the milling process, respectively; f, h, and D stand for feedrate, depth of cut, and diameter of cutter, respectively
Balogun and Mativenga [18]	$E = P_b \cdot t_b + (P_b + P_r) \cdot t_r + P_{\text{air}} \cdot t_{\text{air}} + (P_b + P_r + P_{\text{cool}} + k \cdot v) \cdot t_c$, where P_b, P_r, P_{cool}, and P_{air} represent the basic and ready state powers, coolant pumping power requirements and the average power requirements for a non-cutting approach and retract moves over the component, respectively, t_b, t_r, and t_c are the basic, ready, and cutting times, respectively, t_{air} represents the total time duration of the non-cutting moves; k with units of kJ/cm^3 is the specific cutting energy, v in cm^3/s is the rate of material processing

- Many of the above research works are still preliminary and energy efficiency has not been systematically addressed in process planning and scheduling in a dynamic shop floor. It is critical to develop systematic, adaptive and efficient approaches to address multiple performance criteria and technical constraints such as productivity, surface quality, makespan, and precedence constraints among machining features/operations from both the process level and the system level.

4.3 System Framework

Some essential considerations in process planning and scheduling are [28]:

- Generating optimized process parameters of a part machined on a machine to meet desired functional specifications and achieve good manufacturability;
- Determining the machining feature/operation sequence, set-up plan and schedule according to performance criteria and precedence constraints. Process sequencing means a set of machining features/operations will be sequenced according to some performance criteria such as productivity and quality and constrained by some technical or geometrical precedence constraints among machining features/operations. A set-up can be generally defined as a group of features/operations that are manufactured on a single machine using the same fixture. The scheduling task is to assign the parts and their machining features/operations to specific machines to be executed in different time slots, targeting at good shop floor performance, such as the shortest makespan and the total lowest energy consumption in the shop floor.

In a dynamic machining situation, a part can be manufactured using different process parameters and on different candidate machine systems, which generate different process plans and schedules. In summary, a group of alternative process plans and schedules can be generated using three strategies: machine tool flexibility, process sequencing and setting-up flexibility, and schedule flexibility [29]. Machine tool flexibility refers to the possibility of performing a feature/operation on alternative machine tool systems. Process sequencing and setting-up flexibility corresponds to the possibility of changing the sequence and set-up in which the features/operations are performed. Meanwhile, for a group of parts, alternative schedules can be created based on scheduling flexibility, which relates to the possibility of arranging different schedules to manufacture the features/operations of the parts to achieve the shortest makespan, lowest energy consumption, and/or better performance for other shop floor indicators.

A Gantt chart has been popularly used to represent a schedule of a group of parts, as illustrated in Fig. 4.1. In the Gantt chart, the order in which the parts and their operations are carried out is laid out and the dependencies of the tasks are managed. The *X*-axis of the Gantt chart represents time. Each row in the *Y*-axis represents a machine and the specific arrangement of the operations of the parts on the machine. A machine is comprised of a number of time slots, which can be further classified into idle time slots, preparation time slots for machining operations (further including the set-up time and/or the tool change time), and machining time slots of operations.

Based on the above, in this research, a two-stage optimization approach is proposed, detailed below (illustrated in Fig. 4.2):

(1) The first stage is to optimize the key parameters of a milling process for a part in a single machine to achieve a better energy efficiency and productivity while the surface quality requirement is met. These performance indicators in

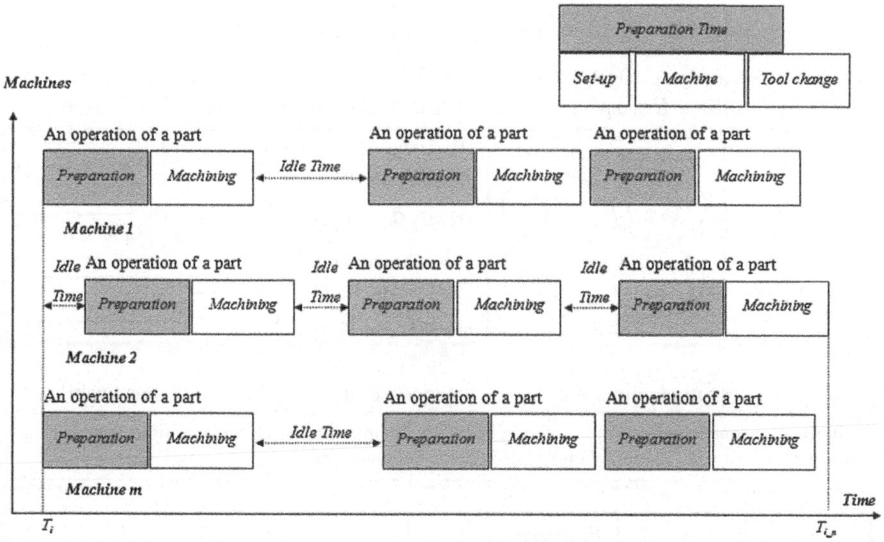

Fig. 4.1 A Gantt chart for scheduling parts and their machining features/operations

machining, i.e., energy efficiency for production, surface quality, and productivity, are incorporated into a constraint-based multiple-objective optimization problem, while critical process parameters affecting the performance indicators, including spindle speed, cutting speed, depth of cut, and width of cut, are taken into account as variables to support the above optimization modeling and processing. The relationships between the variables and the energy efficiency/surface quality could be highly nonlinear. As thus, artificial neural networks (ANNs) are employed in order to present the nonlinear relationship among the variables and performance indicators adaptively and effectively. Intelligent algorithms are applied to identify optimized process parameters for individual parts on individual machines;

(2) Based on the above optimized results of individual machines for individual parts, the second stage of the approach is to identify optimized process sequence, set-up, and schedule with multiple machines for manufacturing multiple parts. In this research, the energy efficiency and makespan of a machine system are integrated as multiple optimization objectives, and precedence constraints among features/operations are considered. Intelligent algorithms are then applied to determine an optimized process plan and schedule. In order to effectively generate a comprehensive search space to support the optimization processes, the aforementioned three strategies, i.e., machine tool flexibility, process sequencing, and setting-up flexibility and scheduling flexibility, are used for the generation of alternative process plans and schedules as a feasible search space to support the above optimization process.

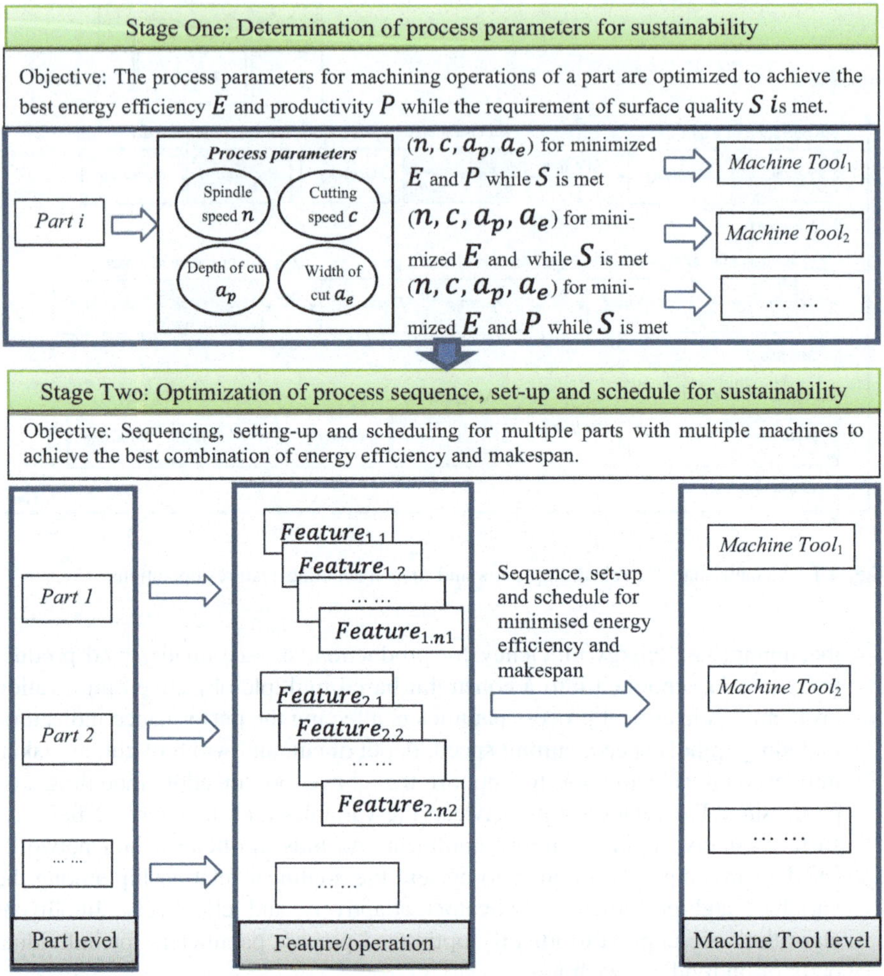

Fig. 4.2 Two stages of sustainable process planning and scheduling optimization

4.4 Process Parameter Identification for Sustainability

Key parameters in milling, such as spindle speed, cutting speed, depth of cut, and width of cut, affect the performance of a milling process, such as energy consumption, surface quality, and productivity. In the following, based on the relationship between the key process parameters and the performance of a milling process, a normalization process and an optimization process have been developed to ensure good machining process in terms of energy efficiency, surface quality, and productivity.

4.4.1 Process of Parameter Identification

Milling, which is a primary process in machining, is considered in this research. The energy performance, surface quality, and productivity of a machining process can be evaluated using three indicators, i.e., energy consumption (*E*), surface roughness (*S*), and machining removal rate (MRR). These indicators interlace each other, and a better performance of one indicator could need trade-off of the other indicators [14]. On the other hand, key parameters of a milling process including spindle speed (*n*), cutting speed (*c*), depth of cut (a_p), and width of cut (a_e) affect these performance indicators significantly.

Given the surface roughness *S* is pre-decided by users as a constraint, optimization of energy consumption *E* and machining removal rate MRR are modeled as a constraint-based multi-objective optimization problem, and the four process parameters (n, c, a_p, a_e) are considered as variables in the optimization problem. The target is to obtain optimized multi-objectives *E* and MRR while the pre-set *S* (denoted as μ) as a constraint is met. Upon the completion of optimization, the values of (n, c, a_p, a_e) within their working ranges, which meet the above optimized objectives and constraint, are identified. In the process, ANNs have been constructed for representing the relationships between (n, c, a_p, a_e) and *S*, and between (n, c, a_p, a_e) and *E*, respectively. The grey relationship analysis approach has been employed to normalize *E* and MRR in the formation of a multiple-objective target function. Several optimization algorithms have been applied for identifying the optimal values of the indictors and process parameters. The process is illustrated in Fig. 4.3, and the details are explained below.

4.4.2 Representation of Energy Consumption (E), Surface Roughness (S) and Machining Removal Rate (MRR), and Their Grey Relational Analysis Processes

To develop an optimization model of a machining process, spindle speed (*n*), cutting speed (*c*), depth of cut (a_p), and width of cut (a_e) are used to represent energy consumption (*E*), surface roughness (*S*), and machining removal rate (MRR). In this research, the units for spindle speed, cutting speed, depth of cut and width of cut, energy consumption, surface roughness, and machining removal rate are rounds/minute (RPM), mm/minute, mm, mm, kilo-Joules (KJs), μm, and mm^3/minute, respectively.

The MRR for a milling process can be calculated below:

$$\mathrm{MRR} = c \cdot a_p \cdot a_e \tag{4.1}$$

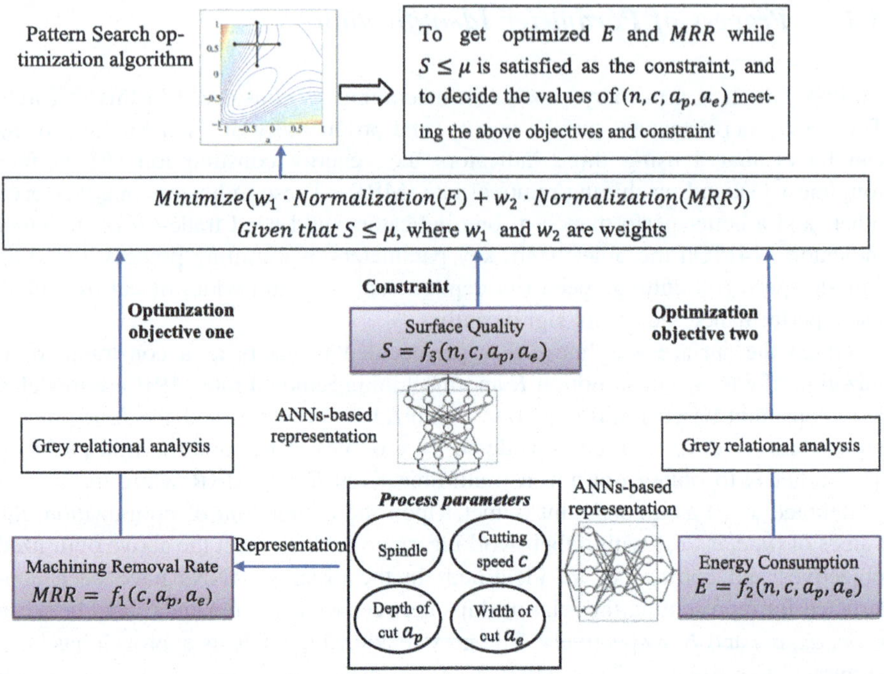

Fig. 4.3 Optimization of milling indicators and process parameters

The relationship between the process parameters and the energy consumption (E) or surface roughness (S) is highly nonlinear and an analytical solution is not easy to extract. In this research, the representation of E or S between the measured data set of (n, c, a_p, a_e) is constructed using a multi-layer feed-forward (MLFF) ANNs trained using a back-propagation (BP) algorithm (shown in Fig. 4.4). ANNs offer several valuable characteristics: (1) The ability to capture and represent complicated input/output relationships; (2) no prior knowledge about the input and output mapping is required for the model development. Unknown relationships are inferred from the data provided for training. Therefore, with ANNs, the fitting function is represented by the networks and does not have to be explicitly defined; and (3) the ability for generalization, meaning they can respond correctly to new data that have not been used for the ANNs model development [28].

The grey relational analysis process [30] consists of two steps. The first step is the normalization of the original sequences in the range between zero and one, and the second step is to calculate the grey relational coefficient to express the relationship between the ideal and actual normalized experimental results. Details are described below.

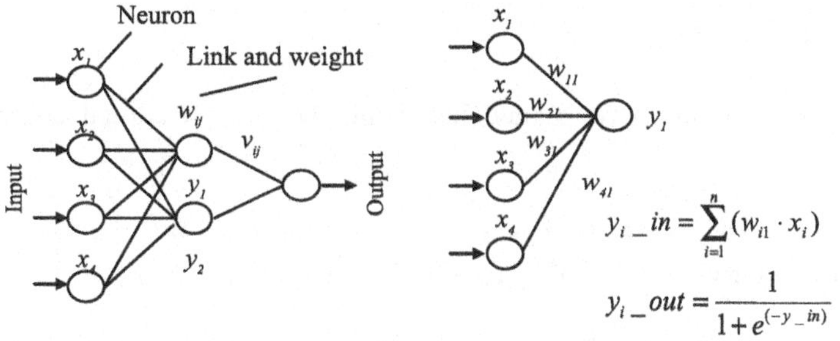

Fig. 4.4 MLFF ANNs for energy consumption and roughness representation

1. Normalization:

Since the final optimization is to find the minimal value of objective function which is the sum of the grey relational coefficients of E and MRR, and the expectancy of energy consumption E is the smaller the better, the original sequence of E_i should be normalized as:

$$Z_i' = \frac{E_i - \min(E_i)}{\max(E_i) - \min(E_i)} \tag{4.2}$$

where E_i $(i = 1, 2, \ldots, n)$ is the energy value of a sample in an experimental set, and n is the number of the experimental set; on the contrast, the expectancy of productivity MRR is the larger the better. The original MRR should be normalized as:

$$Z_i'' = \frac{\max(\text{MRR}_i) - \text{MRR}_i}{\max(\text{MRR}_i) - \min(\text{MRR}_i)} \tag{4.3}$$

where MRR_i $(i = 1, 2, \ldots, n)$ are the MRR values of a sample in an experimental set, and n is the number of the experimental set.

2. Calculation of the grey relational coefficients for E and MRR:

$$\text{GRC}_j = \frac{\Delta_{\min} + w \cdot \Delta_{\max}}{\Delta_j + w \cdot \Delta_{\max}} \tag{4.4}$$

where $\Delta_j = \|Z_0 - Z_j\|$, $\Delta_{min} = \min_{1 \le j \le n} \Delta_j$, $\Delta_{min} = \max_{1 \le j \le n} \Delta_j$, and $w \in [01]$. Usually $w = 0.5$ is used.

4.4.3 Optimization Process

The optimization objective is modeled using the grey relational coefficients as below:

$$\begin{cases} \min(w_1 \cdot \mathrm{GRC}_E + w_2 \cdot \mathrm{GRC}_{\mathrm{MRR}}), & S \leq \mu \\ \min\left(w_1 \cdot \mathrm{GRC}_E + w_2 \cdot \mathrm{GRC}_{\mathrm{MRR}} + \left(\frac{S}{\mu} - 1\right) \times 100\right), & S > \mu \end{cases} \quad w_1 + w_2 = 1 \tag{4.5}$$

The constraint $S \leq \mu$ *(μ the user-defined surface roughness) is modeled in the objective function as a penalty*

Bounds *the upper and lower bounds of input variables* (n, c, a_p, a_e) *are limited by the maximum and minimum values of the measurement samples* $\mathrm{MRR} = f_1(c \cdot a_p \cdot a_e)$, $E = f_2(n, c, a_p, a_e)$, *and* $S = f_3(n, c, a_p, a_e)$

w_1 and w_2 are the user-defined weights for energy consumption and productivity, respectively. For instance, if only energy consumption is concerned, then set $w_1 = 1$ and $w_2 = 0$. Usually both indicators are taken into account by setting balanced weights with $w_1 = 0.5$ and $w_2 = 0.5$. A set of optimization algorithms, including pattern search, genetic algorithm, and simulated annealing algorithm, are applied to this problem. Optimization results show that the pattern search method exhibits a better computational efficiency and a more reliable optimization performance for this case. Thus, the pattern search method is introduced herewith. Pattern search belongs to direct search for solving optimization problems that does not require the gradient of the objective function. It would iterate from search, polling, and expanding/contracting processes until the optimal result is found. The detailed procedure is as follows:

1. Choose an initial vector point x_0 and define the pattern vectors. For a problem with four input variables, there are total eight pattern vectors as: $v_1 = [1 \ \ 0 \ \ 0 \ \ 0] \cdots v_4 = [0 \ \ 0 \ \ 0 \ \ 1]$, $v_5 = [-1 \ \ 0 \ \ 0 \ \ 0] \cdots v_8 = [0 \ \ 0 \ \ 0 \ \ -1]$;
2. Search for a mesh point x_i around x_0 that has a less objective function compared to x_0. The search mesh is generated as $x_i = x_0 + \Delta_i$, where $\Delta_i = \Delta m \cdot v_i$, Δm is the current mesh size, the upper and lower bounds for each of the variables are to be checked;
3. If a better solution x is found, the poll is successful, update the vector point $x_0 = x$ and increase the mesh size: $\Delta m = 2 \cdot \Delta m$; otherwise, keep the original x_0 and reduce the mesh size: $\Delta m = 0.5 \cdot \Delta m$;
4. Check if any of the stop conditions (the mesh size is less than mesh tolerance or the difference between the function value at the previous best point and at the current best point is less than the value of function tolerance) is met, if yes, stop the optimization. Otherwise, go to above Step 2.

4.5 Sustainable Optimization for Machining Systems

There are alternative sequences and set-ups between machining features/operations and a group of machines available as candidate resources for scheduling. In the following, the energy consumption modeling for features/operations in a manufacturing system is built, followed by an optimization process for the model. During the processes, the sequences of machining features/operations are constrained by some technical or geometrical requirements of parts, which are handled in the optimization process by introducing a penalty function [28].

4.5.1 Energy Consumption Modeling

For a machine, its energy power profile is illustrated in Fig. 4.5, which consists of start-up phases, idle/change phases, working phases (operation), and shutdown phases. Hence, the energy consumption of a machine can be separated into the corresponding four segments.

The energy consumption during the idle phase can be formulated as:

$$E(M_i)_{\text{idle}} = P_{\text{idle}}(M_i) \cdot T_{\text{idle}}(M_i) \tag{4.6}$$

where $E(M_i)_{\text{idle}}$ stands for the energy consumption of the idle phase of the *i*th machine; $P_{\text{idle}}(M_i)$ is the power demand of the *i*th machine during the idle phase, which is the sum of the power demand of all the components in the machine and usually a constant value for the particular machine; $T_{\text{idle}}(M_i)$ stands for the total idle time of the *i*th machine, which can be represented as:

$$T_{\text{idle}}(M_i) = \sum_{j=1}^{n} T_{j,j+1}(O_{j,i}, M_i) \tag{4.7}$$

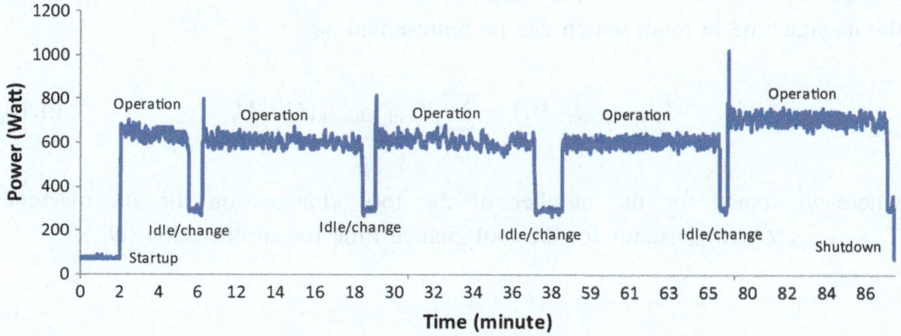

Fig. 4.5 Different phases of energy consumption of operations in a machine

where n stands for the number of the operations denoted as $O_{j,i}$ to be executed on the ith machine; $T_{j,j+1}(O_{j,i}, M_i)$ stands for the idle time between the jth and $(j+1)$th operations to be executed on the ith machine. $T_{j,j+1}(O_{j,i}, M_i)$ can be obtained using the following pseudocodes.

```
If T_{j.end}(O_{j,i}, M_i) < T_{j+1.start}(O_{j+1,i}, M_i) then
    // The j^th operation ends before the (j + 1)^th operation
    // Operation and the machine will be idle
    T_{j,j+1}(O_{j,i}, M_i) = T_{j+1.start}(O_{j+1,i}, M_i) - T_{j.end}(O_{j,i}, M_i)
Else            // There is no waiting
    T_{j,j+1}(O_{j,i}, M_i) = 0
Endif
```

Here, $T_{j.\text{end}}(O_{j,i}, M_i)$ and $T_{j+1.\text{start}}(O_{j+1,i}, M_i)$ stand for the end time of the jth operation and the start time of the $(j+1)$th operation, respectively.

The energy consumption during the working phase can be represented as:

$$E(M_i)_{.\text{working}} = \sum_{j=1}^{n} E_j(O_{j,i}, M_i)_{.\text{working}} \quad (4.8)$$

where $E(O_{j,i}, M_i)_{.\text{working}}$ stands for the energy consumption of the working phase for the ith machine; n stands for the number of the operations denoted as $O_{j,i}$ to be executed on the ith machine; $E_j(O_{j,i}, M_i)_{.\text{working}}$ stands for the energy consumption of the jth operation on the ith machine.

The energy consumption of the tool change phase can be computed as:

$$E(M_i)_{.\text{tool\_change}} = P_{\text{idle}}(M_i) \cdot T_{\text{tool\_change}}(M_i) \quad (4.9)$$

where $E(M_i)_{.\text{tool\_change}}$ represents the energy consumed during the tool change phase; P_{idle} is the idle power needed during the machine waiting phase for this tool change on the ith machine; $T_{\text{tool\_change}}(M_i)$ stands for the total tool change time on the ith machine in total, which can be represented as:

$$T_{\text{tool\_change}}(M_i) = \sum_{j=1}^{m} T_{\text{tool\_change}}(O_{j,i}, M_i) \quad (4.10)$$

where m stands for the number of the tool changes on the ith machine; $T_{\text{tool\_change}}(O_{j,i}, M_i)$ stands for the tool change time for an operation $(O_{j,i})$.

The energy consumption of the set-up phase can be computed as:

$$E(M_i)_{.\text{set-up}} = P_{\text{idle}}(M_i) \cdot T_{\text{set-up}}(M_i) \tag{4.11}$$

where $E(M_i)_{.\text{set-up}}$ represents the energy consumed during the set-up phase; $P_{\text{idle}}(M_i)$ is the idle power needed during the machine waiting phase for this set-up; $T_{\text{set-up}}(M_i)$ stands for the total set-up time on the *i*th machine, which can be represented as:

$$T_{\text{set-up}}(M_i) = \sum_{j=1}^{p} T_{\text{set-up}}\left(O_{j,i}, M_i\right) \tag{4.12}$$

where p stands for the number of the set-up on the *i*th machine; $T_{\text{set-up}}\left(O_{j,i}, M_i\right)$ stands for the set-up time for an operation $(O_{j,i})$ on the *i*th machine.

Based on the energy consumption of the above phases, the total energy consumption of a machine can be represented below:

$$E(M_i) = E(M_i)_{\text{idle}} + E(M_i)_{.\text{working}} + E(M_i)_{.\text{tool\_change}} + E(M_i)_{.\text{set-up}} \tag{4.13}$$

where $E(M_i)$ stands for the total energy consumption of the *i*th machine.

Therefore, if there are n machines to be used in the process planning and scheduling, the overall energy consumed by all the machines to machine all the parts is:

$$E_{total} = \sum_{i=1}^{n} E(M_i) \tag{4.14}$$

Makespan means the maximum interval time spent to machine all the parts. It can be defined in the following:

$$\text{Makespan} = \max(T(M_i)) \tag{4.15}$$

where $T(M_i)$ is the time interval between the stop time of the *i*th machine and the start time of the entire job. It includes the start time and utilization time of the *i*th machine, which includes idle, working, tool change, and set-up phases. That is, $T(M_i)$ can be represented as:

$$T(M_i) = T_{\text{start}}(M_i) + T_{\text{idle}}(M_i) + T_{\text{working}}(M_i) + T_{\text{tool\_change}}(M_i) + T_{\text{set-up}}(M_i) \tag{4.16}$$

4.5.2 Optimization Process

As the two different objective functions, i.e., total energy and makespan, can have very different magnitudes, normalization of the two objective functions is required prior to the optimization of the weight summed objective function. Unlike the optimization of milling parameters in terms of energy consumption E and productivity MRR described in Sect. 4.4.2, which maximum and minimum values are already known, the maximum and minimum values of these two objective functions are unknown before optimization. In this case, a suitable normalization schema that normalizes the objective functions by the differences of objective functions in the Nadir and Utopia points is employed [31]. The Utopia point z_i^U provides the lower bound of the ith objective function and can be obtained by minimizing the ith objective function individually, i.e.,

$$z_i^U = f_i\left(x^i\right) = \min\{f_i(x)\} \tag{4.17}$$

The upper bound is then obtained from the Nadir point z_i^N, which is defined as:

$$z_i^N = f_i\left(x^k\right) = \max_{1 \ll j \leq I}\left\{f_i\left(x^j\right)\right\} \tag{4.18}$$

where I is the total number of objective functions.

This normalization schema may be computationally expensive when the problem dimension is very large. For this research, the time spent on this calculation is acceptable as the number of optimization parameters is not very large. Hence, the energy consumption and timespan are to be normalized individually as:

$$\begin{cases} NE = \left(E_{\text{total}} - z_1^U\right) / \left(z_1^N - z_1^U\right) \\ NT = \left(\text{Makespan} - z_2^U\right) / \left(z_2^N - z_2^U\right) \end{cases} \tag{4.19}$$

The objective function is calculated as weighted sum of the two objectives:

$$\text{Objective: } \min(w_1 \cdot NE + w_2 \cdot NT), \quad w_1 + w_2 = 1 \tag{4.20}$$

Unlike the optimization stage of the process parameters in Sect. 4.3, the pattern search method is not suitable for this type of problem as all the variables need to be optimized are discrete values. The performances of the genetic algorithm and simulated annealing algorithm are then compared and the simulated annealing algorithm is proven to be more reliable in finding the global optimum. As thus, the simulated annealing algorithm is employed here.

The geometric and manufacturing interactions between machining features/operations and the technological requirements in parts are considered to generate some precedence constraints between the machining features/operations. The definitions and classifications of precedence constraints between machining features/operations can be found in [28]. A penalty function for handling the precedence constraints is used in the optimization process.

4.6 Case Studies

Two parts shown in Fig. 4.6 provided by a medium-sized manufacturing company have been used to validate the developed approach. Both parts are imported as STEP files and a developed machining feature recognition processor has been applied to extract machining features [28]. The relevant specifications of the features in each part are listed in Tables 4.4 and 4.5, respectively. The assumptions for process planning and scheduling include the following:

- Parts are independent, and part preemption is not allowed;
- A penalty function is applied to the optimization process to ensure that the sequence of the operations generated for each part complies with precedence constraints;
- All parts and machines are available at time zero simultaneously;
- Each operation can be performed on multiple machines, and each machine can only execute an operation each time;
- Machines are continuously available for production;
- If a machine is broken down, or a new part is inserted, the algorithm can restart and generate new process plans and a schedule efficiently;
- The time for a set-up is assumed to be identical and independent of specific operations. For the milling operations on each machine, the cutter will be kept the same. From a milling operation to a drilling operation, a tool change will be made. The time for a machine change or a tool change is also assumed to be identical and independent of specific operations;
- This research is only for milling process parameter optimization at this moment. For the drilling features/operations in the parts, the energy consumption for the drilling process of each feature/operation is estimated and proportional to the volume of the feature.

Three CNC machines have been used as the candidate machines for this research validation. The first machine is the Hurco vertical machine center, and the other machines are Haas vertical machine centers VF-4 and VF-7.

(a) Part 1 and its recognized machining features

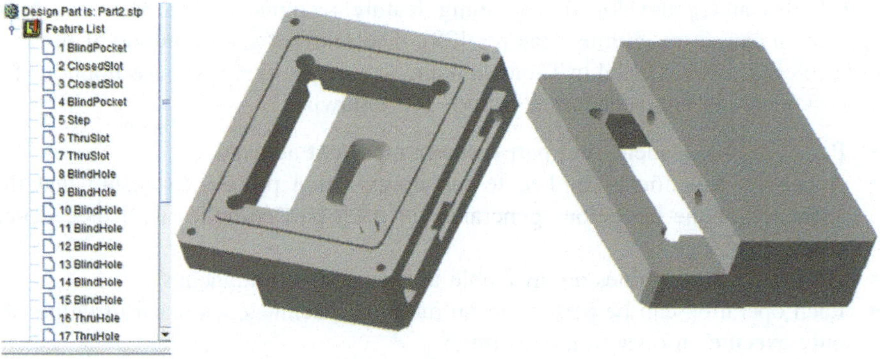

(b) Part 2 and its recognized machining features

Fig. 4.6 Feature lists of test parts with Part 1 having 14 features and Part 2 having 29 features

Table 4.4 Specifications of feature operations in Part 1

Features	Volume (mm^3)
1-BlindPocket	70,800
2-ClosedSlot	1360
3-ThruPocket	3590
4-ThruSlot	1536
5-ThruSlot	1408
6-ClosedSide	6984
7/8/9/10-BlindHole	28.5
11/12-SunkHole	154.3
13-Face	8375
14-Face	8375

Table 4.5 Specifications of feature operations in Part 2

Features	Volume (mm^3)
1-BlindPocket	798
2-ClosedSlot	420
3-ClosedSlot	2190
4-BlindPocket	16,200
5-Step	15,000
6/7-ThruSlot	86.6
8/9-BlindHole	67
10/11/12/13-BlindHole	23.75
14/15/-BlindHole	196.25
16/17-ThruHole	196.25
18/19/20/21/22-Face	2850/1080/4350/710/2850
23/24/28-Face	300/219.5/986
25/26/27/29-Face	40

4.6.1 Monitoring System Deployment

A wireless sensor network system for monitoring three-phase electricity consumption and a cloud-enabled data server to record and share data over the Internet has been developed and deployed in a shop floor. Energy information is measured from machines and transmitted as IPv6 packets to the data server using a wireless transport protocol 6LoWPAN. The sample rate of energy measurement is at 100 samples per second. The above system is illustrated in Fig. 4.7. The hardware photographs and software interfaces are shown in Fig. 4.8. Surface roughness is measured off-line.

4.6.2 Optimization of Milling Process Parameters

As described in Sect. 4.4.1, in order to optimize the milling parameters, ANNs are constructed to represent the relationships between the key milling parameters (n, c, a_p, a_e) and the measured surface roughness S, and between (n, c, a_p, a_e) and measured energy consumption E. To save space, only the measured data set from the Hurco vertical machine center is used here to illustrate the optimization procedure of milling process parameters. The lower and upper bounds and intervals of the milling parameters set for optimization are shown in Table 4.6. To train the ANNs properly, the measured data set is divided into three groups: 4 data as validation data set, 4 data as test data set, and the remaining 19 data as test data set. Three-layer ANNs with 6 hidden neurons are constructed to approximate the

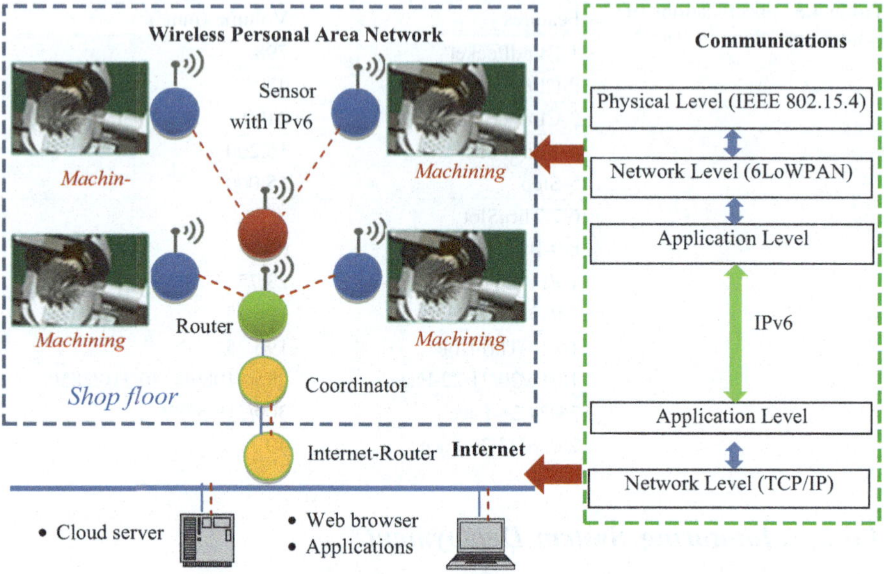

Fig. 4.7 Deployment diagram of the energy monitoring system

measured energy consumption as is shown in Fig. 4.9. The mean squared root error between the measured and predicted energy consumption values is 1.39 KJs, and the maximum error in percentage is 9.82% which occurs in the test set. Similarly, three-layer ANNs with 10 hidden neurons are employed to represent the surface roughness. The comparison of the predicted and measured surface roughness is depicted in Fig. 4.10, with a mean squared root error of 0.018 μm and a maximum error of 6.85%.

The ANNs-predicted energy consumption and calculated MRR are then pre-processed into grey relational coefficients to form the sum-weighted objective function for optimization. The constraint on surface roughness is accounted into the objective function as a penalty function when the ANNs-predicted surface roughness is greater than the value of the user-defined surface roughness.

Three optimization algorithms, i.e., pattern search, genetic algorithm, and simulated annealing algorithm, have been tested with the requirement of surface roughness S set at 2.5 μm. The optimization processes and results are shown in Fig. 4.11 and Table 4.7. It is observed that pattern search is the most stable and efficient among the three algorithms.

Surface roughness is not considered as an optimization objective directly in this work. Instead it is used as a constraint which can provide the flexibility to users to set the desirable surface roughness depending upon the process requirement of roughing, semifinish, and finish during process planning. As thus, the optimized

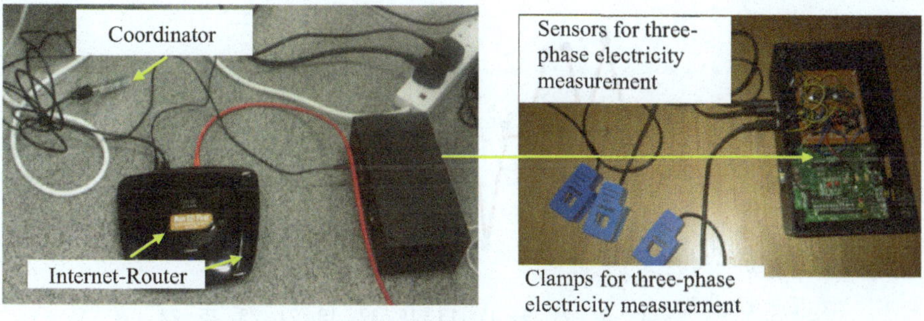

(a) Wireless sense network-based three-phase electricity consumption measurement system

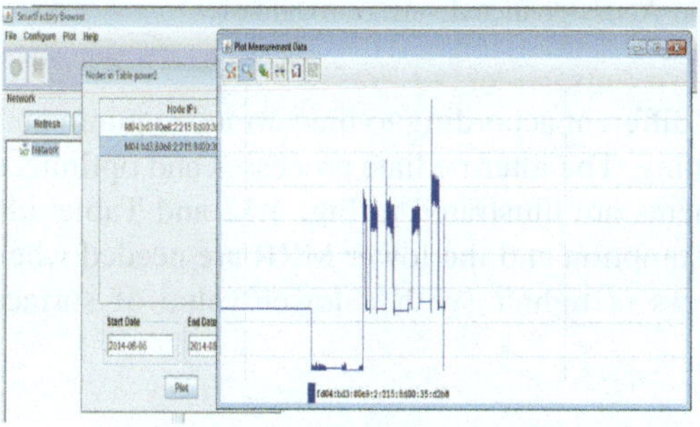

(b) Measured electricity consumption stored in Cloud server

(c) Surface roughness tester (off-line measurement)

Fig. 4.8 Energy monitoring system and surface roughness testing

Table 4.6 Bounds of the milling parameters for optimization

Milling parameters	n (r/min)	c (mm/min)	a_p (mm)	a_e (mm)
Lower bound	1000	200	0.2	5
Upper bound	2000	300	0.4	15
Interval	500	50	0.1	5

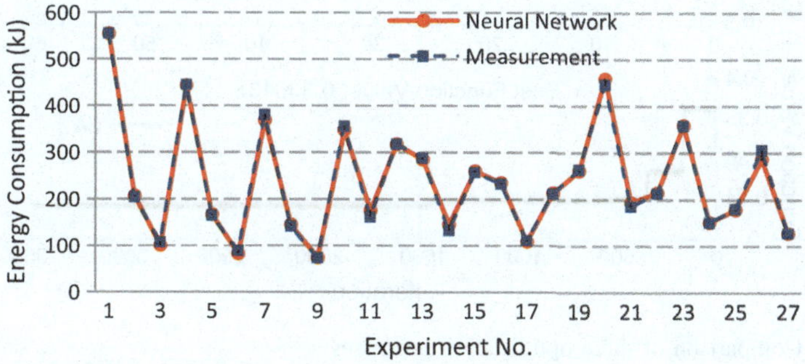

Fig. 4.9 Comparison of the measured and ANNs-predicted energy consumption

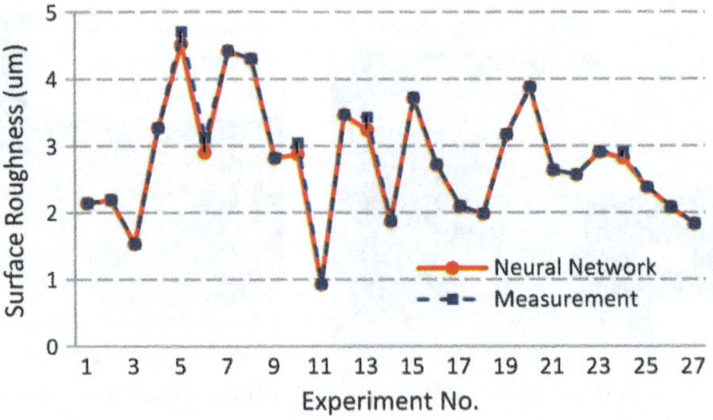

Fig. 4.10 Comparison of the measured and ANNs-predicted surface roughness

results of process parameters will be different according to the various requirements of surface roughness in process planning. The intermediate processes and optimized results for three roughness requirements are illustrated in Fig. 4.12 and Table 4.8. As expected, the greater energy consumption and the lower MRR are needed when the requirement on surface roughness is tighter (with a lower value of surface roughness).

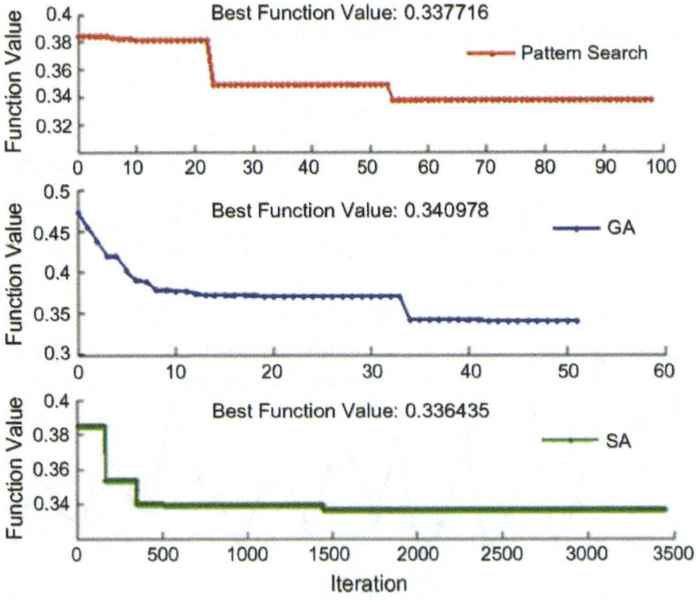

Fig. 4.11 Comparison of three optimization processes

Table 4.7 Result comparison of the three optimization algorithms

Algorithms	Average time (*s*)	Best solutions (3 trials with S $\leq$ 2.5 μm)		
Pattern search	9.23	0.3377	0.3377	0.3377
Genetic algorithm	24.25	0.3376	0.3406	0.3410
Simulated annealing	152.84	0.3364	0.3392	0.3409

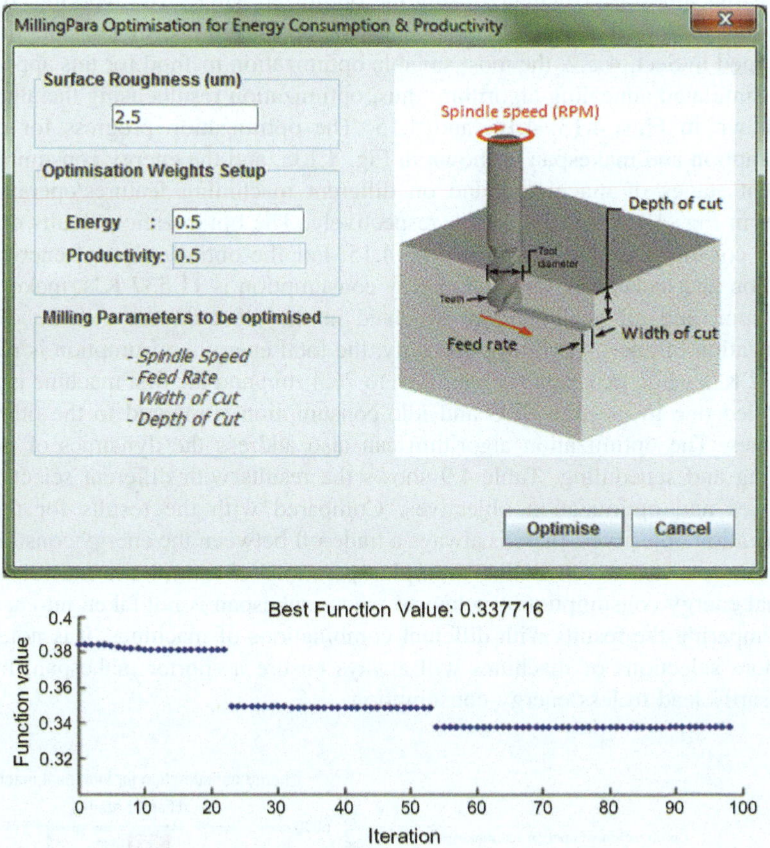

Fig. 4.12 Intermediate results of pattern search

4.6.3 Optimization of Process Sequencing, Setting-Up, and Scheduling

The optimized milling parameters of individual machines according to the roughness requirement are then recorded and the optimized MRR and energy consumption for individual machines are fed into the optimization of process planning and scheduling as inputs. The machining times of individual features/operations on

Table 4.8 Optimized process parameters for different constraints on surface roughness

R_a(μm)	n (r/min)	c (mm/min)	a_p (mm)	a_e (mm)	MRR (mm^3)	E (kJ)
2.5	1103.0	300.0	0.3938	15.0	1771.9	84.276
2.0	1000.0	205.64	0.4000	15.0	1234.0	105.690
1.5	1113.5	200.0	0.3938	15.0	1181.3	116.180

various machines are calculated using the optimized MRR. The working SEC for individual machines can be obtained using the optimized energy consumption. As mentioned in Sect. 4.5.2, the most suitable optimization method for this application is the simulated annealing algorithm; thus, optimization results using the algorithm are shown in Figs. 4.13, 4.14, and 4.15. The optimization progress for energy consumption and makespan is shown in Fig. 4.13a, and the energy consumption at different stages of machining and on different machining features/operations is shown in Fig. 4.13b and Fig. 4.14, respectively. The optimization results only for energy consumption are shown in Fig. 4.15. For the optimization of energy consumption and makespan, the total energy consumption is 11,537 KJs, makespan is 53.97 min, and all machines are involved in the jobs; on the contrast, for the optimization of energy consumption only, the total energy consumption is reduced to 8742 KJs while makespan is increased to 76.4 min and the first machine is not be scheduled due to its high SEC and idle consumption compared to the other two machines. The optimization algorithm can also address the dynamics of process planning and scheduling. Table 4.9 shows the results with different selections of machines and optimization objectives. Compared with the results for the two optimization objectives, there is always a trade-off between the energy consumption and makespan when scheduling multiple operations over multiple machines. The optimal energy consumption is achieved when makespan is not taken into account. By comparing the results with different combinations of machines, it is noted that the more selections of machines will always ensure a shorter makespan, but not necessarily lead to less energy consumption.

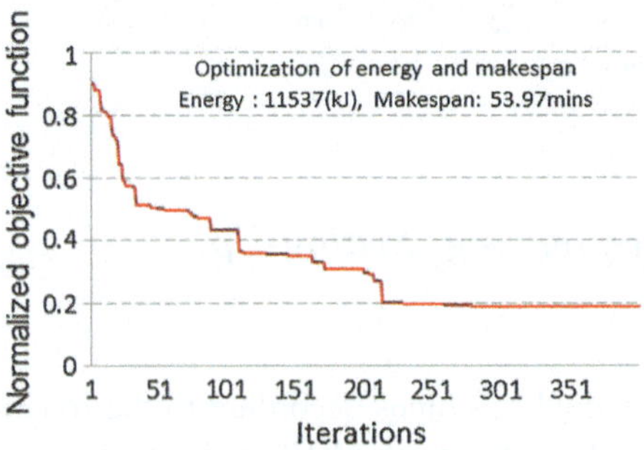

(a) Optimization process

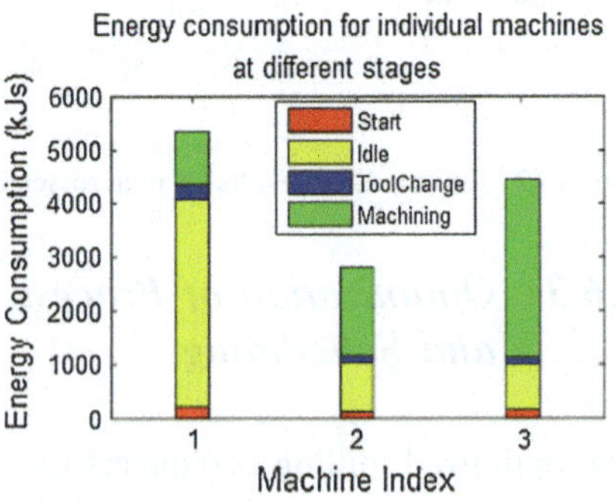

(b) Energy consumption at different stages

Fig. 4.13 Optimization of machine systems in terms of energy consumption/makespan

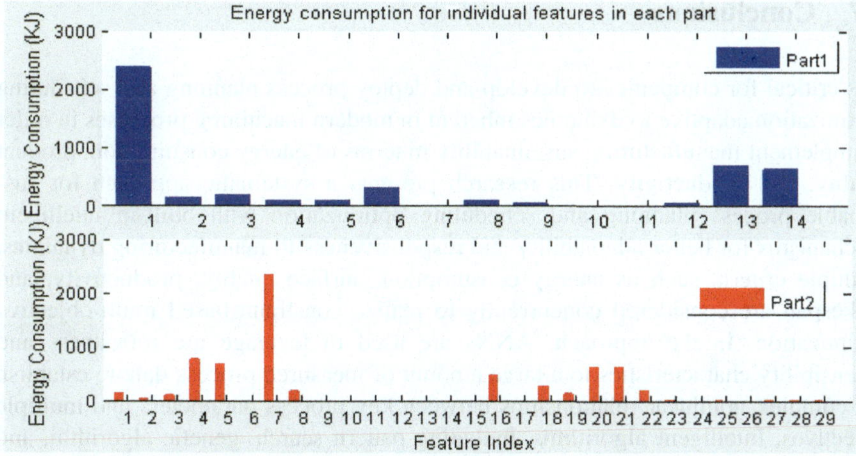

Fig. 4.14 Energy consumption for machining features of the two parts

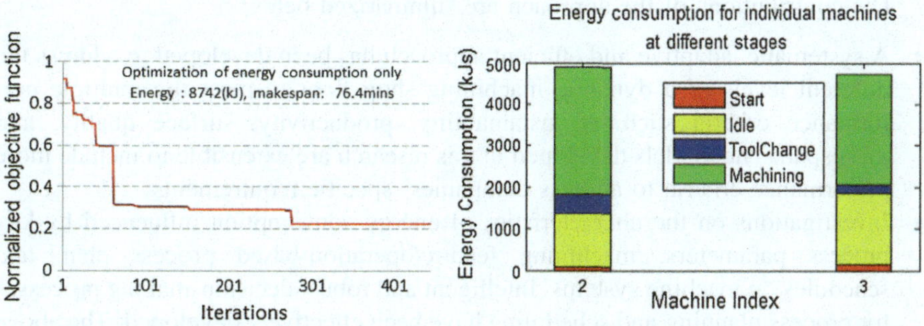

(a) Optimization of process planning and scheduling (b) Energy consumption at different stages

Fig. 4.15 Optimization of process planning and scheduling in terms of energy consumption

Table 4.9 Optimization results under different conditions and objectives

Optimization objective	Available machines	Energy (KJs)	Makespan (minutes)
Energy and makespan	All	11,537	53.97
	2&3	11,376	67.59
	1&3	12,368	65.43
	1&2	12,497	66.84
Energy	All	8742	76.40
	2&3	8742	76.40
	1&3	9453	75.58
	1&2	10,012	78.36

4.7 Conclusions

It is critical for companies to develop and deploy process planning and scheduling optimization adaptive to dynamics inherent in modern machining processes in order to implement manufacturing sustainability in terms of energy consumption, product quality, and productivity. This research presents a systematic approach for sustainable process planning and scheduling optimization with built-in intelligent mechanisms for better adaptability and responsiveness to manufacturing dynamics. Multiple criteria such as energy consumption, surface quality, productivity, and makespan are considered concurrently to realize constraint-based multi-objective optimization. In the approach, ANNs are used to leverage the robustness and extensibility characteristics to a large amount of measured process data to establish the complex nonlinear relationships between key process parameters and multiple objectives. Intelligent algorithms, including pattern search, genetic algorithm, and simulated annealing algorithm, are applied and benchmarked to identify optimized solutions. The developed approach, verified through industrial case studies, shows significant application potential.

The contributions of the approach are summarized below:

- A systematic, adaptive and efficient approach has been developed to address the different levels of a dynamic machining shop floor to meet the multiple performance criteria such as sustainability, productivity, surface quality, and makespan. The models developed in this research are extensible to include more performance criteria to address companies' specific requirements;
- Investigations on the characteristics of energy consumption influenced by key process parameters, machining feature/operation-based process plan, and schedules on machine systems. Intelligent and robust decision-making processes for process planning and scheduling have been effectively developed. The above work paves the way for the approach to be integrated into modern feature-based CAM systems to facilitate the sustainable management of shop floors in companies.

Further investigations and improvements of the research are ongoing, mainly from the following aspects:

- In a shop floor, air-conditioning, ventilation and compressed air equipment and related networks could consume energy significantly. The issue will be investigated in the future research;
- Machining operations deduced from machining features need to be further refined to support the sustainability decision making in more detail;
- Full-scale industrial pilot runs of the system in machining companies in UK, Sweden, Spain, and Germany for demonstration to the entire European industries have been in place.

References

1. Jovane, F., Yoshikawa, H., Alting, L., Boer, C. R., Westkamper, E., Williams, D., et al. (2008). The incoming global technological and industrial revolution towards competitive sustainable manufacturing. *CIRP Annals, 75,* 641–659.
2. Mayers, C. K. (2007). Strategic, financial, and design implications of extended producer responsibility in Europe: A producer case study. *Journal of Industrial Ecology, 11,* 113–131.
3. O'Driscoll, E., & O'Donnell, G. (2013). Industrial power and energy metering—a state-of-the-art review. *Journal of Cleaner Production, 41,* 53–64.
4. Bunse, K., Vodicka, M., Schonsleben, P., Brulhart, M., & Ernst, F. O. (2011). Integrating energy efficiency performance in production management—gap analysis between industrial needs and scientific literature. *Journal of Cleaner Production, 19,* 667–679.
5. Tolio, T., Ceglarek, D., ElMaragphy, H. A., Fischer, A., Hu, S. J., Laperriere, L., et al. (2011). SPECIES—co-evolution of products, processes and production systems. *CIRP Annals—Manufacturing Technology, 59,* 672–693.
6. Wang, L. H., & Shen, W. M. (2010). *Process planning and scheduling for distributed manufacturing*. Berlin: Springer.
7. Duflou, J. R., Sutherland, J. W., Dornfeld, D., Herrmann, C., Jeswiet, J., Kara, S., et al. (2012). Towards energy and resource efficient manufacturing: A process and system approach. *CIRP Annals—Manufacturing Technology, 61,* 587–609.
8. Abele, A., Anderl, R., & Birkhofer, H. (2005). Environmentally-friendly product development—methods and tools. ISBN: 1-85233-903-9 (Chap. 3).
9. Gutowski, T., Dahmus, J., & Thiriez, A. (2006). Electrical energy requirements for manufacturing processes. In *Proceedings of the 13th CIRP International Conference of Life Cycle Engineering. Lueven, May 31st—June 2nd.*
10. Mori, M., Fujishima, M., Inamasu, Y., & Oda, Y. (2011). A study on energy efficiency improvement for machine tools. *CIRP Annals—Manufacturing Technology, 60*(1), 145–148.
11. Newman, S. T., Nassehi, A., Imani-Asrai, R., & Dhokia, V. (2012). Energy efficient process planning for CNC machining. *CIRP Journal of Manufacturing Science and Technology, 5,* 127–136.
12. Hu, S. H., Liu, F., He, Y., & Hu, T. (2012). An on-line approach for energy efficiency monitoring of machine tools. *Journal of Cleaner Production, 27,* 133–140.
13. Camposeco-Negrete, C. (2013). Optimizing of cutting parameters for minimizing energy consumption in turning of AISI 6061 T6 using Taguchi methodology and ANOVA. *Journal of Cleaner Production, 53,* 195–203.
14. Yan, J., & Li, L. (2013). Multi-objective optimization of milling parameters—the trade-offs between energy, production rate and cutting quality. *Journal of Cleaner Production, 52,* 462–471.
15. Winter, M., Li, W., Kara, S., & Hermann, C. (2014). Determining optimal process parameters to increase the eco-efficiency of grinding processes. *Journal of Cleaner Production, 66,* 644–654.
16. Avram, O. I., & Xirouchakis, P. (2011). Evaluating the use phase energy requirements of a machine tool system. *Journal of Cleaner Production, 19,* 699–711.
17. Kong, D., Choi, S., Yasui, Y., Pavanaskar, S., Dornfeld, D., & Wright, P. (2011). Software-based tool path evaluation for environmental sustainability. *Journal of Manufacturing Systems., 30,* 241–247.
18. Balogun, V. A., & Mativenga, P. T. (2013). Modelling of direct energy requirements in mechanical machining processes. *Journal of Cleaner Production, 41,* 179–186.
19. Wang, Q., Liu, F., & Li, C. (2013). An integrated method for assessing the energy efficiency of machining workshop. *Journal of Cleaner Production, 52,* 122–133.
20. Dai, M., Tang, D. B., Giret, A., Salido, M. A., & Li, W. D. (2013). Energy-efficient scheduling for a flexible flow shop using an improved genetic-simulated annealing algorithm. *Robotics and Computer-Integrated Manufacturing, 29*(5), 418–429.

21. Aramcharoen, A., & Mativenga, P. T. (2014). Critical factors in energy demand modelling for CNC milling and impact of toolpath strategy. *Journal of Cleaner Production, 78,* 63–74.
22. Li, W., & Kara, S. (2011). An empirical model for predicting energy consumption of manufacturing processes: A case of turning process. *Proceedings of Institution of Mechanical Engineers, Part B: Journal of Engineering Manufacture, 225,* 1636–1646.
23. Kara, S., & Li, W. (2011). Unit process energy consumption models for material removal processes. *CIRP Annals—Manufacturing Technology, 60*(1), 37–40.
24. Li, W., Winter, M., Kara, S., & Herrmann, C. (2012). Eco-efficiency of manufacturing processes: A grinding case. *CIRP Annals—Manufacturing Technology, 61*(1), 59–62.
25. Diaz, N., Redelsheimer, E., Dornfeld, D. (2011). Energy consumption characterization and reduction strategies for milling machine tool use. In *Proceedings of the 18th CIRP International Conference on Life Cycle Engineering. Braunschweig, May 2nd–4th*, pp. 263–267.
26. Li, L., Yan, J., & Xing, Z. (2013). Energy requirements evaluation of milling machines based on thermal equilibrium and empirical modelling. *Journal of Cleaner Production, 52,* 113–121.
27. He, Y., Liu, F., Wu, T., Zhong, F. P., & Peng, B. (2012). Analysis and estimation of energy consumption for numerical control machining. *Proceedings of Institution of Mechanical Engineers, Part B: Journal of Engineering Manufacture., 226,* 255–266.
28. Li, W. D., Ong, S. K., & Nee, A. Y. C. (2006). Integrated and collaborative product development environment—technologies and implementation. World Scientific Publisher.
29. Li, W. D., & McMahon, C. A. (2007). A simulated annealing-based optimization approach for integrated process planning and scheduling. *International Journal of Computer Integrated Manufacturing, 20*(1), 80–95.
30. Tzeng, C. J., Lin, Y. H., Yang, Y. K., & Jeng, M. C. (2009). Optimization of turning operations with multiple performance characteristics using the Taguchi method and grey analysis. *Journal of Materials Processing Technology, 209,* 2753–2759.
31. Mausser, H. (2006). Normalization and other topics in multi-objective optimization. In *Proceedings of the Fields–MITACS Industrial Problems Workshop*. pp. 89–101.

Chapter 5
Experimental Investigation and Multi-objective Optimization Approach for Low-Carbon Milling Operation of Aluminum

C. Y. Zhang, W. D. Li, P. Y. Jiang and P. H. Gu

Abstract In the past, milling operations have been mainly considered from the economic and technological perspectives, while the environmental consideration has been becoming highly imperative nowadays. In this study, a systemic optimization approach is presented to identify the Pareto-optimal values of some key process parameters for low-carbon milling operation. The approach consists of the following stages. Firstly, regression models are established to characterize the relationship between milling parameters and several important performance indicators, i.e., material removal rate, carbon emission, and surface roughness. Then, a multi-objective optimization model is further constructed for identifying the optimal process parameters, and a hybrid NSGA-II algorithm is proposed to obtain the Pareto frontier of the non-dominated solutions. Based on the Taguchi design method, dry milling experiments on aluminum are performed to verify the proposed regression and optimization models. The experimental results show that a higher spindle speed and feed rate are more advantageous for achieving the performance indicators, and the depth of cut is the most critical process parameter because the increase of the depth of cut results in the decrease of the specific carbon emission but the increase of the material removal rate and surface roughness. Finally, based on the regression models and the optimization approach, an online platform is developed to obtain in-process information of energy consumption and carbon emission for real-time decision making, and a simulation case is conducted in three different scenarios to verify the proposed approach.

C. Y. Zhang · P. Y. Jiang (✉)
State Key Laboratory for Manufacturing Systems Engineering,
Xi'an Jiaotong University, Xi'an, People's Republic of China
e-mail: pjiang@xjtu.edu.cn

W. D. Li
Faculty of Engineering, Environment and Computing,
Coventry University, Coventry, UK

P. H. Gu
College of Engineering, Shantou University, Shantou, People's Republic of China

W. Li and S. Wang (eds.), *Sustainable Manufacturing and Remanufacturing Management*, https://doi.org/10.1007/978-3-319-73488-0_5

Keywords Specific carbon emission · Multi-objective optimization
Dry milling · NSGA-II · Online analysis platform

5.1 Introduction

With the aggravation of global warming and quick increase of energy cost, research to develop energy-efficient and low-carbon emission technologies for the manufacturing industry, which consume significant raw materials and energy, has been becoming paramount. In the USA, the manufacturing sector was responsible for 22% of energy consumption in 2006, and the associated energy costs were about $50 billion [1]. Manufacturing results in substantial stress on the environmental concerns [2]. Research has been actively carried out to improve the sustainability in manufacturing, such as sustainable production scheduling [3], better workshop management for less energy consumption, and machine parameter optimization for energy-efficient machining processes [4].

For machining, research has shown that energy savings up to 6–40% could be obtained based on the optimum choice of cutting parameters, tools, and optimum tool path design [4]. Therefore, machining parameters optimization leading to energy saving and minimized carbon emission in manufacturing workshops is imperative.

In machining processes, the most commonly used optimization criteria are material removal rate (MRR), surface roughness (SR), cutting force, tool life, and power consumption [5]. Although several optimization approaches have been proposed to reduce the environmental impacts of machining processes, most of them are qualitative analysis methods, such as gray relational analysis [6], response surface methodology (RSM) [7], and factor effect analysis [8]. Through establishing the regression and optimization models, a systemic approach is proposed to analyze and optimize machining parameters quantitatively and achieve a better eco-efficiency which means lower manufacturing costs, better production rate, and less carbon emission. Furthermore, an online platform for carbon emission analysis is developed to realize prompt decision making during the above processes.

The rest of this study is organized as follows. The related research is reviewed in Sect. 5.2. In Sect. 5.3, regression models of milling processes are constructed to characterize the relationship between the milling parameters and the environmental/productivity/quality objectives firstly. Then, a mathematical optimization model is constructed and a hybrid Non-dominated Sorting Genetic Algorithm-II (NSGA-II) is proposed to identify the optimal milling parameters. Section 5.4 shows the experimental work for establishing the regression models. The regression analysis and optimization analysis are carried out on the basis of the experimental results in Sect. 5.5. An online platform for carbon emission analysis is developed, and a simulation case is shown to illustrate the feasibility of the method in Sect. 5.6. Finally, some conclusions are made in Sect. 5.7.

5.2 Research Background

5.2.1 Energy Reduction of Machine Tools

Reducing the machining energy of machine tools can significantly improve the environmental performance of manufacturing process [9]. Therefore, several researchers have focused on the energy monitoring and reduction for machine tools. By presenting a detailed description of different test procedures based on standardized workpieces, Behrendt et al. [10] proposed a novel and coherent method to assess energy consumption of machine tools. Hu et al. [11] developed a new online energy efficiency monitoring approach without using any torque sensor or dynamometer to minimize the implementation cost and difficulty. Kara and Li [12] presented an empirical model to characterize the relationship between energy consumption and process variables for material removal processes and tested and validated the model on a number of turning and milling machine tools. In addition, some studies were conducted from the viewpoint of machine tool components and internal energy dissipation units. Through measuring the power consumption of a machining center under different conditions, a new acceleration control method was developed to reduce energy consumption by synchronizing spindle acceleration with the feed system [13]. Newman et al. [4] presented a framework to validate the introduction of energy consumption in the objectives of process planning for computer numerical control (CNC) machining on the basis of the state-of-the-art in process planning and energy consumption in manufacturing research. In addition, a model for the optimization of machining parameters was presented for the minimum energy consumption in a multi-pass turning operation [14], and the model takes into account finishing and roughing passes separately for the energy optimization followed by the dual optimization of the energy functions for a combination of one finishing pass and multiple roughing passes. In order to obtain the optimum machining parameters, Kant and Sangwan [15] provided a multi-objective predictive model for the minimization of power consumption and surface roughness in machining, using gray relational analysis coupled with principal component analysis and response surface methodology. From the above literature, it can be seen that energy modeling and qualitative analysis of machine tools from different viewpoints have drawn much attention, while another important aspect, i.e., the optimization of cutting parameters and quantitative analysis for energy consumption reduction, has not well researched. Therefore, more efforts need to be made to search quantitative methods for the energy conservation and carbon emission reduction of machine tools.

5.2.2 *Low-Carbon-Oriented Modeling of Machining Processes*

With the purpose of analyzing machining process and reducing its environmental impact, models have been developed to reveal the relationship between machining parameters and some performance indicators, as shown in Table 5.1. Choudhury and Appa Rao [16] established a tool life estimation equation from experimental data and the adhesion wear model. Lalwani et al. [17] established a linear model to fit the variation of cutting forces with feed rate and depth of cut by conducting machining experiments based on RSM and the sequential approach. Moreover, Zain et al. [18] established a predicted model of the SR to show its relationship with the decision variables (cutting speed, the feed per tooth, the axial depth of cut, the radial depth of cut, and machining tolerance).

In addition, some researchers conducted machining experiment and regression analysis to minimize energy consumption and carbon emission. Campatelli et al. [19] focused on the efficiency of the machining centers and developed a quadratic regression model through an experimental approach to evaluate and optimize the process parameters in order to minimize the power consumption in a milling process performed on a modern CNC machine. An orthogonal array, signal to noise (S/N) ratio, and analysis of variance (ANOVA) were employed to analyze the

Table 5.1 Modeling methods of machining processes

Focus	Authors	Machine performances	Machining process	Model or methodology
Technical performance	Choudhury and Appa Rao [16]	Tool life	Milling	Experimental methods
	Lalwani et al. [17]	Cutting forces	Turning	RSM, the sequential approach
	Zain et al. [18]	SR	Milling	Artificial neural network
Power or energy consumption	Campatelli et al. [19]	Power consumption	Milling	RSM
	Camposeco-Negrete [20]	Energy consumption and SR	Turning	Orthogonal array, S/N, and ANOVA
	Bhattacharya et al. [21]	Surface finish and power consumption	Turning	Taguchi technique, ANOVA
	Winter et al. [22]	SR, cost and carbon footprint	Grinding	Regression analysis and sensitivity analysis
	Bhushan [23]	Power consumption and tool life	Turning	RSM, desirability function approach
	Rajemi et al. [24]	Energy footprint and tool life	Turning	$T_{\text{opt}-E} = \left(\frac{1}{\alpha} - 1\right)\left(\frac{P_0 t_3 + y_E}{P_0}\right)$

effects and contributions of depth of cut, feed rate, and cutting speed on the energy consumption [20]. Bhattacharya et al. [21] outlined an experimental study to investigate the effects of cutting parameters on surface finish and power consumption during high-speed machining of AISI 1045 steel by employing the Taguchi techniques and ANOVA. Winter et al. [22] presented a generic regression model to describe and analyze the influence of grinding process parameters in conjunction with different cutting fluids on SR, cost, and carbon footprint and applied the sensitivity analysis to reveal the trends of each process parameter in relation to the preference of technological, economic, and environmental objectives. Bhushan [23] conducted experimental investigations to establish relationships between cutting speed, feed rate, depth of cut and nose radius and power consumption and tool life in CNC turning of 7075 Al alloy 15 wt% SiC composite by using the RSM. However, these works considered the power consumption as environmental target which did not reflect the real energy consumption of machining process directly because the energy consumption is also related to processing time except power consumption. Although Rajemi et al. [24] modeled the total energy of machining a component by a turning process and optimized it to derive a minimum carbon footprint requirement, the carbon emission of cutting tools and cutting fluids was not considered [25].

5.2.3 Mathematical Optimization of Machining Processes

To obtain the optimal cutting parameters and reduce the carbon emission, mathematical optimization approaches were used to identify the optimal or close to the optimal solution of a given task regarding constraints and a set of given functions. The tasks can be generally classified as single-objective or multi-objective optimization, as shown in Table 5.2. For the former, the aim is to solve a single-objective function by identifying the minimum or maximum value. Nalbant et al. [26] used the Taguchi method to find optimal cutting parameters for SR in turning. Wibowo and Desa [27] presented a technique by using the hybridization of kernel principal component analysis (KPCA)-based nonlinear regression and genetic algorithms (GAs) to estimate the optimum values of the three parameters (namely radial rake angle, speed, and feed rate) such that the estimated SR was as low as possible. In order to get the minimum energy consumption, an experimental study to optimize cutting parameters during turning of AISI 6061 T6 under roughing conditions was presented [20].

However, single-objective approaches are limited in identifying the optimal cutting parameters, because several objectives are sometimes contradictory and must be simultaneously optimized. Hence, multi-objective approaches for cutting parameters optimization which consider multi-objectives were developed. Quiza Sardiñas et al. [28] constructed a multi-objective optimization model to obtain the optimal tool life and operation time and used micro-GAs to obtain the non-dominated points. Pawade and Joshi [29] applied a new effective approach,

Table 5.2 Mathematical optimization comparison of machining processes

Focus	Authors	Objectives	Optimization methods
Single-objective optimization	Nalbant et al. [26]	SR	Taguchi method
	Wibowo and Desa [27]	SR	KPCA, nonlinear regression, and GA
	Camposeco-Negrete [20]	Energy consumption	RSM
Multi-objective optimization	Quiza Sardiñas et al. [28]	Tool life and operation time	Micro-GAs
	Pawade and Joshi [29]	SR and cutting forces	Taguchi gray relational analysis
	Somashekhar et al. [6]	MRR, overcut, SR	ANOVA, gray relational analysis
	Kuram et al. [30]	Specific energy, tool life, and SR	D-optimal method
	Yan and Li [8]	Cutting energy, MRR, SR	Weighted gray relational analysis, RSM, and SQP
	Winter et al. [22]	SR, cost, and carbon footprint	Geometric programming algorithm
	Bhushan [23]	Power consumption and tool life	Desirability function analysis
	Campanelli et al. [7]	Ablation depth, MRR, SR	RSM (contour plot methodology)

named the Taguchi gray relational analysis to experimental results in order to optimize the high-speed turning of Inconel 718 with consideration to multiple performance measures. A new approach for the optimization of the micro-wire electric discharge machining process with multiple performance characteristics is attempted based on the statistical-based ANOVA and gray relational analysis [6]. Kuram et al. [30] investigated the effects of cutting fluid types as a function of three milling factors (cutting speed, depth of cut, and feed rate) on process responses (specific energy, tool life, and SR). Yan and Li [8] presented a multi-objective optimization method based on the weighted gray relational analysis and RSM and optimized the cutting parameters in milling process by using the sequential quadratic programming (SQP) algorithm. Winter et al. [22] presented an approach to identify the process parameters and developed Pareto-optimal solutions for advancing the eco-efficiency of grinding operations, including SR, cost, and carbon footprint. Based on the contour plot methodology, a multi-objective statistical optimization was performed for improving the machining productivity and surface quality of laser milling [7]. It is the fact that most of these studies are limited to qualitative analysis of some optimization objectives by using gray relational analysis, desirability function analysis, sensitivity analysis, RSM, etc. Moreover, many studies transformed multi-objective problems into single-objective problems and employed traditional mathematical programming methods to solve the problems. However, few researchers have used Pareto-optimal methods or intelligent algorithms to optimize machining parameters which are more effective.

5.3 Approach

5.3.1 Workflow of the Approach

Figure 5.1 presents the approach to identify the optimal milling parameters for better quality, higher productivity, and lower carbon emission. Firstly, owing to the stochastic nature of milling process, regression models are constructed to characterize the relationship between the milling parameters and the respective objectives. Secondly, based on the regression models, an optimization model is established and a hybrid NSGA-II is adopted to identify the optimal milling parameters. Thirdly, experiments based on the Taguchi design method are designed to identify the levels of experimental variables with the minimal amount of experiments. Fourthly, the milling experiments are performed according to the experimental plan and the regression models are derived from the experimental results with the statistical analyses software SPSS®, and single-objective analysis and multi-objective optimization are carried out to obtain the Pareto frontier of milling parameters. Finally, an online platform is developed to obtain in-process information about the energy consumption and carbon emission to support above real-time decision making, and a simulation case is conducted in three different scenarios to verify the proposed

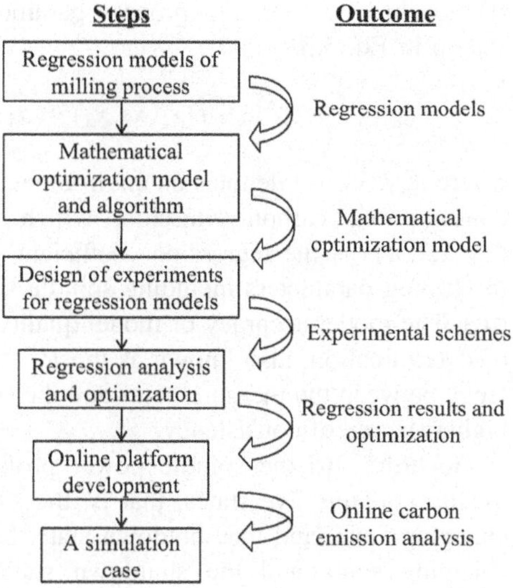

Fig. 5.1 Schematic of the solving approach

method. Although Palanikumar et al. [31] applied similar statistical models and NSGA-II to optimize the cutting conditions of glass fiber-reinforced plastic composites, they did not consider the energy consumption and carbon emission.

5.3.2 Regression Models of Milling Process

1. Production rate

The MRR (in mm^3/min), which is the most commonly used optimization criterion of production rate in milling processes, can be computed by Eq. 5.1.

$$\mathrm{MRR} = d * f * a_p \tag{5.1}$$

where d is the cutting tool diameter in mm, f means feed rate in mm/min, and a_p represents the depth of cut in mm.

2. Environmental impact

The power of a milling process is determined by the milling force and milling velocity of machines [32], which can be calculated by Eq. 5.2.

$$P_c = 42.4 * 10^{-5} k_p d_0^{-0.3} a_{se} f_z^{0.75} a_{sp}^{1.1} z n_0^{0.8} \tag{5.2}$$

where k_p, d_0, $a_{se}(a_{sp})$, f_z, z, and n_0 represent correction factor, cutter diameter in mm, cutting depth in mm, feed rate in mm/min, number of teeth, and spindle speed in r/min, respectively.

According to Eq. 5.2, a generic regression model is developed to describe the relationship between the process parameters and the environmental impact, as shown in Eq. 5.3.

$$f(x_1, x_2, x_3) = \alpha_1 * x_1^{\alpha_2} * x_2^{\alpha_3} * x_3^{\alpha_4} \tag{5.3}$$

where $f(x_1, x_2, x_3)$ denotes an environmental impact such as cutting power, energy consumption, carbon emission, which will be described in detail later. α_i (i = 1,2,3,4) is the regression coefficient, and x_i (i = 1,2,3) represents one of the machining parameters including spindle speed (n), feed rate (f), and cutting depth (a_p). The model accuracy or model quality can be ascertained using the coefficient of determination, also known as the R^2 value. The R^2 value describes the consistency between the measurements and the statistical model. The higher R^2 value, the higher degree of consistency.

According to the spindle power profile of a machine, a machining process mainly contains five states, that is, the startup state, idle state, cutting state, tool changing state, and the shutdown state. Since the power of the startup state, tool changing state, and the shutdown state has nothing to do with the milling parameters, only the cutting power (P_c) and air cutting power (P_a) were chosen to analyze the relationship between machining power and milling parameters, as illustrated in Eqs. 5.4 and 5.5.

$$P_c = \beta_1 * n^{\beta_2} * f^{\beta_3} * a_p^{\beta_4} \tag{5.4}$$

$$P_a = \theta_1 * n^{\theta_2} * f^{\theta_3} \tag{5.5}$$

where β_i (i = 1,2,3,4) and θ_i (i = 1,2,3) denote regression coefficients.

In addition, the specific carbon emission (SCE [$kgCO_2$-e/cm^3]) was used to evaluate the environmental impact of different machining processes, as shown in Eq. 5.6. Here, the carbon emission contains two parts: one from machine tools (SCE_{energy}) and another one from cutting tools (SCE_{tool}) [25]. For the former, only the carbon emission of a machine tool due to the electrical energy consumption was taken into account, while the carbon emission from the production and procurement of the machine tool was not considered because this part of carbon emission cannot be influenced by the machining parameters. For the latter, the carbon emission of cutting tools is calculated by comparing machining time with tool life as the production of cutting tools consumes energy and each cutting tool has a lifetime. Notably, these life cycle analysis of cutting tools only considered the production of cutting tools and the disposal phase was excluded due to the absent data; thus, the carbon emission due to cutting tools shown in this study may be underestimated. In order to improve the quality of the analysis, the entire life cycle of the cutting tools was suggested to be considered in the future research.

$$\mathrm{SCE} = \mathrm{SCE}_{\mathrm{energy}} + \mathrm{SCE}_{\mathrm{tool}} = \eta_1 * n^{\eta_2} * f^{\eta_3} * a_p^{\eta_4} \tag{5.6}$$

where η_i (i = 1, 2, 3, 4) denotes regression coefficients.

3. Product quality

The SR (R_a [um]), which was widely used to assess product quality, was selected to evaluate the production quality target. It was found that the cutting parameters n, f, and a_p have a strong effect on SR [33].

In order to characterize the relationships between the SR and the above process parameters, the RSM method was chosen due to its adaptability in applications where several input variables (independent variables) potentially influence some performance measure or quality characteristic of products or processes [8]. Usually, the first-order model of RSM is ineffective because it includes only the main effect of the variables. Here, the second-order model of RSM was adopted based on its flexibility. A general form is shown in Eq. 5.7.

$$R_a = \gamma_0 + \sum_{i=1}^{3} \gamma_i * x_i + \sum_{i \le j}\sum \gamma_{ij} * x_i * x_j \tag{5.7}$$

where γ_i (i = 0, 1, 2, 3) and γ_{ij} are the regression coefficients and x_i (i = 1, 2, 3) represents one of the machining parameters.

The regression coefficients of models in Eqs. 5.4–5.7 will be obtained through the experimental work in Sect. 5.4, and the regression results will be shown in Sect. 5.5.1.

5.3.3 *Mathematical Optimization Model*

Considering the eco-efficiency of milling processes, an optimization model is established in which the production rate MRR, specific carbon emission SCE_{total}, and surface roughness R_a are chosen to represent the production target, environmental target, and quality target, respectively, as shown in Eq. 5.8. In addition, the total cutting power P_c stands for the real-time machining power reflecting the state of runtime machine. High cutting power can incur the greater vibration of machine tools, or bigger cutting tool wear, so there should be an upper bound limit for the cutting power. Based on the above regression models, the optimization model is shown as follows:

Objectives:

$$\begin{cases} \mathrm{SCE} = \eta_1 * n^{\eta_2} * f^{\eta_3} * a_p^{\eta_4} \\ \mathrm{MRR} = d * f * a_p \\ R_a = \gamma_0 + \sum_{i=1}^{3} \gamma_i * x_i + \sum_{i \le j} \sum \gamma_{ij} * x_i * x_j \end{cases} \tag{5.8}$$

Constraints:

$$\beta_1 * n^{\beta_2} * f^{\beta_3} * a_p^{\beta_4} \le \bar{P}_c \tag{5.9}$$

$$0 < n \le n^{\max} \tag{5.10}$$

$$0 < f \le f^{\max} \tag{5.11}$$

$$0 < a_p \le a_p^{max} \tag{5.12}$$

$$d > 0, x_i > 0, x_j > 0, \beta_1 > 0, \eta_1 > 0, \gamma_0 > 0 \tag{5.13}$$

where $\bar{P}_c$ denotes the upper limit of the cutting power which can ensure the machine tool in normal operation state. According to the actual processing capacity of a machine tool, $n^{\max}$, $f^{\max}$, and $a_p^{\max}$ represent the maximum of spindle speed, feed rate, and depth of cut, respectively.

To solve the above multi-objective optimization problem, a hybrid NSGA-II algorithm is proposed to identify the optimal milling parameters, as shown in Fig. 5.2. NSGA-II can get the Pareto frontier of solutions through non-domination

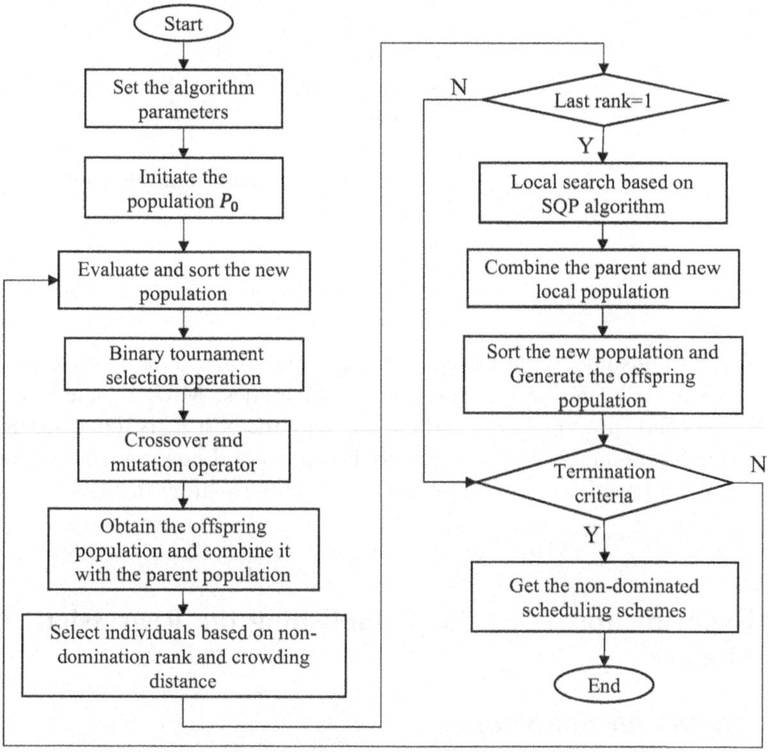

Fig. 5.2 Flowchart of the hybrid NSGA-II

sorting and crowding distance calculation, which allows the operator to choose the appropriate solution according to specific needs. The main components of the proposed algorithm are summarized below:

Step 1: Set the algorithm parameters like number of population, maximum number of generations, crossover, and mutation probabilities.

Step 2: Generate the initial population P_0 randomly within the range of parameters.

Step 3: Evaluate the objective functions (i.e., SCE, R_a, MRR), and regard the constraint (P_c) as an additional objective function to conduct the non-dominated sorting, as shown in Fig. 5.3. Then, sort them with the assigned non-domination level number and the value of crowding distance.

Step 4–6: Perform selection, crossover, and mutation operation [34].

Step 7: When ranks of all chromosomes in the parent population equal one, divide the population into three sub-populations and perform local search for each sub-population based on SQP algorithm. For example, for the first population, the SQP algorithm is used to obtain the best

Fig. 5.3 Non-dominated sorting of the constrained NSGA-II algorithm

```
// Non-dominated sorting of the constrained NSGA-II algorithm
Choose any chromosome C1 and C2;
If (Pc of chromosome C1) ≤ P̄c and (Pc of chromosome C2) > P̄c
    C1 dominates C2;
Else if (Pc of C1) ≤ P̄c and (Pc of C2) ≤ P̄c
    If SCE, Ra, MRR of C1 are all better than that of C2
        C1 dominates C2;
    End if
End if
```

chromosome with the lowest SCE and form a new sub-population. Similarly, the best chromosome with highest MRR is selected for the second sub-population and the one with the best R_a is found for the third sub-population. Then, combine the parent and new population, and sort them based on non-domination rank and crowding distance.

5.4 Experimental Work for Establishing the Regression Models

5.4.1 Experimental Setup

The experimental environment and measurement equipment are shown in Fig. 5.4. The experiments were performed on a CNC micromachining center (Manix CNC MM-250S3, Fig. 5.4a) with 1.2 kW motor rated power and maximum spindle speed of 6400 rpm. The power demand of the milling process was acquired by using the Janitza power analyzer UMG 604 (Fig. 5.4b), and SR was measured by the surface roughness tester TR300 (Fig. 5.4c). Since the main propose of this experiment was to obtain the total power consumption of the micromachining center, the power analyzer was connected with the main input wire of the machine. The used power analyzer with a temporal resolution of 10 ms was configured to record the total active power of the MM-250S3. The real-time power data from the power analyzer was recorded through an online platform for energy consumption analysis and process planning, which will be introduced in Sect. 5.6.1. Since many problems such as health and environmental issues are identified with the use of flood cutting fluids in machining processes, considerable attention has been given to reduce or completely omit the cutting fluids and meet the demands for environment-friendly cutting processes [35]. Therefore, the dry milling of aluminum is researched in this study. A 7.8-mm diameter, 4 flutes carbide tool was employed for the dry cutting of an 80 mm × 80 mm × 80 mm aluminum block.

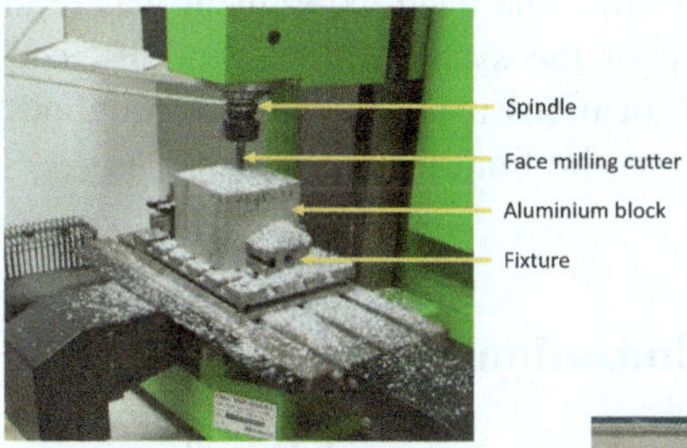

(a) Manix CNC MM-250S3

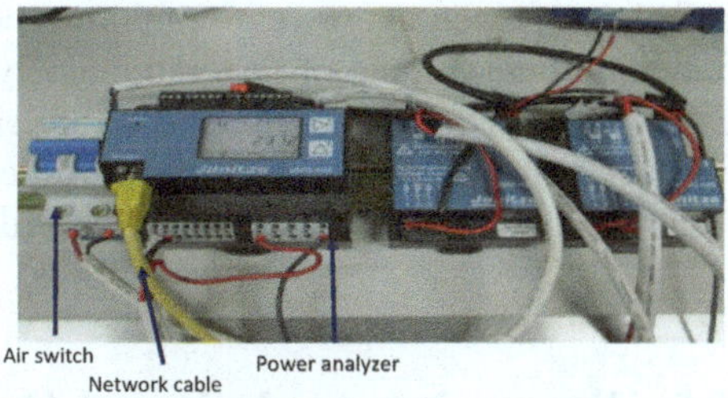

(b) Janitza UMG-604 power analyser

(c) Surface roughness tester TR300

Fig. 5.4 Experimental environment and measurement equipment

5.4.2 *Design of Experiments*

Taking the actual processing capacity of the MM-250S3 into consideration, the milling parameters were set up in the recommended ranges and the tool wear did not deteriorate significantly according to preliminary tests. As mentioned before, the spindle speed n (r/min), feed rate f (mm/min), and depth of cut a_p (mm) were chosen due to their major influence on the milling process. The variances of n, f, and a_p were customized according to the machine tool's parameter range. The cutting parameters and their levels are shown in Table 5.3. In order to reduce the times of experiments, the Taguchi design method of experiments was adopted. Since each parameter had four levels, the standard orthogonal array L16(4^5) was chosen. But only three columns in the L16(4^5) were used to obtain the experimental data because there were only three parameters in this experiment. As mentioned before, multiple independent experimental data was measured throughout the experiments, including processing time, air cutting power,

Table 5.3 Design of experiments

Parameters	Range	Level 1	Level 2	Level 3	Level 4
n (r/min)	1000–4000	1000	2200	3000	4000
f (mm/min)	4–16	4	8	12	16
a_p (mm)	0.4–1.6	0.4	0.8	1.2	1.6

machining power, energy consumption, and SR. Each measurement was taken after removing unit volume of material, namely 1 cm^3, and each experiment was replicated twice in order to reduce the influence of the system errors. For SR, each measurement was taken from three different locations using the surface roughness tester, and the average values were recorded as the final result.

5.5 Regression Analysis and Optimization

5.5.1 Regression Analysis Based on Experimental Results

After carrying out the above experiments, all results of different combinations of milling parameter are shown in Table 5.4. As aforementioned, the objectives of the mathematical optimization can be expressed as productivity, environmental, and quality target functions. Based on the experimental results, a nonlinear regression analysis was performed to derive each target function via the IBM SPSS Statistics 19. The regression coefficients and R^2 values of machining power and SCE are listed in Table 5.5. A summary of ANOVA results for the regression models is presented in Table 5.6, and it can be clearly seen that the models achieve a great accuracy because of a high R^2 value.

Based on the experimental data in Table 5.4, the second-order polynomial regression model of the SR was developed by using the IBM SPSS Statistics 19 software, as shown in Eq. 5.14. The ANOVA for R_a is presented in Table 5.7, and

Table 5.4 Experimental results of different milling parameter combination

No.	n	f	a_p	P_a	P_c	MRR	SCE	R_a
1	1000	4	0.4	381.756	474.95	12.48	2.931	0.134
2	1000	8	0.8	382.3	502.39	49.92	0.755	0.189
3	1000	12	1.2	384.397	532.73	112.32	0.346	0.234
4	1000	16	1.6	387.319	559.13	199.68	0.200	0.318
5	2200	4	1.2	504.056	661.4	37.44	1.177	0.181
6	2200	8	1.6	506.442	687.93	99.84	0.452	0.21
7	2200	12	0.4	504.63	615.73	37.44	1.128	0.152
8	2200	16	0.8	498.654	656.9	99.84	0.440	0.231
9	3000	4	1.6	578.283	793.03	49.92	0.989	0.231
10	3000	8	1.2	581.997	759.16	74.88	0.641	0.277
11	3000	12	0.8	583.974	733.53	74.88	0.627	0.275
12	3000	16	0.4	585.892	696.65	49.92	0.911	0.216
13	4000	4	0.8	687.115	832.69	24.96	2.041	0.239
14	4000	8	0.4	681.655	797.86	24.96	1.985	0.241
15	4000	12	1.6	686.775	922.19	149.76	0.364	0.292
16	4000	16	1.2	699.73	894.15	149.76	0.357	0.295

Table 5.5 Regression coefficients of Pa, Pc, and SCE

P_a		P_c		SCE	
β_1	18.46	θ_1	36.421	η_1	1.0577
β_2	0.432	θ_2	0.375	η_2	0.222
β_3	0.007	θ_3	0.021	η_3	−1.01
β_4	/	θ_4	0.099	η_4	−0.965

Table 5.6 Analysis of variance for Pa, Pc, and SCE

P_a	Factor	DOF	SS	MS	F	Sig. F
	Regression model	2	0.7250	0.3625	387.13	2.6E−12
	Error	13	0.0122	0.0009	–	–
	Total	15	0.7372	–	–	–
	S = 0.0306		R-Sq = 98.35%		R-Sq(adj) = 98.09%	
P_c	Factor	DOF	SS	MS	F	Sig. F
	Regression model	3	0.5969	0.1990	246.68	4.81E−11
	Error	12	0.0097	0.0008	–	–
	Total	15	0.6065	–	–	–
	S = 0.028		R-Sq = 98.40%		R-Sq(adj) = 98.01%	
SCE	Factor	DOF	SS	MS	F	Sig. F
	Regression model	3	8.2985	2.7662	6118.44	2.28E−19
	Error	12	0.0054	0.0005	–	–
	Total	15	8.3039	–	–	–
	S = 0.021		R-Sq = 99.93%		R-Sq(adj) = 99.92%	

Table 5.7 ANOVA for Ra

R_a	Factor	DOF	SS	MS	F	Sig. F
	Regression model	6	0.0349	0.0058	59.27	1.56E−5
	Error	9	0.0053	0.0006	–	–
	Total	15	0.0402	–	–	–
	S = 0.024		R-Sq = 93.11%		R-Sq(adj) = 88.52%	

it can be observed that the coefficient of determination R-Sq (adj) for the regression model of R_a is equal to 0.885, which indicates that the model has good compatibility to the experimental data. Therefore, this regression model based on the Taguchi method and RSM is suitable for establishing prediction models.

$$\begin{aligned} R_a = {} & 0.04 + 4.615 * 10^{-5} * n - 0.003 * f \\ & + 0.147 * a - 2.417 * 10^{-5} * n * a + 0.007 * f * a - 0.05 * a^2 \end{aligned} \tag{5.14}$$

5.5.2 Single-Objective Analysis

In order to investigate the contribution and effects of milling parameters on the different objectives including SCE, R_a, P_a, and P_c, the surface plots and contour plots were created to perform single-objective analysis.

1. Environmental impact analysis

The environmental impact is presented in Fig. 5.5 and shows that the SCE changes over the depth of cut a_p and the feed rate f, with four fixed values for the cutting speed n, namely 1000, 2200, 3000, and 4000 r/min.

It can be observed that the SCE decreases with the increase of f and a_p simultaneously and f and a_p have a similar effect on the SCE. In particular, the SCE declines obviously when f and a_p are small relatively. If f >11 mm/min and a_p >1.1 mm, the SCE changes very little, which means that f = 11 mm/min and a_p = 1.1 mm are the critical points for carbon emission reduction. The influence of

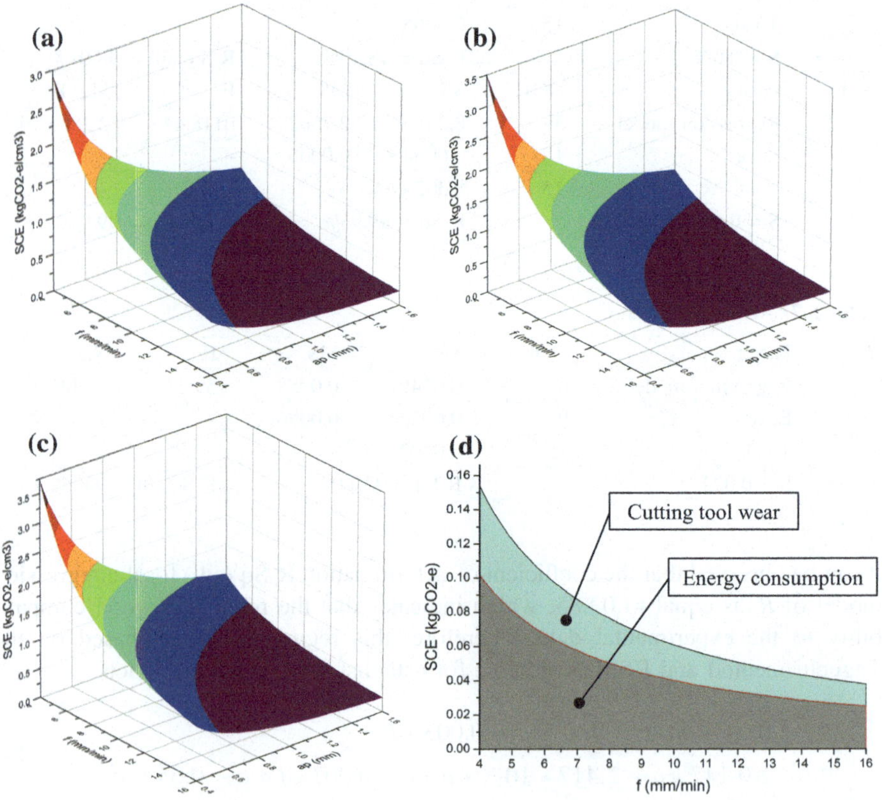

Fig. 5.5 Specific carbon emission analysis. **a** n = 1000 r/min, **b** n = 2200 r/min, **c** n = 3000 r/min, **d** n = 4000 r/min and a_p = 1.0 mm sectional view

n on SCE is not obvious, especially when f and a_p are large relatively. Therefore, compared with the cutting speed, the feed rate and depth of cutting are more important for SCE. From the specific carbon emission sectional view in Fig. 5.5d, the carbon emission of energy consumption decreases significantly due to the reduction of processing time; however, the carbon emission of cutting tools decreases not obviously, which shows that the reduction of carbon emission mainly comes from the energy consumption for the chosen parameters.

2. Product quality analysis

Figure 5.6 presents the response surfaces of the empirical regression model for the product quality impact, i.e., SR of the milling process. The impact is also presented over the depth of cut a_p and the feed rate f, with four fixed values for the cutting speed n, namely 1000, 2200, 3000, and 4000 r/min.

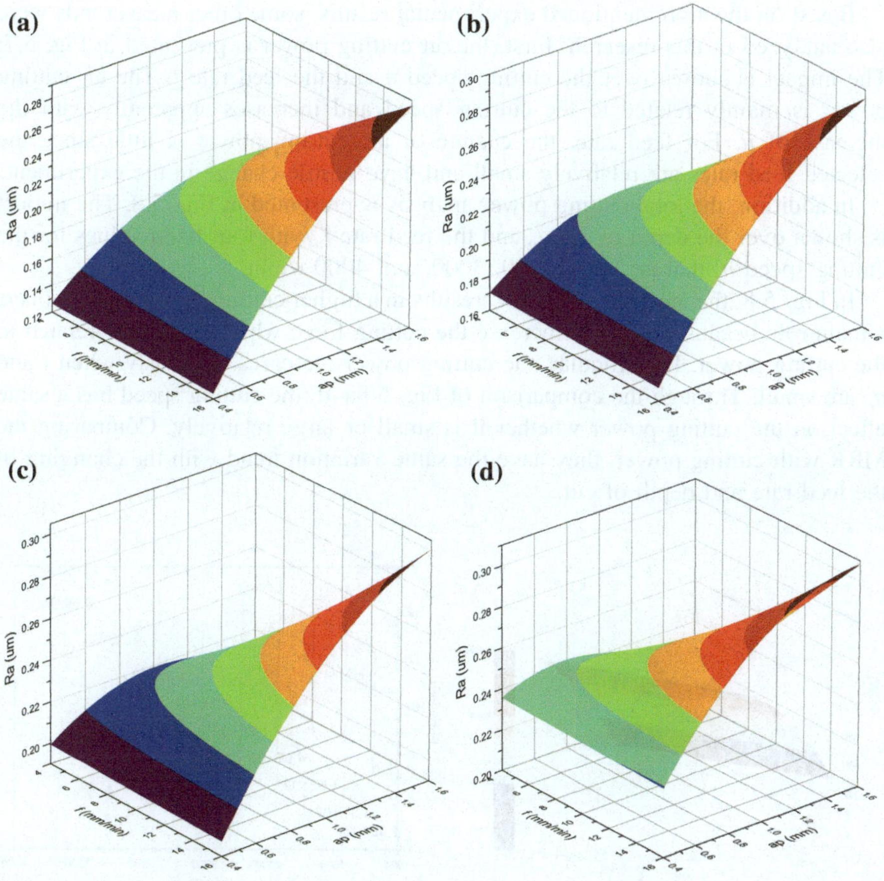

Fig. 5.6 Surface roughness analysis. **a** n = 1000 r/min, **b** n = 2200 r/min, **c** n = 3000 r/min, **d** n = 4000 r/min

From Fig. 5.6, it can be clearly seen that the increase of a_p and f leads to the increase of the measured SR, and a_p has a more significant impact due to the superposition of geometrical and kinematical effects on the milling process. In particular, the influence of f is not obvious when a_p is small, and the increase of f will cause the changing of SR if a_p >0.8 mm. Similarly, when f <6 mm/min, the increase of a_p will cause little change of SR, which means there is a critical region (f <6 mm/min or a_p <0.8 mm) in which the part has a good quality and the SR changes little due to the increase of f and a_p. Conversely, the influence of the cutting speed is obvious only within the critical region. Overall, when f and a_p are small (f <6 mm/min or a_p <0.8 mm), the cutting speed will have more influence on the SR, but f and a_p will affect the SR obviously beyond the critical region.

Moreover, by comparing SR and MRR, a_p and f have an opposite effect on them, so that the optimal SR and MRR cannot be obtained simultaneously.

3. Other measurands

Based on the aforementioned experimental results, some other measurands were also analyzed in this research. First, the air cutting power is presented in Fig. 5.7. The impact is shown over the cutting speed n and the feed rate f. The air cutting power is mainly related to the cutting speed and increases apparently with the increase of n. For feed rate, the change of air cutting power is little since the selected feed rates are relatively small and have a little change in the experiment.

In addition, the total cutting power analysis is presented in Fig. 5.8. The impact is shown over the depth of cut a_p and the feed rate f, with four fixed values for the cutting speed n, that is, 1000, 2200, 3000, and 4000 r/min.

In Fig. 5.8, the increase of f and a_p results in a higher cutting power, and a_p plays a main role because they will increase the cutting force which is directly related to the cutting power. In particular, the cutting power increases obviously when f and a_p are small. Through the comparison of Fig. 5.8a–d, the cutting speed has a same effect on the cutting power whether it is small or large relatively. Comparing the MRR with cutting power, they have the same variation trend with the changing of the feed rate and depth of cut.

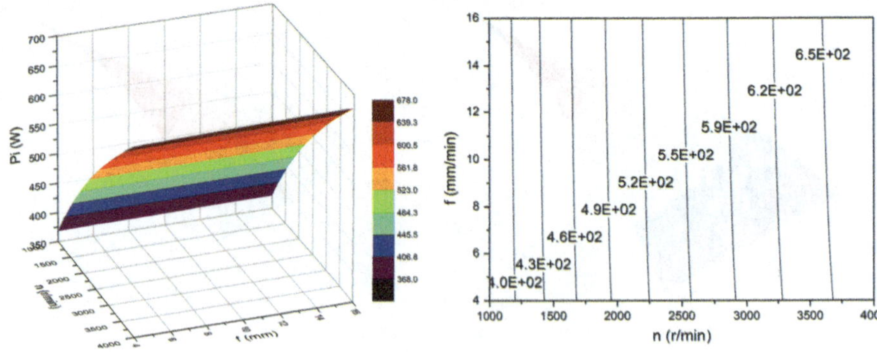

Fig. 5.7 Air cutting power analysis (left-side: response surface and right-side: contour plot)

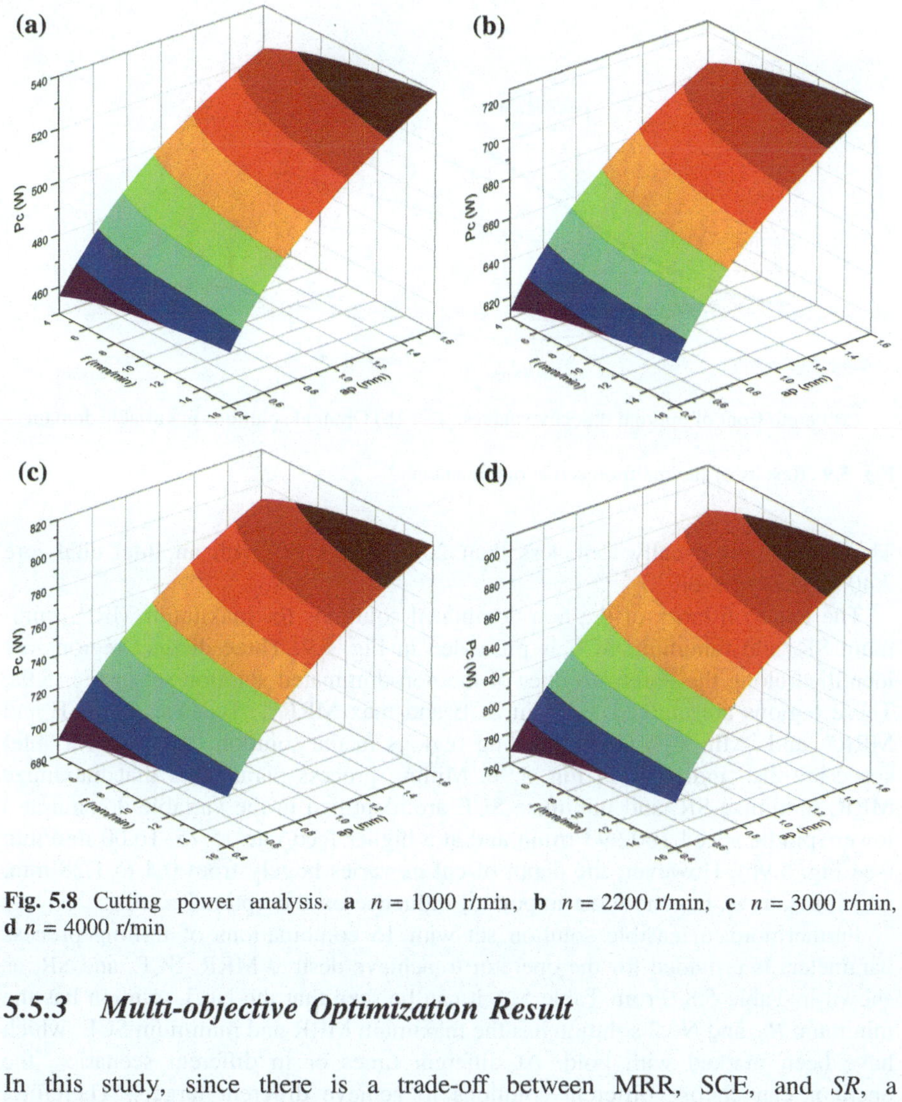

Fig. 5.8 Cutting power analysis. **a** n = 1000 r/min, **b** n = 2200 r/min, **c** n = 3000 r/min, **d** n = 4000 r/min

5.5.3 Multi-objective Optimization Result

In this study, since there is a trade-off between MRR, SCE, and *SR*, a multi-objective optimization becomes necessary. Finding the optimal process parameters to achieve the desired level of response (maximum MRR, minimum SCE, or minimum *SR*) can be performed.

The multi-objective optimization model is described in Sect. 5.3.3. According to the actual operation of the milling machine, the total power is constrained to be less than or equal to 530 W. The simulations were run by using the hybrid NSGA-II with a population of 80 chromosomes and a maximum number of 500 iterations. After obtaining the best milling parameter combinations, the Pareto frontier was plotted in a three-dimensional objective space for viewing (shown in Fig. 5.9a).

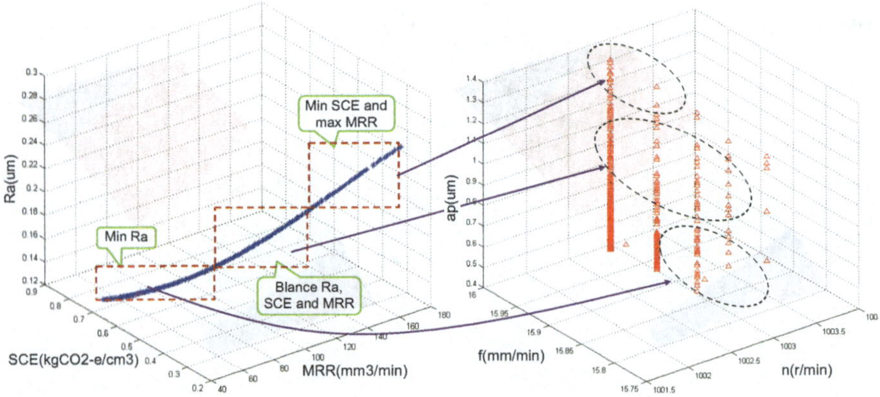

(a) Pareto front of optimal objective values; (b) Optimal solutions in variable domain

Fig. 5.9 Results of the multi-objective optimization

The simulations usually took less than 20 min in a PC with an Intel dual-core 2.40 GHz processor.

The Pareto frontier of the non-dominated solutions for maximum MRR, minimum SR, and minimum SCE is presented in Fig. 5.9. Three distinct regions are identified along the Pareto frontier of the non-dominated solution set in Fig. 5.9a. These regions are marked as "Min SCE and max MRR", "Balance R_a, SCE and MRR", and "Min R_a". Corresponding regions in the solution (decision variable) space are also indicated in Fig. 5.9b. Milling process parameters that maximize MRR, minimize SR, and minimize SCE are identified in the variable domain at a lower spindle speed 1002.93 r/min and at a higher feed rate 15.75–16.00 mm/min (see Fig. 5.9b). However, the depth of cut a_p varies hugely from 0.4 to 1.28 mm, which means a_p has the most important influence on the optimal results.

Furthermore, a feasible solution set with 15 combinations of milling process parameters is provided for the operator to achieve desired MRR, SCE, and SR, as shown in Table 5.8. From Table 5.8, it can be seen that the No.1 solution has the minimum R_a, and No.2 solution has the maximum MRR and minimum SCE, which have been marked with bold. At different times or in different scenarios, the operator can choose different solutions to achieve different targets. Therefore, compared to other traditional multi-objective optimization algorithms such as desirability analysis [23] and weighted gray relational analysis [8], the multi-objective optimization model based on the constrained NSGA-II can get a Pareto-optimal set which includes all possible optimal solutions and the operator can make the final decision up to the practical situation and specific demands.

In addition, since convergence performance is an important criterion to evaluate optimization algorithms, many methods are proposed to assess it. Generational

Table 5.8 A set of non-domination solutions

No.	n	f	a_p	MRR	SCE	R_a
1	1002.93	15.99	0.40	49.92	0.722	**0.124**
2	1002.93	15.99	1.32	**164.99**	**0.228**	0.261
3	1002.93	15.99	1.02	127.48	0.292	0.226
4	1002.93	15.81	0.95	116.62	0.319	0.215
5	1002.93	15.86	0.51	62.74	0.579	0.144
6	1002.93	15.81	0.79	97.91	0.377	0.193
7	1002.93	15.99	1.11	138.79	0.269	0.238
8	1002.93	15.99	1.29	160.80	0.233	0.258
9	1002.93	15.99	1.09	135.52	0.275	0.234
10	1002.93	15.86	0.66	81.70	0.449	0.171
11	1002.93	15.99	0.41	51.76	0.697	0.127
12	1002.93	15.86	1.14	140.48	0.266	0.240
13	1002.93	15.99	0.64	79.96	0.458	0.168
14	1002.93	15.86	0.63	77.89	0.470	0.166
15	1002.93	15.99	0.42	52.91	0.682	0.129

distance (GD) is widely used for the assessment [36] which has the following representation:

$$GD = \sqrt{\sum_{i=1}^{n} d_i^2 / n} \tag{5.15}$$

where n is the number of the solutions in the current Pareto front, and d_i stands for the Euclidean distance between ith solution in the current Pareto front and the nearest solution in the reference set. And the GD with higher value means worse convergence performance to the reference set.

In order to compare the performance of original NSGA-II and the proposed hybrid approach, both approaches are implemented ten times for the low-carbon optimization model which is discussed in Sect. 5.3.3, and the results are listed in Table 5.9. The simulation results show that the proposed hybrid NSGA-II algorithm has better convergence performance than the original NSGA-II.

Table 5.9 Generational distance of the original NSGA-II and this approach

/	Generational distance									
	1	2	3	4	5	6	7	8	9	10
Original NSGA-II	0.047	0.048	0.055	0.060	0.056	0.053	0.060	0.058	0.042	0.060
Proposed algorithm	0.037	0.034	0.050	0.050	0.044	0.041	0.058	0.039	0.029	0.046

5.6 Online Platform Development and Simulation Case

5.6.1 Online Platform for Carbon Emission Analysis and Optimization

Since there are many kinds of data about carbon emission which need to be analyzed, such as machining power, air cutting power, and energy consumption, we require a platform to satisfy the demand of data collection and analysis. Meanwhile, the real-time data needs to be gathered to validate and amend the proposed models because different parts and processes may influence the regression models and optimization results. Therefore, an online platform for carbon emission analysis was developed to analyze the carbon emission and optimize the process parameters. Moreover, it can provide the function of early warning of fault through the monitoring and simple analysis of the processing power, which can reduce accidents during machining processes. Also, it is simple and convenient for field operation since mobile devices such as smart phones can access the platform.

The schematic diagram of the online platform is shown in Fig. 5.10. Firstly, the power sensor receives the data of power of the milling process in real time. The platform can analyze the original power data for making statistics related to machining power, air cutting power, energy consumption, and total carbon emission and further optimize process parameters. Then, it deposits the results into the database which will be passed to the operator through the Internet. The operator can monitor the carbon emission information and process planning results via his/her handheld tablets or PCs.

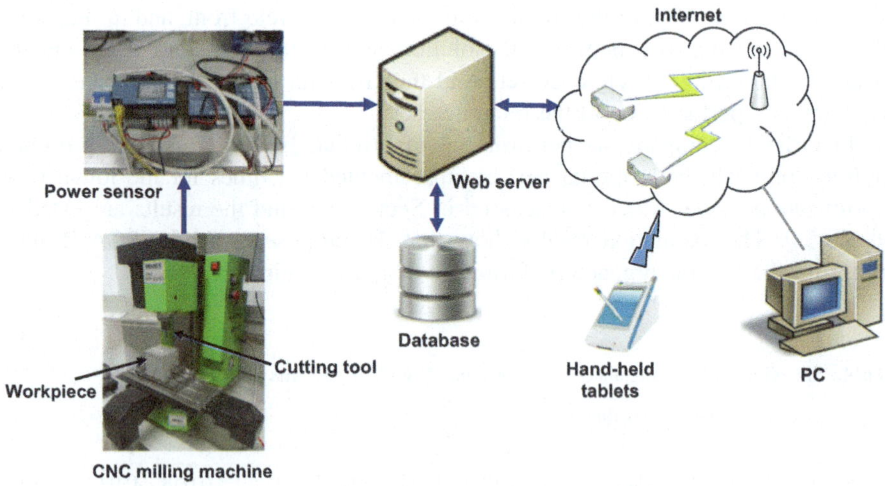

Fig. 5.10 Schematic diagram of the online platform

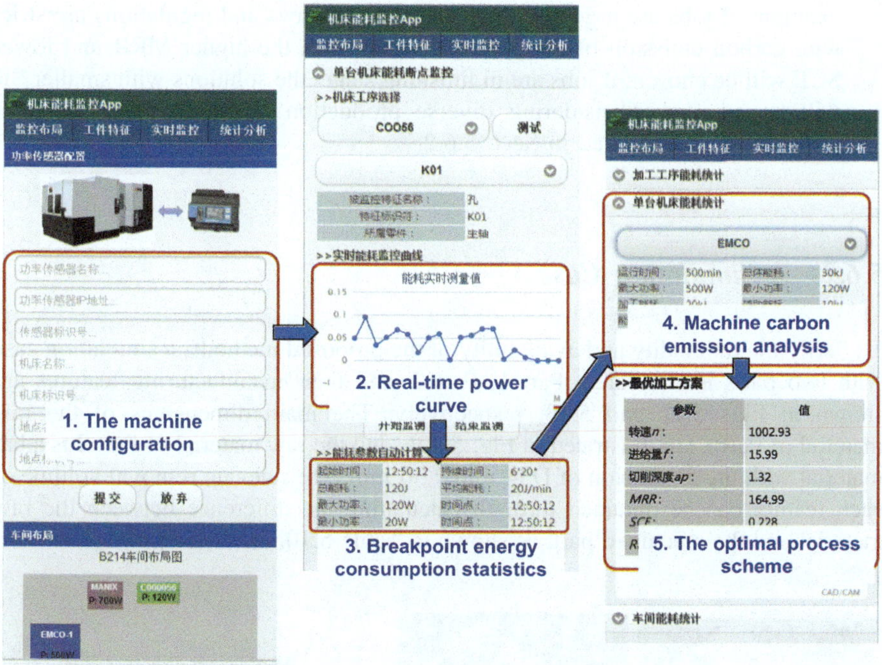

Fig. 5.11 Operation procedures of the online platform

Based on the above schematic diagram, the operation procedure of the platform mainly contains five steps, as shown in Fig. 5.11:

(1) Machine configuration: It mounts sensors to machines for performance monitoring.
(2) Real-time power curve: When the machine starts, the real-time power curve will be plotted and the frequency of data collection is three in one second.
(3) Breakpoint energy consumption statistics: When the machining process is finished and the sensor is stopped, the real-time power curve will end and several parameters will be calculated automatically, such as processing time, total energy consumption/carbon emission, average energy consumption, the average power.
(4) Machine carbon emission analysis: Based on the statistics, the analysis module can analyze the relationship between cutting power, energy consumption, carbon emission, MRR, etc., and milling parameters, namely the regression models.
(5) Optimization and real-time decision making: Based on the analysis of the new data, the regression models will be amended to reduce the error. Then, the optimization process will be performed again to obtain the new and accurate parameters. In accordance with the new Pareto-optimal results, the operator will change the machining parameters according to their specific requirement. For

example, if jobs are urgently demanded and the laws and regulations are strict with carbon emission of the plant, solutions with the higher MRR and lower SCE will be chosen; if jobs are in finishing stage, the solutions with smaller SR will be adopted. Considering diverse production occasions, their different scenarios are considered in Sect. 5.6.2.

5.6.2 A Simulation Case

To verify the rationality and availability of the proposed methods, a simulation case with two parts is simulated. Part 1, Part 2, and their manufacturing features are shown in Figs. 5.12 and 5.13, respectively. The main dimensions of the raw material of Part 1 are illustrated in Fig. 5.12a, and the raw material of Part 2 is a bar material with the dimension of D34 mm * 30 mm. The relevant removal volume of each feature can be obtained through calculating the difference between the raw material and the machined part, as shown in Table 5.10. Since the proposed model

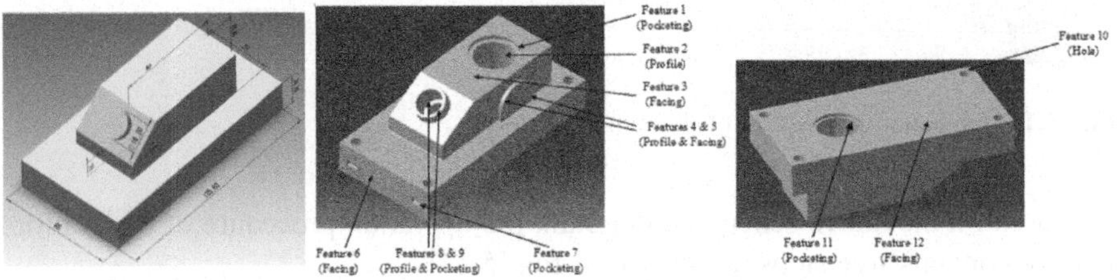

(a) Raw material of Part 1 (b) Features of Part 1

Fig. 5.12 Raw material and features of Part 1

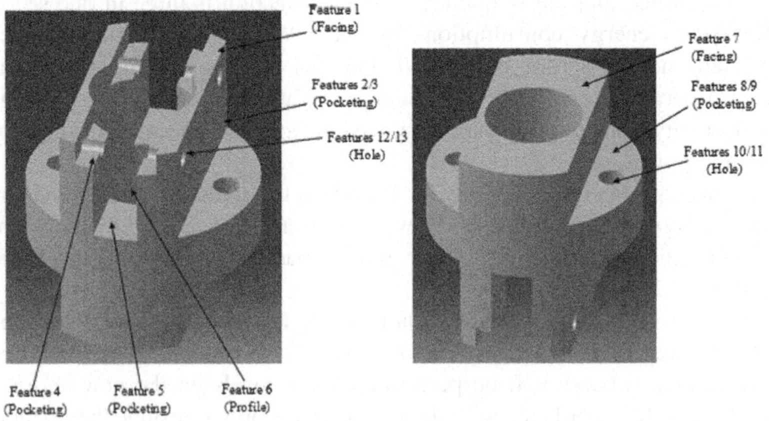

Fig. 5.13 Features of Part 2

Table 5.10 Removal volume of each feature of Part 1 and Part 2

Part 1		Part 2	
Features	Volume (mm^3)	Features	Volume (mm^3)
1	760.3	1	960.0
2	8906.0	2/3	2900.0
3	3562.4	4	802.4
4	1022.5 × 2	5	1245.3
5	917.6 × 2	6	1102.8
6	2968.7	7	960.0
7	55.8	8/9	2780.5
8	235.9		
9	502.7		
11	760.3		
12	14,925.0		

mainly takes the milling process into account, the features of holes are not considered in this study, i.e., feature 10 of Part 1, feature 10/11 and feature 12/13 of Part 2. The parts will be machined on the CNC MM-250S3 milling machine. Considering the different importance of MRR, SR, and SCE under diverse production conditions, three different scenarios are considered as follows:

Scenario 1: MRR and SCE are the main concerns.
Scenario 2: SR is mainly concerned.
Scenario 3: MRR, SCE, and SR are equally important.

For the Scenario 1 and Scenario 2, the No.2 and No.1 processing schemes in Table 5.8 are appropriate, respectively, while the No.6 is suitable for the Scenario 3 since its MRR, SCE, and R_a are 97.91 mm^3/min, 0.377 $kgCO_2$-e/cm^3, and 0.193 um which are all medium. The operator can browse the optimization schemes through accessing the online platform.

Through the proposed empirical modeling and optimization methodology, the processing results of Part 1 can be predicted without the actual processing, as shown in Table 5.11. Except the completion time, the carbon emission, the SR, and further measurands are presented to describe the influence of the process parameters on the energy consumption factors, including air cutting power and cutting power. Also, these empirical modeling methods and processing results are beneficial to the low-carbon design of products.

After analyzing the results, it can be clearly seen that the completion time of Scenario 1 is the shortest, and its carbon emission is also less than other scenarios. However, its product quality is poorer, whose SR is 0.261 um. If the machining scheme of Scenario 2 is adopted, Part 1 can obtain the best machining quality, but its completion time is the longest and its carbon emission is 26.67 $kgCO_2$-e, which is the highest. Comparing with Scenario 1 and Scenario 2, Scenario 3 is a compromise choice, whose results of completion time, carbon emission, and SR are all medium. Overall, the increase of depth of cut will lead to the decrease of

Table 5.11 Simulation results of Part 1

Results	Scenario 1	Scenario 2	Scenario 3	Traditional optimization	Another scheme [8]
Cutting speed (r/min)	1002.93	1002.93	1002.93	1002.28	1000
Feed rate (mm/min)	15.99	15.99	15.81	15.78	16
Depth of cut (mm)	1.32	0.40	0.79	1.05	1.6
Completion time (min)	221.57	732.32	373.38	282.31	183.08
Carbon emission ($kgCO_2$-e)	8.34	26.39	13.78	10.6	6.92
SR (um)	0.261	0.124	0.193	0.228	0.286
Air cutting power (W)	372.6	372.6	372.5	372.4	372.1
Cutting power (W)	529.7	470.7	503.4	517.7	539.3

completion time and carbon emission, but will result in a clear opposing impact on the SR.

In addition, for the three scenarios, the difference of their air cutting power is very small, while the cutting power will decrease with the increase of the depth of cut. Due to the constraint of the maximum cutting power, the cutting powers in these three scenarios are all less than 530 W.

In order to highlight the difference between traditional optimization method and this eco-efficiency method, the optimal milling parameters determined by traditional method were also obtained, as shown in Table 5.11. As previously mentioned, the traditional optimization problem means cutting parameter optimization based on traditional optimization objectives such as MRR, SR, and cutting force, which does not take environmental impact into consideration. In this study, the traditional objective optimization of milling parameters was executed using the similar algorithm. Moreover, MRR and SR were employed as the traditional optimization objectives, and it is assumed that the two objectives are equally important (i.e., weight = 1:1). As noted from Table 5.11, the results of the traditional method are similar to that of the Scenario 1, but the total carbon emission of the latter decreased 21.3%. It is obvious that the results of the proposed method are obtained after the trade-offs between MRR, SR, and carbon emission. Since considering environmental impact shifts the balance to the carbon emission optimization, the SR becomes a little larger when the carbon emission decreases compared to the traditional optimization result. However, this problem could be solved if the constraint of SR is considered.

Furthermore, referring to Yan and Li [8], the machining result by using their optimal milling parameters is also listed in Table 5.11. Although its completion time and carbon emission are optimal, its SR increases from 0.261 to 0.286 comparing to the result of Scenario 1. Meanwhile, it does not consider the cutting power constraint. In a word, their method only considered one situation, but the actual requirement may be varying, and this developed methodology can provide many alternatives for the operator. In other words, the operator can adopt different processing parameters from the non-dominated solutions according to various processing requirements and different processing stages which fit for dynamic manufacturing.

The simulation results of Part 2 are shown in Table 5.12. The similar conclusion can be drawn from Part 2. Comparing Part 2 with Part 1, it can be seen that the total completion time of Part 2 is shorter, so as for its carbon emission. The main reason is that the total volume of Part 2 is much smaller than Part 1.

5.6.3 Discussions

As summarized in Sect. 5.2.3, there are many multi-objective optimization methodologies in manufacturing technologies, such as the Taguchi gray relational analysis for high-speed turning [29], gray relational analysis for micro-wire electric discharge machining [6], and contour plot methodology for laser milling [7]. Through the analysis above, it can be observed that the proposed systemic optimization method has two advantages with respect to other multi-objective optimization models:

(1) Most of these optimization methodologies focus on the qualitative analysis of some optimization objectives, which can only reflect the influence trend of the different parameter combinations on the objectives. The proposed method in this study is an accurate method for the parameter optimization.
(2) Considering the conflict among multiple objectives, the proposed optimization method can generate different Pareto-optimal results, and the operator can choose the suitable parameters according to different requirements.

However, there are also some drawbacks in the proposed method if applied in other manufacturing technologies. Firstly, the universality of the method needs to be validated in different machine tools and different manufacturing technologies because only a simulation case was studied. So various experiments and applications need to be performed in the future work to improve the universality of the proposed method. Then, the multi-objective decision method needs to be established to help the operator to choose a better parameter combination.

Table 5.12 Simulation results of Part 2

	Scenario 1	Scenario 2	Scenario 3
Completion time (min)	65.16	215.36	109.80
Carbon emission (kgCO_2-e)	2.45	7.76	4.05
SR (um)	0.261	0.124	0.193

5.7 Conclusions

Carbon emission reduction in manufacturing industry is imperative. In this study, a systemic optimization approach is presented to identify the values of some key process parameters leading to low-carbon milling operation. By considering production rate, carbon emission, and product quality concurrently, regression models are constructed to characterize the relationship between environmental/productivity/quality objectives and milling parameters. Then, a multi-objective optimization model is further constructed for identifying the optimal process parameters, where the MRR is maximized, and the carbon emission and SR are minimized simultaneously. After several dry milling experiments of different combinations of milling parameters, the regression models are derived and they have a great reliability for depicting behavior of tested milling processes because of a high R^2 value. The hybrid NSGA-II is adopted to solve the optimization model, and the Pareto frontier of the non-dominated solutions are obtained. Finally, based on the regression models and the optimization approach, an online platform is developed to obtain in-process information about the energy consumption and carbon emission for real-time decision making.

Some conclusions are drawn as follows:

(1) The Pareto frontier of non-dominated solutions show that when the optimal spindle speed is 1002.93 r/min, feed rate is 15.99 mm/min, and depth of cut ranges from 0.4 to 1.28 mm; the biggest effect on the objectives is achieved. The increase of depth of cut results in the decrease of SCE and the increase of MRR and SR.
(2) The simulation case shows that in the optimal solutions MRR has a positive correlation with carbon emission, and there is an opposing relationship between MRR and SR.
(3) Comparing to other existing methods, the results of the simulation case indicate that the proposed method can obtain multiple eco-efficient milling schemes through only one calculation which is more efficient for dynamic manufacturing.

Comparing to existing process models and optimization methods for manufacturing process, this research derives some regression models for characterizing milling processes. As a general recommendation, empirical process models need to be developed for other cutting tools, workpiece materials, cutting fluids, and machine tools. The analysis of cutting tools impact also requires improvements by considering its entire life cycle, since it is one of the main contributors regarding the environmental impacts. Therefore, a more accurate analysis can be achieved in the future.

References

1. EIA U., Annual Energy Review. (2011). U.S. Energy Information Admin: Washington DC, USA, pp. 3–4.
2. Duflou, J. R., Sutherland, J. W., Dornfeld, D., Herrmann, C., Jeswiet, J., Kara, S., et al. (2012). Towards energy and resource efficient manufacturing: A processes and systems approach. *CIRP Annals—Manufacturing Technology, 61*(2), 587–609.
3. Bruzzone, A., Anghinolfi, D., Paolucci, M., & Tonelli, F. (2012). Energy-aware scheduling for improving manufacturing process sustainability: A mathematical model for flexible flow shops. *CIRP Annals—Manufacturing Technology, 61*(1), 459–462.
4. Newman, S. T., Nassehi, A., Imani-Asrai, R., & Dhokia, V. (2012). Energy efficient process planning for CNC machining. *CIRP Journal of Manufacturing Science and Technology, 5*(2), 127–136.
5. Gopalsamy, B. M., Mondal, B., & Ghosh, S. (2009). Taguchi method and ANOVA: An approach for process parameters optimization of hard machining while machining hardened steel. *Journal of Scientific & Industrial Research, 68*(8), 686–695.
6. Somashekhar, K. P., Mathew, J., & Ramachandran, N. (2011). Multi-objective optimization of micro wire electric discharge machining parameters using grey relational analysis with Taguchi method. *Proceedings of the Institution of Mechanical Engineers, Part C: Journal of Mechanical Engineering Science, 225*(7), 1742–1753.
7. Campanelli, S. L., Casalino, G., & Contuzzi, N. (2013). Multi-objective optimization of laser milling of 5754 aluminum alloy. *Optics & Laser Technology, 52*(1), 48–56.
8. Yan, J., & Li, L. (2013). Multi-objective optimization of milling parameters—the trade-offs between energy, production rate and cutting quality. *Journal of Cleaner Production, 52*(1), 462–471.
9. Vijayaraghavan, A., & Dornfeld, D. (2010). Automated energy monitoring of machine tools. *CIRP Annals—Manufacturing Technology, 59*(1), 21–24.
10. Behrendt, T., Zein, A., & Min, S. (2012). Development of an energy consumption monitoring procedure for machine tools. *CIRP Annals—Manufacturing Technology, 61*(1), 43–46.
11. Hu, S., Liu, F., He, Y., & Hu, T. (2012). An on-line approach for energy efficiency monitoring of machine tools. *Journal of Cleaner Production, 27*(1), 133–140.
12. Kara, S., & Li, W. (2011). Unit process energy consumption models for material removal processes. *CIRP Annals—Manufacturing Technology, 60*(1), 37–40.
13. Mori, M., Fujishima, M., Inamasu, Y., & Oda, Y. (2011). A study on energy efficiency improvement for machine tools. *CIRP Annals—Manufacturing Technology, 60*(1), 145–148.
14. Arif, M., Stroud, I. A., & Akten, O. (2014). A model to determine the optimal parameters for sustainable-energy machining in a multi-pass turning operation. *Proceedings of the Institution of Mechanical Engineers, Part B: Journal of Engineering Manufacture, 228*(6), 866–877.
15. Kant, G., & Sangwan, K. S. (2014). Prediction and optimization of machining parameters for minimizing power consumption and surface roughness in machining. *Journal of Cleaner Production, 83,* 151–164.
16. Choudhury, S. K., & Appa, Rao I. (1999). Optimization of cutting parameters for maximizing tool life. *International Journal of Machine Tools and Manufacture, 39*(2), 343–353.
17. Lalwani, D. I., Mehta, N. K., & Jain, P. K. (2008). Experimental investigations of cutting parameters influence on cutting forces and surface roughness in finish hard turning of MDN250 steel. *Journal of Materials Processing Technology, 206*(1), 167–179.
18. Zain, A. M., Haron, H., & Sharif, S. (2010). Prediction of surface roughness in the end milling machining using artificial neural network. *Expert Systems with Applications, 37*(2), 1755–1768.
19. Campatelli, G., Lorenzini, L., & Scippa, A. (2013). Optimization of process parameters using a response surface method for minimizing power consumption in the milling of carbon steel. *Journal of Cleaner Production, 66*(1), 309–316.

20. Camposeco-Negrete, C. (2013). Optimization of cutting parameters for minimizing energy consumption in turning of AISI 6061 T6 using Taguchi methodology and ANOVA. *Journal of Cleaner Production, 53*(1), 195–203.
21. Bhattacharya, A., Das, S., Majumder, P., & Batish, A. (2009). Estimating the effect of cutting parameters on surface finish and power consumption during high speed machining of AISI 1045 steel using Taguchi design and ANOVA. *Production Engineering, 3*(1), 31–40.
22. Winter, M., Li, W., Kara, S., & Herrmann, C. (2014). Determining optimal process parameters to increase the eco-efficiency of grinding processes. *Journal of Cleaner Production, 66*(1), 644–654.
23. Bhushan, R. K. (2013). Optimization of cutting parameters for minimizing power consumption and maximizing tool life during machining of Al alloy SiC particle composites. *Journal of Cleaner Production, 39*(1), 242–254.
24. Rajemi, M. F., Mativenga, P. T., & Aramcharoen, A. (2010). Sustainable machining: Selection of optimum turning conditions based on minimum energy considerations. *Journal of Cleaner Production, 18*(10–11), 1059–1065.
25. Narita, H., Desmira, N., Fujimoto, H., Environmental burden analysis for machining operation using LCA method. In *Manufacturing Systems and Technologies for the New Frontier, 2008, Springer*, pp. 65–68.
26. Nalbant, M., Gökkaya, H., & Sur, G. (2007). Application of Taguchi method in the optimization of cutting parameters for surface roughness in turning. *Materials and Design, 28* (4), 1379–1385.
27. Wibowo, A., & Desa, M. I. (2012). Kernel based regression and genetic algorithms for estimating cutting conditions of surface roughness in end milling machining process. *Expert Systems with Applications, 39*(14), 11634–11641.
28. Quiza, Sardiñas R., Rivas, Santana M., & Alfonso, Brindis E. (2006). Genetic algorithm-based multi-objective optimization of cutting parameters in turning processes. *Engineering Applications of Artificial Intelligence, 19*(2), 127–133.
29. Pawade, R. S., & Joshi, S. S. (2011). Multi-objective optimization of surface roughness and cutting forces in high-speed turning of Inconel 718 using Taguchi grey relational analysis (TGRA). *The International Journal of Advanced Manufacturing Technology., 56*(1–4), 47–62.
30. Kuram, E., Ozcelik, B., Bayramoglu, M., Demirbas, E., & Simsek, B. T. (2013). Optimization of cutting fluids and cutting parameters during end milling by using D-optimal design of experiments. *Journal of Cleaner Production, 42*(1), 159–166.
31. Palanikumar, K., Latha, B., Senthilkumar, V. S., & Karthikeyan, R. (2009). Multiple performance optimization in machining of GFRP composites by a PCD tool using non-dominated sorting genetic algorithm (NSGA-II). *Metals and Materials International, 15*(2), 249–258.
32. China D.O.M.I. (1997). Mechanical engineering handbook: Machine building technology and equipment (II). 1997. Beijing, China, China Machine Press.
33. Fang, X. D., & Safi-Jahanshahi, H. (1997). A new algorithm for developing a reference-based model for predicting surface roughness in finish machining of steels. *International Journal of Production Research, 35*(1), 179–199.
34. Deb, K., Pratap, A., Agarwal, S., & Meyarivan, T. (2002). A fast and elitist multiobjective genetic algorithm: NSGA-II. *IEEE Transactions on Evolutionary Computation, 6*(2), 182–197.
35. Davoodi, B., & Tazehkandi, A. H. (2013). Experimental investigation and optimization of cutting parameters in dry and wet machining of aluminum alloy 5083 in order to remove cutting fluid. *Journal of Cleaner Production, 68*(1), 234–242.
36. Yang, L., Deuse, J., & Jiang, P. (2013). Multi-objective optimization of facility planning for energy intensive companies. *Journal of Intelligent Manufacturing, 24*(6), 1095–1109.

Chapter 6
Cyber-Physical System and Big Data-Enabled Scheduling Optimization for Sustainable Machining

Y. C. Liang, X. Lu, S. Wang and W. D. Li

Abstract Modern manufacturing is challenged by customized, high-variety and low-volume job orders, dynamic ambient working conditions in shop floors, and stricter requirements on sustainability. Data-driven approach for manufacturing planning, control, and management will be a new research trend to tackle the challenge. In this research, based on in-process monitoring on energy consumption of machining processes and data analytics technologies, an innovative Cyber-physical system (CPS) and Big Data-enabled scheduling optimization system for sustainable computer numerical control (CNC) machining has been developed. This system is augmented with intelligent mechanisms for enhancing adaptability to condition dynamics in machining shop floors. The system consists of scheduling and re-scheduling functions. For scheduling, an artificial neural networks (ANNs)-based algorithm has been designed to establish energy models of components machined in a shop floor according to current working conditions of CNC machines. Based on the energy models, a fruit fly optimization (FFO) algorithm has been applied to generate a multi-objective optimized schedule. For re-scheduling, another ANNs-based algorithm has been developed to monitor the energy consumption of components during machining in real time. Scheduling optimization will be triggered to generate an updated schedule if there are significantly varying working conditions and re-scheduling adjustments are necessary. The system has been validated through deployment into a European machining company and industrial case studies to demonstrate technical innovations and the great potential of applicability in practice.

Keywords Cyber-physical system · Big Data · Energy efficiency Scheduling optimization

---

Y. C. Liang · X. Lu · S. Wang · W. D. Li (✉)
Faculty of Engineering, Environment and Computing,
Coventry University, Coventry, UK
e-mail: weidong.li@coventry.ac.uk

W. Li and S. Wang (eds.), *Sustainable Manufacturing and Remanufacturing Management*, https://doi.org/10.1007/978-3-319-73488-0_6

6.1 Introduction

Modern manufacturing industries are characterized by customized, high-value, low-volume job orders as well as increasing sustainability requirements, keener competition for zero-defect manufacturing in quality and faster delivery of manufactured products. In a manufacturing shop floor, there are various dynamic changes of job priority and ambient working conditions, unexpected delay, dimensional and geometric deviations of components during manufacturing, etc. This requires companies to configure their processes efficiently so as to adapt to customized jobs and ambient working conditions during manufacturing lifecycles. In the meantime, manufacturing processes are energy intensive making the sector one of the primary energy consumption and carbon footprint sources. Manufacturing processes in factories, in which motors, compressors and machine tools need to be powered and adequately heated, contribute to over 24% of total European energy consumption [1]. Therefore, effective implementation of manufacturing sustainability is increasingly prevalent for companies. The roadmap research of EU intelligent manufacturing toward 2020, conducted by an international consortium consisting of active and leading researchers from Europe, Japan, Korea, and the USA (FP7 project IMS2020 [2]), indicated that energy efficiency and carbon emission reduction of manufacturing should be enhanced during manufacturing process planning and management. Energy-efficient regulations and incentives should take effect and the manufacturing industry is required to undergo a full-scale transformation toward sustainability.

Scheduling is a key enabler in manufacturing shop floors to minimize lifecycle cost, enhance adaptability to customized manufacturing, and improve manufacturing sustainability [3, 4]. Data-driven approach is becoming a new trend in manufacturing scheduling and control. Different from conventional manufacturing management approaches that are based on pre-defined manufacturing conditions, data-driven manufacturing is enabled by smart sensors, Cyber-physical system (CPS) and in-process Big Data analytics to address varying working conditions and customized manufacturing processes for optimization based on current conditions.

In this chapter, an innovative CPS and Big Data-enabled scheduling optimization system for sustainable computer numerical control (CNC) machining is presented. CPS, Big Data and intelligent mechanisms have been developed to enhance the system's adaptability to customized jobs, dynamic machining conditions and sustainability requirements. The system consists of scheduling and re-scheduling functions as follows:

- For scheduling, based on innovative CPS design and its integration with CNC machines, energy data are continuously collected in real time to reflect the current working conditions of machines. An artificial neural networks (ANNs)-based algorithm is then used to establish the energy models of components machined in the shop floor. Based on the energy models, a fruit fly optimization (FFO) algorithm is applied to optimize multi-objective schedules;

- For re-scheduling, another ANNs-based algorithm is applied to monitor the energy consumption patterns of machining components during machining processes in real time. Optimization is applied to generate an updated re-scheduled plan if there are significantly varying conditions and re-scheduling adjustments are needed.

The innovative characteristics of the system are from the following aspects:

- Scheduling optimization is supported by CPS and energy consumption Big Data collected and analyzed in real time. Knowledge of manufacturing conditions is accumulated and updated during machining processes. Prior experiments required to establish the energy models of customized machining processes for scheduling optimization are therefore minimized, and scheduling efficiency and effectiveness are greatly improved.
- Dynamic working conditions during machining lifecycles are efficiently addressed. ANNs are designed not only for establishing energy models to support scheduling, but also for monitoring energy patterns to support re-scheduling optimization according to dynamic conditions during machining lifecycles;
- Innovative design of CPS, Big Data, ANNs, and optimization algorithm are integrated for systematic and effective implementation of manufacturing intelligence. The system has been validated through various real-world industrial case studies in a European machining company to demonstrate the significant potentials of applicability in practice.

The rest of the chapter is organized as follows: In Sect. 6.2, a literature survey on energy modeling and machining scheduling optimization, CPS and Big Data for intelligent manufacturing systems is presented. In Sect. 6.3, the system framework is presented. In Sects. 6.4, 6.5, and 6.6, design of the ANNs, optimization algorithm, and CPS system are discussed, respectively, in detail. Case studies and system validation are given in Sect. 6.7. In Sect. 6.8, conclusions are drawn.

6.2 Literature Survey

Energy-efficient scheduling for machining has been mainly investigated from two aspects: energy modeling to support scheduling, and effective algorithms and strategies for energy-efficient scheduling optimization. To effectively support decision making for energy-efficient scheduling, research on CPS and Big Data for manufacturing has been conducted in recent years. Comprehensive review and research frameworks can be found from [5, 6]. Some latest related works on the above aspects are summarized below.

6.2.1 Energy Modeling for Machining

Sustainability is an important aspect to be considered in machining optimization nowadays. It is important to build accurate energy models for sustainability optimization. Fang et al. [7] developed a general multi-objective mixed integer programming formulation. An energy model has been established based on an empirical machining model, including part width, feed per tooth, machining speed, and specific machining energy. He et al. [8] built an energy assessment framework for a machining workshop based on CNC codes. The framework consists of four layers, i.e., workshop layer, task layer, manufacturing unit layer, and machine tool layer. For each layer, the major element that affects the energy consumption mostly was modeled. Yan and Li [9] proposed a thermal equilibrium and empirical approach for energy consumption modeling during machining processes under various machining parameters, such as material removal rate, idling power, machine tool-specific coefficients, and standby power. Winter et al. [10] utilized a sensitivity analysis method to analyze the energy performance of a grinding process, in which energy consumption is affected by key grinding parameters, including grinding depth, grinding speed, and dressing speed. Wang et al. [4] built multi-level models for energy consumption from two levels. In the machining level, ANNs were employed to estimate energy consumption and surface roughness based on the spindle speed, machining speed, depth of cut and width of cut. Furthermore, an energy model in a shop floor level was developed, in which start-up phases, idle/change phases, working phases, and shutdown phases were considered. Yan et al. [11] designed a multi-level model to optimize energy from both machining parameters and shop floor levels. It requires off-line experiments to build the energy model of machining processes based on qualitative analysis and gray relational analysis. A model in a shop floor consists of processing energy consumption, setup energy consumption, transportation energy consumption, standby energy consumption, and overhead energy consumption. The above works are summarized in Table 6.1.

6.2.2 Energy-Efficient Machining Scheduling

Based on energy models, optimization algorithms need to be applied to improve the energy efficiency of machining processes. A number of research works focus on the investigation of better optimization algorithms to improve the energy efficiency of machining processes. Tang et al. [12] developed an improved particle swarm optimization approach to address dynamic scheduling under unexpected disruptions to reduce energy consumption and makespan simultaneously. Liu et al. [13] developed a novel multi-objective genetic algorithm based on NSGA-II to minimize

Table 6.1 Energy models for machining shop floor

Works	Input	Optimization targets	Research methods
Fang et al. [7]	Machining width, feed per tooth, machining speed, and specific machining energy	Makespan, peak power demand, and carbon footprint	Empirical models and case studies of machining cast iron plates with slots
He et al. [8]	CNC codes	Energy consumption for spindle, axis feed, tool changes, coolant pump, and fixed energy-consuming units of CNC machines	Empirical models for spindle, axis feed, tool changes, coolant pump, and fixed energy-consuming units
Yan and Li [9]	Material removal rate, idle power, machine tool-specific coefficients, and standby power	Energy consumption model	Thermal equilibrium and empirical
Winter et al. [10]	Machining depth, machining speed, and dressing speed	Energy consumption	Sensitivity analysis method
Wang et al. [4]	Spindle speed, machining speed, depth of cut, and width of cut; Number of machines and the number of jobs to be processed	Surface quality, energy consumption, and machining removal rate; Energy consumption for idle, working, tool change, and setup	ANNs to establish a model for surface quality and energy consumption Empirical models for idle, working, tool change, and setup
Yan et al. [11]	Material removal rate, spindle speed; Number of machines and the number of jobs to be processed	Idle power and operation power, energy consumption for processing setup, transportation, standby, and overhead	Off-line experiments for gray relational analysis Empirical models for processing, setup, transportation, standby, and overhead

the total non-processing electricity consumption and total weighted tardiness. The process provides a function for parent and children combination and elitism to improve optimization further. Yan et al. [11] designed a multi-level energy model, utilized gray relational analysis to optimize machining parameters and a genetic algorithm to optimize the makespan and energy consumption. Based on a model related to real-time monitored data, Xu et al. [14] designed an enhanced Pareto-based bees algorithm to optimize energy consumption and productivity. Salido et al. [15] developed a memetic algorithm to minimize energy consumption under makespan constraints within the re-schedule zone. The above works are summarized in Table 6.2.

Table 6.2 Energy-efficient machining scheduling optimization algorithms

Works	Optimization targets	Algorithms
Tang et al. [12]	Energy consumption and makespan	Improved particle swarm algorithm
Liu et al. [13]	Energy consumption and tardiness	NSGA-II based genetic algorithm
Yan et al. [11]	Makespan and energy consumption	Gray relational analysis, genetic algorithm
Xu et al. [14]	Energy consumption and productivity	Enhanced Pareto-based bee algorithm
Salido et al. [15]	Energy consumption	A memetic algorithm

6.2.3 *CPS and Big Data for Sustainable Manufacturing*

CPS and Big Data technologies have been increasingly used for sustainable manufacturing implementation in real time, and the latest sensors, communication technologies, and Big Data analytics are leveraged. Dai et al. [16] developed an integrated Big Data platform in cloud for data flow-based analysis. Nagorny et al. [17] developed a CPS to support reasoning-based control, monitoring and management functions for manufacturing devices in a shop floor. When Big Data are accumulated, effective information management infrastructures, such as Hadoop Distributed File System (HDFS), MapReduce, YARN, HBase, HiveQL, and NoSQL, become effective tools for storage, management, processing, interpretation, and visualization of the data [18]. Big Data technologies for the Engine Health Monitoring Unit (EHMU) were developed in Rolls-Royce Plc. to collect real-time Big Data from working engines, systems, and factory lines to monitor and optimize system performance and manufacturing quality. In Raytheon Corp, a Big Data technology was implemented as a smart factory to manage information from different data sources, such as sensors, simulations, and all other manufacturing records in the factory [19]. Prabhu [20] investigated Big Data collection by using a CPS. The collected data can be modeled as a set (device ID, time, event ID) supported by sensors and actuators. Chaplin et al. [21] developed a method for the integration of legacy CNC controllers and decentralization, context-awareness, and data distribution services. Liu and Jiang [22] designed a CPS for achieving intelligent manufacturing by collecting, processing, and visualizing Big Data in a micro-manufacturing system laboratory. The above works are summarized in Table 6.3.

Based on the analysis, the following research requirements are identified:

- The current research on scheduling optimization is mainly based on empirical modeling and prior/off-line experiments. This will limit the efficiency of scheduling for customized optimization requirements and online adjustments based on ambient working conditions;

Table 6.3 CPS and Big Data for manufacturing applications

Works	System characteristics
Dai et al. [16]	Integrating Big Data platform in cloud for dataflow-based analysis
Nagorny et al. [17]	CPS devices, information collection for reasoning-based control, monitoring, and management functions
Big Data-Startups [18]	Big Data for engine health monitoring unit
Noor [19]	Big Data technology to achieve smart factories to manage information from different data sources
Prabhu [20]	CPS as a set model with device ID, time, event ID, which can be modified
Chaplin et al. [21]	Integration of legacy CNC controllers with decentralization, context-awareness, data distribution services
Liu and Jiang [22]	CPS for achieving intelligent manufacturing establishment for Big Data collection, processing, and visualization

- It is critical to develop an adaptive, cost-effective, CPS, and Big Data-enabled system to efficiently optimize multi-objective schedules for customized machining. Based on the system, changing conditions during customized manufacturing lifecycles will be updated quickly so as to achieve the best system performance during machining lifecycles;
- It will be significant if developed systems are proved through system deployment in factories and using various industrial real-world case studies for system validation.

6.3 System Framework

6.3.1 System Functions

In the past, scheduling for CNC machining processes in a shop floor was developed based on pre-defined machining conditions. Generated scheduled plans are represented in a relatively rigid format [23]. On the other hand, machining in shop floors has become increasingly customized. Machining customization is characterized by upcoming disturbance, disruption, and uncertainty (e.g., dynamic changes of job priority, unexpected delay, aging, or degrading of tooling and machines) [24]. It is essential to update scheduling flexibly when machining conditions are changed in a shop floor to ensure the effectiveness of optimized schedules for the dynamic working conditions. Real-time monitoring, analysis, and optimization based on working conditions are crucial to achieve adaptive scheduling optimization. The current practice of relevant data collection in shop floors is mainly manually based.

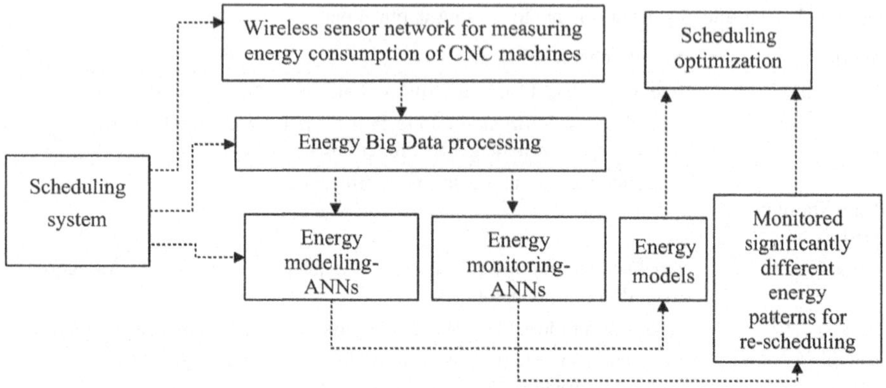

Fig. 6.1 Functions of the system

Due to large quantities and diverse product models, this data collection process is tedious, time-consuming, and error-prone. To address the challenge, in this research, a novel CPS and Big Data-enabled machining scheduling optimization system has been developed. The system, which is shown in Fig. 6.1, consists of the following functions:

- A wireless sensor network has been designed and integrated with CNC machines as CPS for measuring the energy consumption of CNC machines to support scheduling optimization. Electricity measurement sensors are mounted onto CNC machines;
- A Big Data processing infrastructure has been developed for collecting, storing, processing, and visualizing real-time energy data from CNC machines;
- The system consists of scheduling and re-scheduling functions. For scheduling, an ANNs-based algorithm (i.e., energy modeling-ANNs) has been designed to establish the energy models of components machined in a shop floor. Based on these energy models, a scheduling optimization algorithm (i.e., scheduling optimization) will be triggered to generate an optimal schedule with the target of minimized energy consumption, shortest makespan, etc. and addressing working conditions efficiently. An ANNs-based algorithm (i.e., energy monitoring-ANNs) has been developed to monitor the energy patterns of machining components. Scheduling optimization will be triggered to generate an updated scheduling plan whenever there are significantly different energy patterns identified which may indicate machine replacements/tool wear for re-scheduling adjustments. More details of the above processes are explained in the following section.

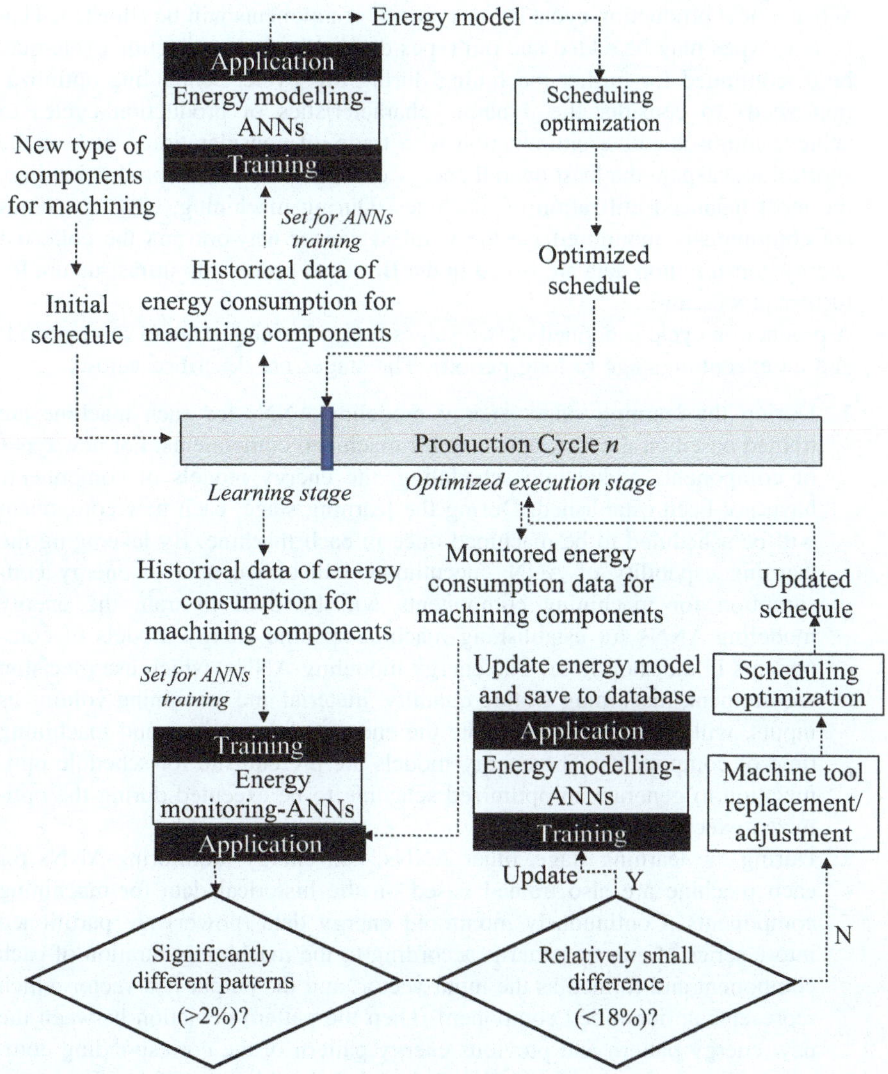

Fig. 6.2 Process of using the system for scheduling optimization

6.3.2 System Flow

The detailed working process of the system is shown in Fig. 6.2. Explanations are given below:

- A customized machining process can be managed as a series of production cycles. During a production cycle, types of components for production are certain while quantities of each component for production could be varying.

When a new production cycle starts, types of components will be adjusted. That is, new types may be added and old types during the last production cycle may be discontinued for further machining during this cycle. Scheduling optimization needs to consider the dynamic characteristics of production cycles to achieve multi-objective optimization with trade-off considerations, such as the shortest makespan, the least overall energy consumption for the production line, the most balanced utilization of machines. During machining, CNC machines are continuously monitored via the wireless sensor network and the collected energy consumption data are stored in the Big Data processing infrastructure for further processing;

- A production cycle is defined as two stages, i.e., a learning stage (a short period) and an execution stage (a long period). The stages are described below:

1. During the learning stage, energy modeling-ANNs for each machine are trained based on the historical data for machined components. For new types of components added into scheduling, the energy models of components have not been established. During the learning stage, each new component will be scheduled to be machined once in each machine. By leveraging the learning capability of ANNs, accumulated historical data of energy consumption for machining components will be used to train the energy modeling-ANNs for establishing machine-specific energy models of components in the shop floor. The energy modeling-ANNs, which use precision requirement, machining feature quantity, material and machining volume as inputs, will be trained to estimate the energy consumptions and machining time of components. The energy models are prerequisite for schedule optimization to generate an optimized schedule to be executed during the optimized execution stage;
2. During the learning stage, other ANNs, i.e., energy monitoring-ANNs for each machine are also trained based on the historical data for machining components. Continuously monitored energy data (power) are partitioned into a series of energy patterns according to the machining duration of each component and formed as the input vector, and the output is a vector which represents an identified component. Then the pattern deviation between the new energy pattern and previous energy pattern of the corresponding component is calculated (refers to Formula 6.1 defined later). If the difference is within a small range, energy modeling-ANNs will be fine-tuned (reasonable aging or degrading conditions of machines, tool wear, etc.); otherwise there are severe aging or degrading conditions of machines/cutters, so that machines/cutters should be replaced or maintained/temporarily excluded from scheduling. Under the circumstance, scheduling optimization will be triggered to generate an updated schedule plan. The initial values of difference ranges (2 and 15%) are obtained from experiment results for a simple component, which will be discussed in Sect. 6.4.2, and these values would be updated during machining stage when more energy data are acquired for individual components.

- As thus, the learning and optimization processes are driven by real-time energy Big Data collected during machining. The experiments required to investigate the energy consumption performance of components for CNC machines to support scheduling optimization can be carried out during machining processes. Therefore, the time and cost required for scheduling decision making can be significantly reduced and dynamic working conditions can be efficiently addressed.

6.4 ANNs for Energy Modeling and Monitoring

In the system, two ANNs (i.e., energy modeling-ANNs and energy monitoring-ANNs) have been designed to establish energy models for components machined in a shop floor and monitor energy consumption patterns in real time, respectively. The design and functions of the two ANNs are depicted in the following sections.

6.4.1 Energy Modeling-ANNs

Figure 6.3 shows the relationship between inputs and output of ANNs. In ANNs, calculated weights and bias through a training algorithm are utilized to calculate y_{i_in} and y_{i_out}. ANNs provide several distinguishing characteristics [25]: (1) the capability to capture and perform complex input and output relationships; (2) no prior knowledge regarding input and output is needed to develop learning models. The trained model can enhance the knowledge database and the newly learned knowledge can refine ANNs. A fitting function is performed by the ANNs; (3) the capability to update ANNs when new data are used.

To support scheduling optimization, machine-specific energy models for components to be machined should be established. The design of the energy modeling-ANNs is illustrated in Fig. 6.4.

The inputs of the energy modeling-ANNs are precision requirement, machining features, material and machining volume. The outputs are energy consumption and machining time. The historical information from the learning stage is collected to train the ANNs. The training procedure of the ANNs is shown in Fig. 6.5.

Various training algorithms have been proposed for ANNs modeling in various application areas. According to the users' guide of the MATLAB Neural Network Toolbox™ 7, generally, Levenberg-Marquardt is the fastest. BFGS Quasi-Newton is fast but it is less efficient for a large network. Scaled conjugate gradient and resilient backpropagation are better choices for pattern recognition and require less memory for a large network. Bayesian regularization generally has the good generalization performance for difficult data. Variable learning rate backpropagation is

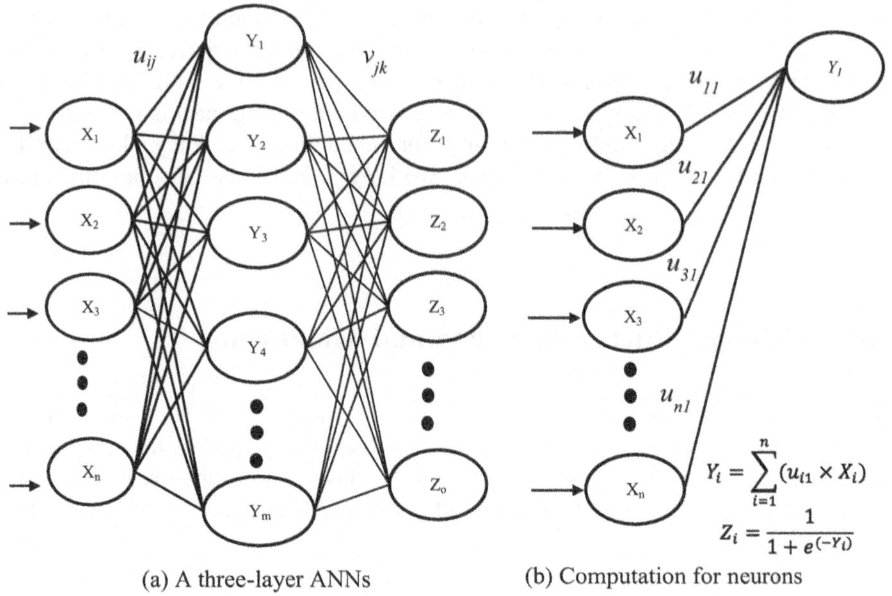

Fig. 6.3 Design of multi-layer ANNs

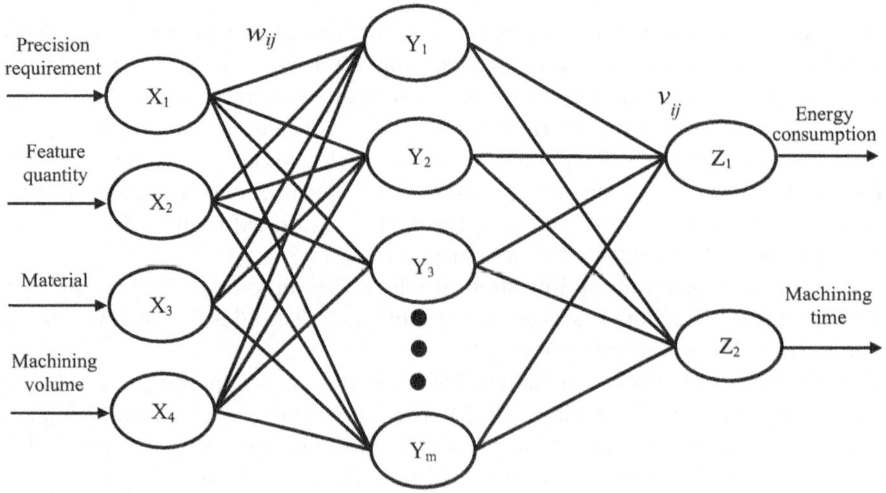

Fig. 6.4 Design of the energy modeling-ANNs

usually the slowest algorithm with similar storage requirements to resilient back-propagation, but it is still useful for some situations as it is preferred to converge slowly for some problems. In this research, nine different popular training algorithms have been utilized during the learning stage considering that different

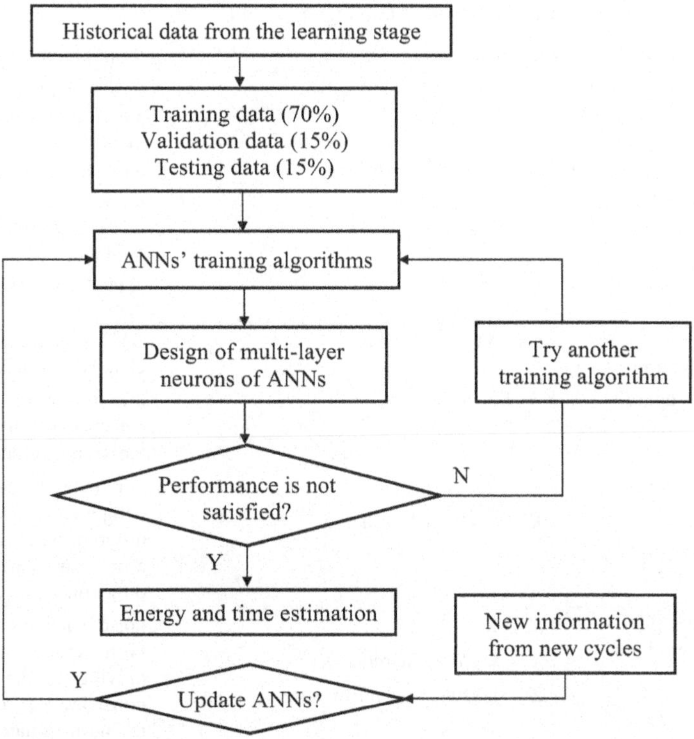

Fig. 6.5 Flow chart for the energy modeling-ANNs' training

training algorithms perform different accuracy and simulation time for different scenarios. The best performance algorithm will then be selected at the learning stage. Table 6.4 shows the details of the training algorithms employed.

6.4.2 Energy Monitoring-ANNs

Identical component operations should generate similar energy consumption patterns with slight deviations. Abnormal energy patterns indicate significant condition changes of machines and/or tooling; thus, a need for machine maintenance is necessary, leading to re-scheduling optimization. The detection of abnormal conditions during machining processes is divided into two steps: the energy monitoring-ANNs and the deviation calculation of energy patterns. The design of energy monitoring-ANNs, which are to identify the machining component during machining processes, is illustrated in Fig. 6.6 and Table 6.5. To train energy monitoring-ANNs with energy profiles, continuously monitored energy data (power) are partitioned into a series of energy patterns according to the machining

Table 6.4 Information of neural network training algorithms

Training algorithm	Equation	Symbol explanation
Levenberg-Marquardt	$jj = jX \times jX$ $je = jX \times E$ $dX = -(jj + I \times \mu)/je$	X is weight and bias, E are all errors, I is the identity matrix, and μ is an adaptive value
BFGS Quasi-Newton	$X = X + a \times dX$ $dX = -H/gX$	dX is the search direction, a is coefficient, gX is the gradient, and H is an approximate Hessian matrix
Resilient backpropagation	$dX = \Delta X. \times \text{sign}(gX)$	ΔX are all initialized elements to delta0
Scaled conjugate gradient	$X = X + a$	a is to minimize the performance time along the search direction
Scaled conjugate gradient	$X = X + a \times dX$ $dX = -gX + dX\_\text{old} \times Z$	Parameter Z can be computed in several different ways, dX_old is a previous search direction
Fletcher-Powell conjugate gradient	$X = X + a \times dX$ $dX = -gX + dX\_\text{old} \times Z$ $Z = \text{normnew}_{\text{sqr}}/\text{norm\_sqr}$	norm_sqr represents the norm square of the previous gradient and normnew_sqr represents the norm square of the current gradient
Polak-Ribiére conjugate gradient	$X = X + a \times dX$ $dX = -gX + dX\_\text{old} \times Z$ $Z = ((gX - gX_{\text{old}})' \times gX)$ $/\text{norm\_sqr}$	gX_old represents the gradient on the previous iteration
One-step secant	$X = X + a \times dX$ $dX = -gX + Ac \times X_{\text{step}}$ $+ Bc \times dgX$	X_step and dgX are the changes for weights and gradient for the previous iteration, respectively; Ac and Bc are the coefficients
Variable learning rate backpropagation	$dX = mc \times dX\text{prev} + lr \times mc \times d\text{perf}/dX$	mc represents a momentum constant, dXprev represents the change to the weight or bias for the last iteration, lr represents the learning rate, and $d$perf is the derivative of the performance

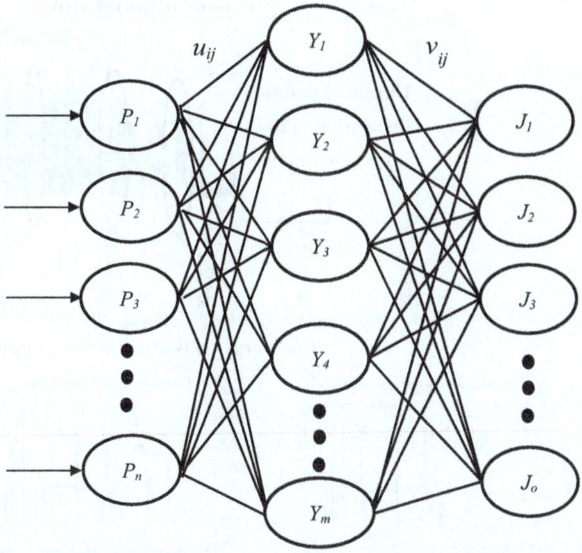

Fig. 6.6 Design of the energy monitoring-ANNs

Table 6.5 Input and output of the energy monitoring-ANNs

Input vector	Output vector
Point 1 in the energy pattern	Component category 1 [1, 0, 0, …, 0]
Point 2 in the energy pattern	Component category 2 [0, 1, 0, …, 0]
Point 3 in the energy pattern	Component category 3 [0, 0, 1, …, 0]
… …	… …
Point *n* in the energy pattern	Component category *o* [0, 0, 0, …, 1]

duration of each component. The monitored power consists of several stages, e.g., idle, machining, machine start-up/shutdown. The data partition process is based on the power range to concentrate on the data of the machining process. When the power is above a given threshold defining the working range for machining, the energy profiles of a component are partitioned from the monitored Big Data. An illustrative example is given in Fig. 6.7. The input is a vector of an extracted energy pattern during production, and the output is a vector representing a component category for the input energy pattern. The vector length of the input n is the maximum length of power readings (maximum number of a machining process of a component). For a component with a smaller number of durations compared to the maximum number of duration, 0 will be added at the end of the pattern to standardize the vector lengths of the patterns to be the same to facilitate ANNs' processing. In terms of output, 0 is the total number of component types. For instance, if the output is for Component 1, the output will be [1 0 0 … 0] and the vector length of the output is the total number of component types.

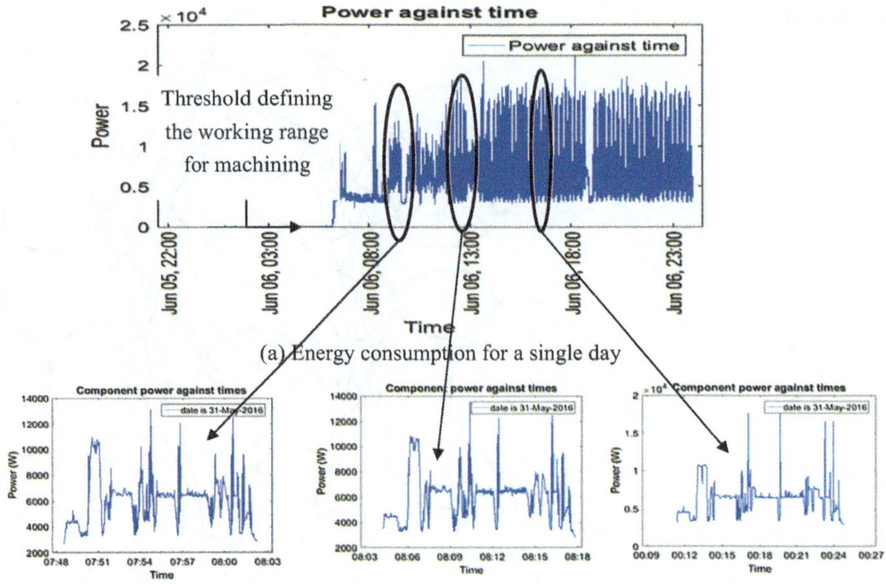

(a) Energy consumption for a single day

(b) Three energy patterns for three individual components partitioned from the daily energy

Fig. 6.7 Examples of energy partition to individual energy patterns for individual components

The abnormal condition is then detected by calculating the deviation of the current energy pattern with the previous energy pattern during machining process for the identified component. The deviation (Δ) of energy consumption between current production and previous one is defined below:

$$\Delta = \frac{E_i - E_j}{E_j} \times 100\% \tag{6.1}$$

where E_i is the energy consumption of the current production and E_j is the energy consumption of previous production.

If the deviation is within a certain range ($\Delta_{max} > \Delta > o'_{min}$) it means the energy modeling-ANNs need to be updated. If the deviation exceeds a large threshold ($\Delta > o'_{max}$) it means the pattern is significantly different from its corresponding previous pattern; this could indicate an abnormal condition. Based on that, the conditions of machines and tools will be investigated. If pattern deviations are caused by severe aging or degrading conditions of tooling and/or machines, the machines or tools will be replaced. Scheduling optimization is triggered to generate an updated schedule to address the current working conditions. The initial value of o'_{min} is set as 2% which is a trade-off of modeling-ANNs accuracy and update frequency, the initial o'_{max} is set as 18% which is obtained from the following

(a) Component under normal condition

(b) Component under abnormal condition

Fig. 6.8 Machined component under normal and abnormal conditions

experiment. Meanwhile, the values for individual components would be updated during machining stage when more energy data are acquired.

The part in Fig. 6.8 was repeatedly machined on a CNC machine for many times. The energy consumption and surface roughness for all machined parts are measured and compared. It is noted that when the machine tool is in severe wear condition, the energy deviation of the two consecutive parts is 21.67% (Fig. 6.9), and the corresponding average surface roughness is 1.15 and 3.22 μm, respectively. For all other parts the deviation is far below 18% and surface roughness is below 1.15 μm. It can be concluded that when the deviation is greater than 18%, the machine tool is quite possibly degrading or there is an abnormal condition in the machine. To this end, the initial o'_{max} is set as 18%.

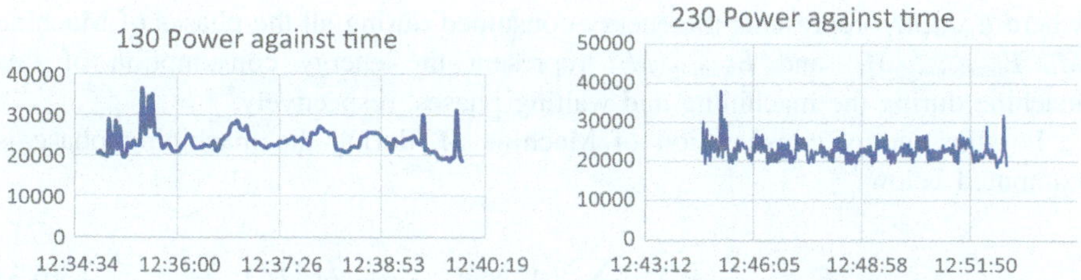

(a) Power profile under normal condition

(b) Power profile under abnormal condition

Fig. 6.9 Power profiles for machining under normal and abnormal conditions (energy consumptions are 2.03 and 2.47 KWh, respectively, and the deviation is 21.67%)

6.5 Optimization for Shop Floor Scheduling

In a shop floor, it is not unusual to see the machines are not managed properly leading to longtime standby of machines. As a result, a large amount of energy is wasted and there is extra labor cost associated with the waste as well. Therefore, scheduling optimization based on energy modeling-ANNs has been implemented for sustainable machining.

6.5.1 Computation of Energy, Makespan, and Machine Utilization Level

In this research, the conditions/assumptions for scheduling optimization are defined as following:

- A shop floor consists of a set of CNC machines: Machine $M = \{M_1, M_2, M_3, \ldots, M_n\}$, to machine a set of components $J = \{J_1, J_2, J_3, \ldots, J_m\}$;
- Each component has to be finished in a single machine when it is available;
- Each component will use different machining time and energy consumption by using different machines;
- The preparation time is assumed to be constant in the same machine;
- There are no sequencing constraints for machining components;
- The machine start-up and shutdown energy are negligible.

The energy consumption of a machine is from machining and waiting phases:

$$E_{\text{total}}(M_i) = E_{\text{machining}}(M_i) + E_{\text{waiting}}(M_i) \tag{6.2}$$

where $E_{\text{total}}(M_i)$ represents the energy consumed during all the phases of Machine M_i. $E_{\text{machining}}(M_i)$ and $E_{\text{waiting}}(M_i)$ represent the energy consumption of this machine during the machining and waiting phases, respectively.

For the energy consumption of Machine M_i during the machining phase is computed below:

$$E_{\text{machining}}(M_i) = \sum_{j=1}^{m} \left(A_{ij} \times E_{\text{machining}}(M_i, J_j)\right) \tag{6.3}$$

where A_{ij} represents whether Machine Mi needs to be machining for Component J_j. $E_{\text{machining}}(M_i, J_j)$ represents the machining energy consumption of Component J_j by Machine M_i. m is the total number of components to be machined. A_{ij} can be defined as below:

$$A_{ij} = \begin{cases} 1 & \text{Componet } J_j \text{ is machined by } M_i \\ 0 & \text{Componet } J_j \text{ is not machined by } M_i \end{cases} \tag{6.4}$$

For the energy consumption of Machine M_i during the waiting phase is computed below:

$$E_{\text{waiting}}(M_i) = \sum_{j=1}^{m} \left(A_{ij} \times E_{\text{waiting}}(M_i, J_j)\right) \tag{6.5}$$

where $E_{\text{waiting}}(M_i, J_j)$ represents the energy consumption of the machining waiting time of Machine M_i for Component J_j.

The total energy consumption for all the machining jobs by all the machines can be calculated below:

$$E_{\text{total}} = \sum_{i=1}^{n} E_{\text{total}}(M_i) \tag{6.6}$$

where E total represents total energy consumption in all machines. n is the number of total machines.

Similarly, the time consumption for each machine during machining can be calculated as below:

$$T_{\text{total}}(M_i) = T_{\text{machining}}(M_i) + T_{\text{waiting}}(M_i) \tag{6.7}$$

where $T_{\text{total}}(M_i)$ represents the total time consumption during all the phases of Machine M_i. $T_{\text{machining}}(M_i)$ and $T_{\text{waiting}}(M_i)$ represent the time demand of this machine during all the machining and waiting phases, respectively.

To calculate the time used during the whole production time: makespan, which is the maximum production time for all components in all machines, can be computed below:

$$\text{Makespan} = \underset{j=1}{\overset{n}{\text{Max}}} \left(T_{\text{total}}(M_i)\right) \tag{6.8}$$

The balanced utilization of machines in a shop floor is defined below:

$$\mu = \frac{\sum_{i=1}^{n} T_{\text{total}}(M_i)}{n} \tag{6.9}$$

$$\text{Utilisation\_level} = \sqrt{\sum_{i=1}^{n} \left(T_{\text{total}}(M_i) - \mu\right)^2} \tag{6.10}$$

6.5.2 Scheduling Optimization Modeling and Normalization

In this research, minimization of energy consumption, makespan, and balanced utilization level of machines are considered. As the three different objective functions can have very different magnitudes, normalization of the objective functions is required prior to the optimization of the weight summed objective function. Since the maximum and minimum values of these three objective functions are unknown before optimization, a suitable normalization schema that normalizes the objective functions in the Nadir and Utopia points is employed [26]. The Utopia point z_i^U provides the lower bound of the ith objective function and can be obtained by minimizing the ith objective function individually, i.e.,

$$z_i^U = f_i\left(x^i\right) = \min\{f_i(x)\} \tag{6.11}$$

The upper bound is then obtained from the Nadir point z_i^N, which is defined as:

$$z_i^N = f_i\left(x^k\right) = \max_{1 \ll j \leq I}\{f_i\left(x^j\right)\} \tag{6.12}$$

where I is the total number of objective functions.

This normalization schema may be computationally expensive when the problem dimension is very large. For this research, the time spent on this calculation is acceptable as the number of optimization parameters is not very large. Hence, the energy consumption, makespan, and utilization level are to be normalized individually as:

$$\begin{cases} NE = (E_{\text{total}} - z_1^U)/(z_1^N - z_1^U) \\ NT = (\text{Makespan} - z_2^U)/(z_2^N - z_2^U) \\ NU = (\text{Utilisation} - z_3^U)/(z_3^N - z_3^U) \end{cases} \tag{6.13}$$

The fitness function is calculated as weighted sum of the three objectives below:

$$\text{Fitness} : \min(w_1 \cdot NE + w_2 \cdot NT + w_3 \cdot NU), w_1 + w_2 + w_3 = 1 \tag{6.14}$$

6.5.3 Optimization Algorithm

In this chapter, a latest evolutional optimization algorithm, i.e., FFO (fruit fly optimization), has been developed and improved for scheduling optimization. FFO is a relatively new optimization algorithm [27]. It provides multiple fruit fly groups for parallel search during the evolution process, which is competitive compared to evolutional and other main-stream optimization algorithms due to the local optima avoidance. The algorithm mimics the search behavior of fruit fly. In this research, the classic procedure of the algorithm is applied for schedule optimization during the learning stage of a production cycle. Meanwhile, a mutation operation has been added into the algorithm for scheduling adjustments.

Scheduling optimization during the learning stage

During the search process, the initial group of fruit flies are swarm centers and subgroups generated around each fruit fly are employed for simulating the leadership hierarchy. In addition, two main steps of searching, smell-based search and vision-based search, are implemented:

- Smell-based search: Sub-populations are randomly generated around each initial fruit fly, and the fitness of each fruit fly is calculated.
- Vision-based search: The fruit fly centers are determined with best fitness.

In the FFO algorithm, initial swarm centers are generated randomly. Sub-populations are generated randomly around the swarm centers. However, the initial swarm centers could be far from the target, which will waste computing resource and potentially miss the best solution. In order to improve the quality of the initial swarm centers, a bigger initial group population will be randomly generated and the best m fruit fly will be selected as the initial swarm centers.

The flow of the FFO algorithm is shown in Fig. 6.10.

The process of applying the algorithm for machining scheduling optimization is depicted below:

1. Create an initial fruit fly population i, which is composed of a matrix containing the information of components and corresponding CNC machines for machining. The matrix below illustrates an example to establish a fly (in the example, assuming there are 3 types of CNC machines and 10 components in total) (Fig. 6.11).
2. Select the best m fruit fly as initial swarm centers and set the maximum iterations for the optimization $T_{\max}$.
3. Smell-based search: Generate one random sub-population around each swarm centers. The matrix below illustrates an example to establish a fly in sub-population. Two components will be randomly exchanged, and one CNC machine type will be randomly switched to another CNC machine. The fitness of each fly is calculated for vision-based search (Fig. 6.12).
4. Vision-based search: Replace current swarm centers with the best fruit fly if the fitness is improved. Also, in order to avoid local optimal result, the probability of accepting worse result is adopted in order to achieve global optimal result when Eq. (6.15) satisfies:

$$e > \text{rand} \tag{6.15}$$

$$e = \exp(-\text{abs}(dC))/\text{Time} \tag{6.16}$$

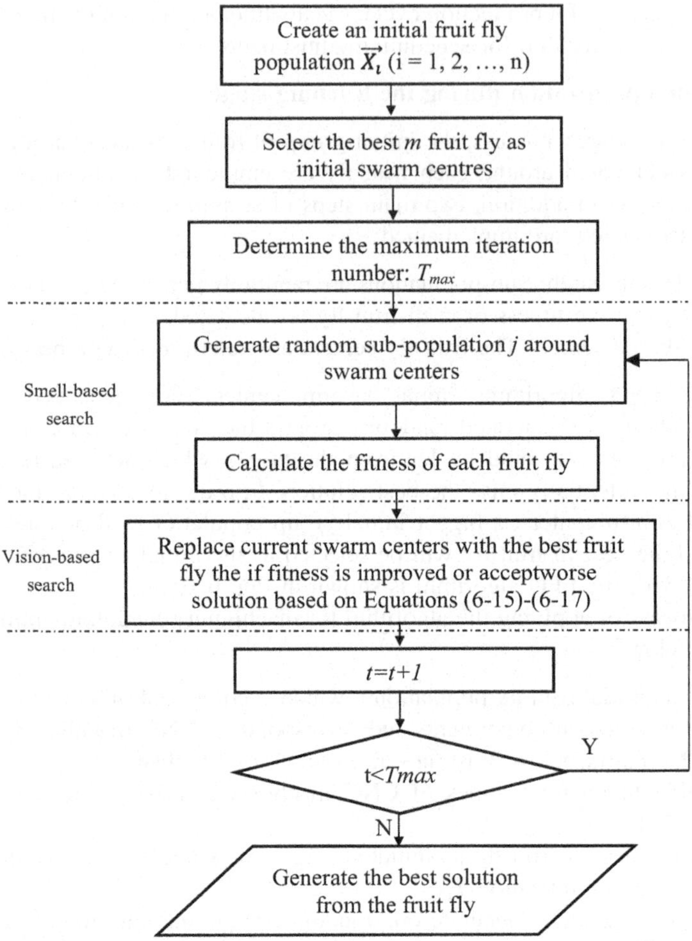

Fig. 6.10 Flow chart of the optimization algorithm

Fig. 6.11 Modeling of a fly for the FAO algorithm

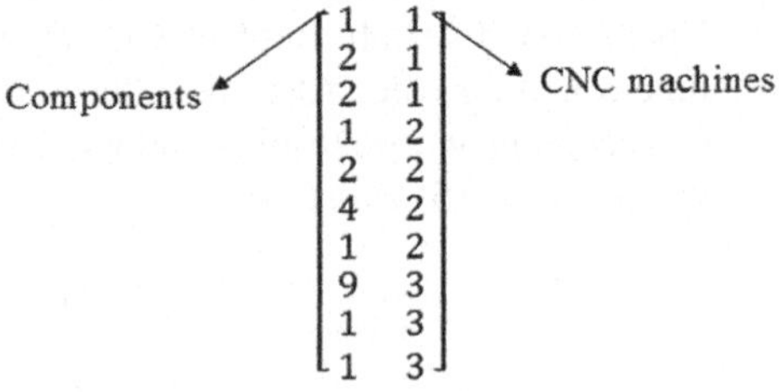

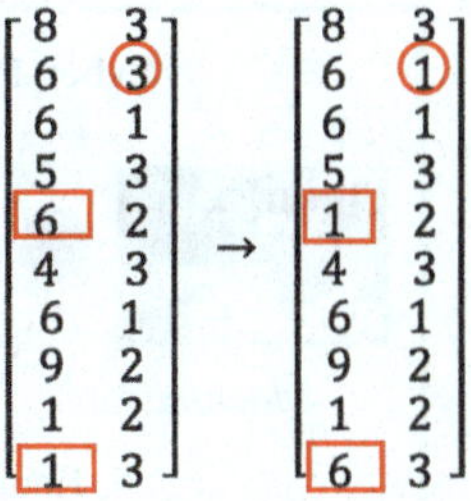

Fig. 6.12 Modeling of a fly in sub-population for the FAO algorithm

$$dC = \text{fitness}\,2 - \text{fitness}\,1 \tag{6.17}$$

where e represents the coefficient that determines whether the new result is accepted, rand is the random number between 0 and 1, dC represents the difference between the new makespan and old one, and Time is the number of current iteration.

5. Repeat Steps 3–4 until reaching the maximum iterations. The best solution in the fly population is selected.

Re-scheduling optimization during the optimized execution stage

Once an optimized schedule is generated through the above process, the solution is applied during the optimized execution stage. In this stage, energy monitoring-ANNs are used for monitoring and identifying significant energy pattern change if any. If the energy pattern change leads to a need for machine maintenance or tool change, a re-scheduling is triggered. As thus, the FFO algorithm is improved and a mutation operation is added so as to make the adjustments on the schedule flexibly. The process is to monitor the machining data during production. If the energy deviation $\Delta > o'_{max}$, re-apply the FFO algorithm to optimize the entire scheduling.

6.6 CPS and Big Data Infrastructure

6.6.1 Design of CPS for Collecting and Monitoring Energy Big Data

Wireless sensor networks (WSNs) are configured and sensors are mounted onto CNC machines as a CPS system. The system is operated through radio communication through recent technologies—IEEE 802.15.4 and WSNs producers' own communication protocols (e.g., NXP). As illustrated in Fig. 6.13, based on the latest WSN technology and the Internet of Things protocols (6LoWPAN (IPv6 over

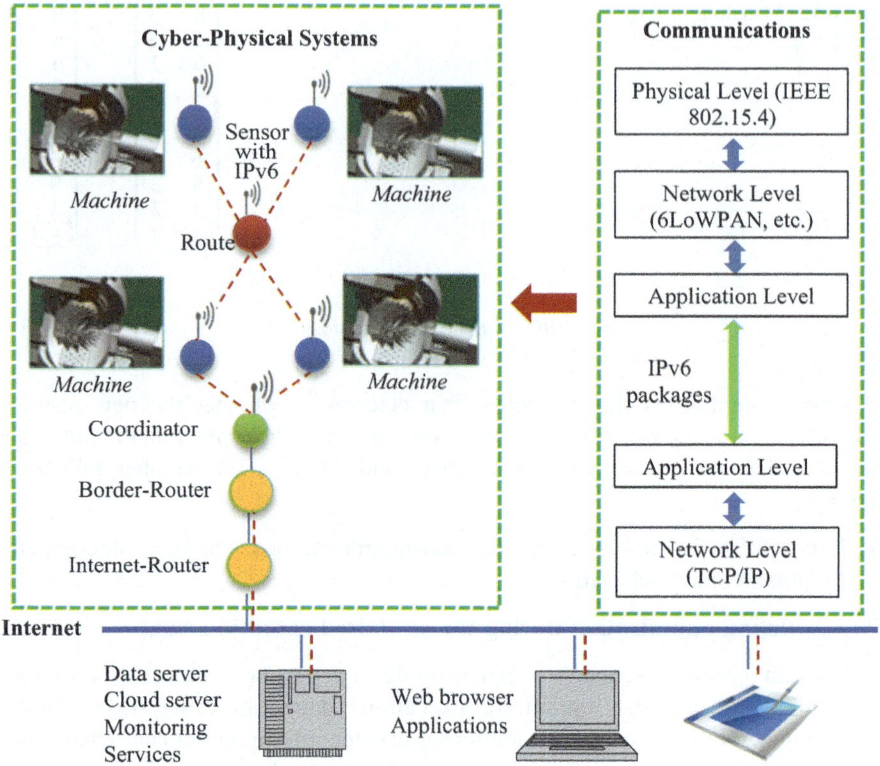

Fig. 6.13 Design of the system for energy Big Data collection

Low power Wireless Personal Area Networks), energy consumption is collected and the data are used to support online decision making of scheduling optimization with sustainability, economic and technical considerations. Three-phase current and voltage are measured utilizing current and voltage sensors. Measured data are transmitted through 2.4 GHz Wi-fi to a coordinator which is connected to Internet-router. Energy data are stored in Hadoop for Big Data processing.

6.6.2 Design of Big Data Infrastructure

Monitored energy data can be defined as Big Data, which are characterized by high volume (e.g., more than 10G volume for six-month monitoring), variability (e.g., time, machine IDs, and current readings), and velocity (e.g., 9 current data samples in a second for 3-machine monitoring) [28]. For Big Data processing, Hadoop

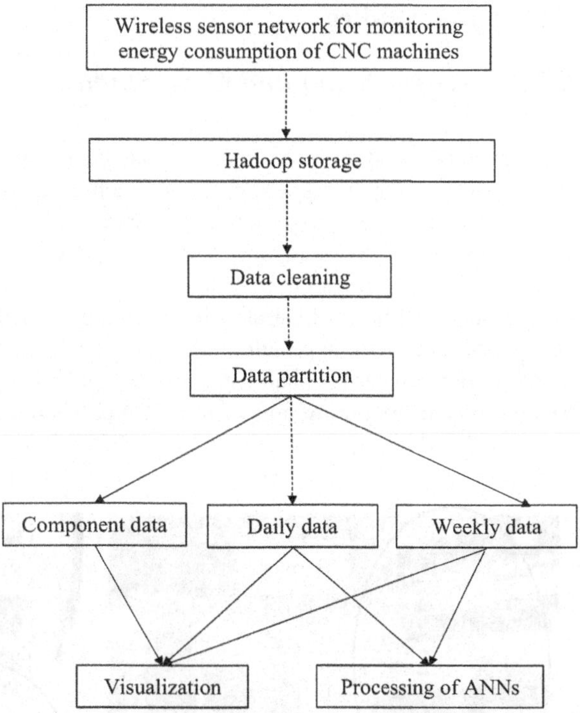

Fig. 6.14 Design of the Big Data infrastructure for energy Big Data processing

Distributed File System (HDFS), MapReduce, and YARN are useful tools for distributed data storage, data processing acceleration, and resource management to deliver consistent operations, security, and data governance tools across clusters, respectively. In this research, the Hadoop Hive system, which integrates the above tools and functions, is used to store and present the monitored energy Big Data. Data partition is used to split the data on component-, daily-, weekly- or monthly basis through MapReduce for data parallel processing. With the parallel storage and data processing functions, the Big Data are processed efficiently which cannot be handled by traditional data platform such as MySQL.

Due to the huge amount of data accumulation, there might be issues when importing data into Hadoop for processing: (1) data duplication due to data accumulation in sensor nodes; (2) data missing due to Wi-fi signal blockage in the shop floor. If there is data duplication or data missing, it is necessary to carry out data cleaning to remove duplicated data and make sure that data sets have the same length to train the ANNs and process data using the ANNs. The relevant process is illustrated in Fig. 6.14.

6.7 Case Studies

6.7.1 System Setup and Case Studies

The developed system in this research has been deployed in a machining company in UK. The company specializes in machining high-precision components for automotive, aerospace, and tooling applications. For case studies, a production line, consisting of 3 CNC machines (MX520, MAZAK, and HAAS) and accessory equipment, has been monitored and analyzed for scheduling optimization. Part of the production line is illustrated in Fig. 6.15, and the specifications are listed in Table 6.6. The system is automated with Lang Eco Towers and robot arms for loading raw materials and storing completed components after machining. Electricity sensors are mounted on the CNC machines. Energy data are transferred

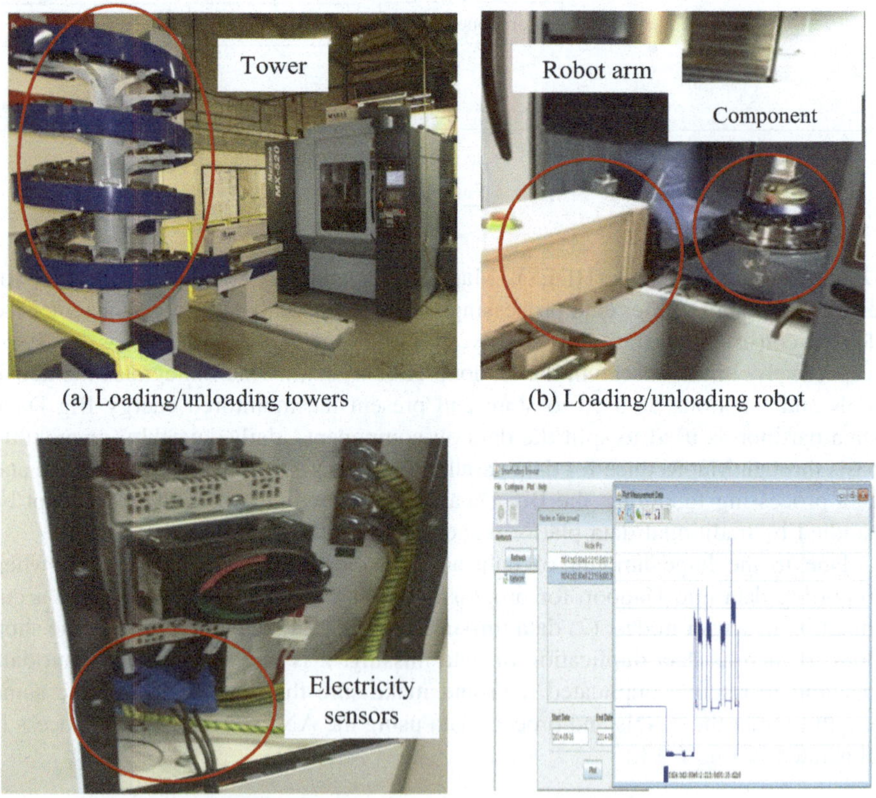

(a) Loading/unloading towers (b) Loading/unloading robot

(c) Power measurement (d) Power measurement software

Fig. 6.15 Loading/unloading towers, robot feeding, and energy measurement of CNC machines

Table 6.6 Specifications of the monitored CNC machines and accessory equipment

Machines and accessory equipment	Specifications			
	Max. travel (mm)	Loading capacity (kg)	Rapid traverse: X/Y/Z (m/min)	Spindle speed rev/min
MX520	630/560/510	150	40	12,000–20,000
MAZAK VTC-800/20SR	2,000/800/720	–	50	18,000
HAAS VF-2TR	762/406/508	36.3	25.4	8100
	Handling weight (kg)	Max. workpiece size (mm)		
Lang eco towers	450	350 × 200 × 200		
Robot arms	30	350 × 200 × 200		

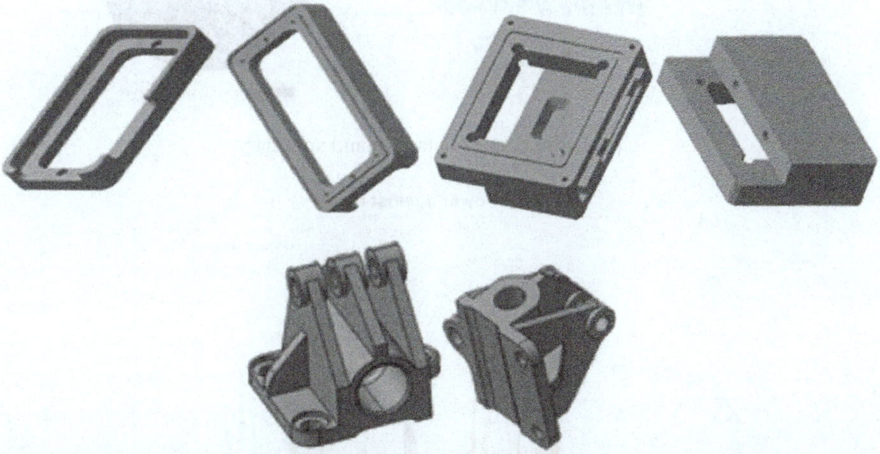

Fig. 6.16 Three example components

to the Hadoop Big Data server through Wi-fi signals in the shop floor. For each CNC machine, the collection rate of energy data is 3 readings/s. Some machined components are shown in Fig. 6.16.

6.7.2 Energy Data Measurement

The energy consumption data can be seen from Table 6.7. Each machine uses 3-phase electricity. The voltage is 220 V and power factor is 0.82. As shown in Fig. 6.17, daily, weekly, and longer period data plots can be generated and visualized from the system.

Table 6.7 Samples of energy Big Data

Time	Current 1 (A)	Current 2 (A)	Current 3 (A)	Voltage (V)	Power (W)
03/10/2016 00:00:00	0.25098	0.25098	0.50196	220	181.107
03/10/2016 00:00:01	0.14091	0.18788	0.09394	220	76.260
03/10/2016 00:00:02	0.14091	0.09394	0.14091	220	67.787
03/10/2016 00:00:03	0.09394	0.18788	0.09394	220	67.787
03/10/2016 00:00:04	0.18788	0.18788	0.04697	220	76.260
03/10/2016 00:00:05	0.14091	0.04697	0.09394	220	50.840

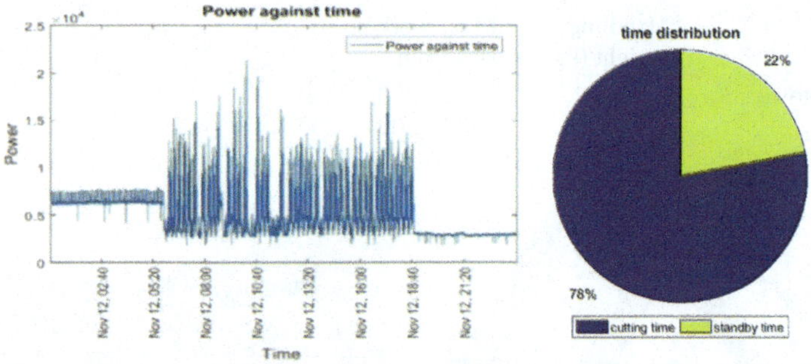

(a) Daily energy data plot and summary

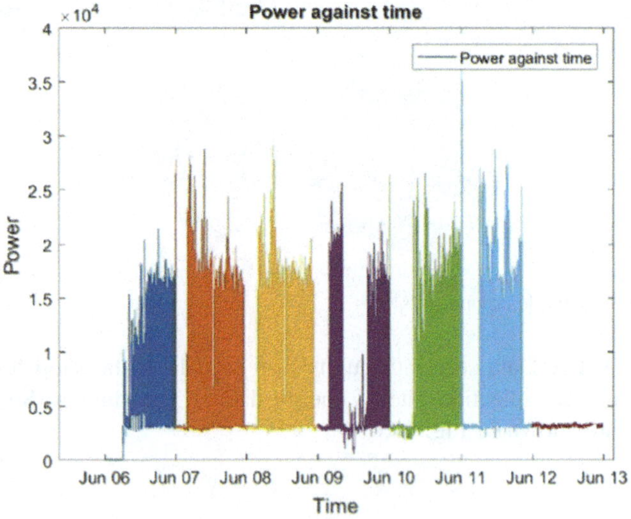

(b) Weekly energy data plot

Fig. 6.17 Examples of energy data plots and summary

6.7.3 Energy Modeling-ANNs

During each production cycle, energy data are collected, and energy models are established by the energy modeling-ANNs. Each machine is associated with an energy modeling-ANNs. The inputs of the ANNs include precision, feature quantity, and machining volume, and the outputs are the predicted energy consumption and machining time to support scheduling optimization. Table 6.8 provides the results of 13 types of components and relevant energy consumption and time generated by the energy modeling-ANNs.

Training results based on different ANNs training algorithms and ANNs structures are shown in Table 6.9. Levenberg-Marquardt with 20 hidden neurons has the best performance of prediction accuracy for energy consumption and machining time. Meanwhile, iterations and training time are relatively short. Therefore, Levenberg-Marquardt with 20 hidden neurons is selected for energy modeling-ANNs.

6.7.4 Energy Monitoring-ANNs

Thirteen types of components with total 78 energy pattern samples are utilized to train the energy monitoring-ANNs. Figure 6.18 shows 3 different samples of 3 components for training in MX520, MAZAK, and HAAS, respectively. Once the ANNs have been trained, the ANNs are used for monitoring energy consumption to identify abnormal patterns. An example is shown in Fig. 6.19, a new pattern for Component 1 with a deviation of 19.24% is identified, which indicates a possible severe tool wear or machine degradation. Through observation, it was found out that the tool was wear at the moment and the quality of the component did not satisfy the roughness requirement. A tool replace is needed, and re-scheduling adjustment was therefore made.

6.7.5 Optimization

The optimization simulation has been programmed using the MATLAB 2016 programming language. Fruit fly optimization (FFO), genetic algorithm (GA), and simulated annealing (SA) have been utilized for comparison. Simulations have been run for 10 times for each algorithm for comparing average results. The optimized multi-objectives are weighted energy consumption, makespan, and utilization level.

Regarding the parameter set, the FFO algorithm has three key parameters to achieve optimum optimization performance: the swarm center population i, the size of sub-population j, and worse result acceptance factor e. According to [29]

Table 6.8 Input and output of the energy modeling-ANNs

Components	Precision (μm)	Feature quantity	Machining volume(cm^3)	Machining time (min)			Energy consumption (KWh)		
				MX520	MAZAK	HAAS	MX520	MAZAK	HAAS
1	2.6	7	242.3	16.5	13	27	0.405	0.443	0.565
2	2.6	11	248.7	13.5	25	37	0.812	0.546	0.932
3	2.6	19	486.6	33	28	38	2.122	1.472	2.886
4	2.6	11	353.8	21.5	24	33	1.521	1.56	1.824
5	2.6	6	462.15	33	32	43	2.203	2.533	2.912
6	2.6	4	237.3	12	22	23	0.962	1.488	1.563
7	2.6	5	392.45	27	23	32	2.423	2.13	2.733
8	2.6	2	216.1	18.5	24	11	0.863	0.845	0.996
9	2.6	9	190	10	18	17	1.135	0.729	1.765
10	2.6	20	218.2	8	19	22	1.002	0.582	1.322
11	2.6	14	260.1	9	13	16	2.201	1.685	2.531
12	2.6	14	333.2	20	26	35	1.967	1.634	2.127
13	2.6	10	295	24	23	17	2.199	2.519	2.329

Table 6.9 Accuracy comparison with different training algorithms and number of neurons

Descriptions	Training time (s)	Iterations	Hidden neurons	RMSE
Levenberg-Marquardt	1.865	78	10	2.167
	1.654	54	20	1.832
Bayesian regularization backpropagation	1.851	86	10	1.998
	2.188	99	20	2.394
BFGS Quasi-Newton	1.809	85	10	2.312
	1.914	85	20	1.897
Resilient backpropagation	1.739	79	10	2.076
	1.745	60	20	2.005
Scaled conjugate gradient	1.632	70	10	1.983
	1.869	79	20	1.933
Scaled conjugate gradient	1.81	86	10	2.123
	1.798	69	20	2.532
Fletcher-Powell conjugate gradient	1.915	104	10	2.612
	1.62	57	20	2.721
Polak-Ribiére conjugate gradient	1.594	63	10	2.661
	1.875	91	20	1.886
One-step secant	1.82	109	10	2.558
	1.83	90	20	1.962
Variable learning rate backpropagation	3.52	62	10	2.009
	4.22	51	20	2.127

regarding the impact of each parameter, the size of sub-population is the most significant parameter. Therefore, the sub-population is the critical parameter.

Figure 6.20 shows the optimization results by FFO, GA, and SA. All the algorithms are executed for 1000 iterations. Table 6.10 shows the compared optimization information.

In the company, the initial energy consumption, makespan, and machine utilization level in shop floor for the same production are 120 KWh, 750, and 250.47 min, respectively. FFO generates the best optimization results for energy consumption, makespan, and utilization level with a good simulation time of 2.03 s.

All optimization results are shown in Table 6.10. Overall, this FFO system can achieve energy consumption saving, makespan reduction, and utilization level improvement for 38.26, 30.61, and 91.83%, respectively. The utilization level optimized by FFO is less than 20.5 min, which means all the three machines relatively equally engaged during the production process. Furthermore, it can be seen from the figures that the best optimized results for FFO can be achieved within 90 iterations, which is faster than the convergence of GA and SA. Therefore, it can be proved that the FFO has good robustness for this scheduling problem.

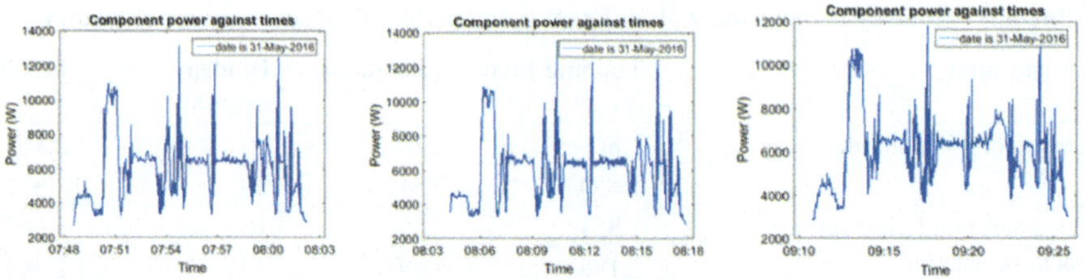

(a) 3 samples for Component 1

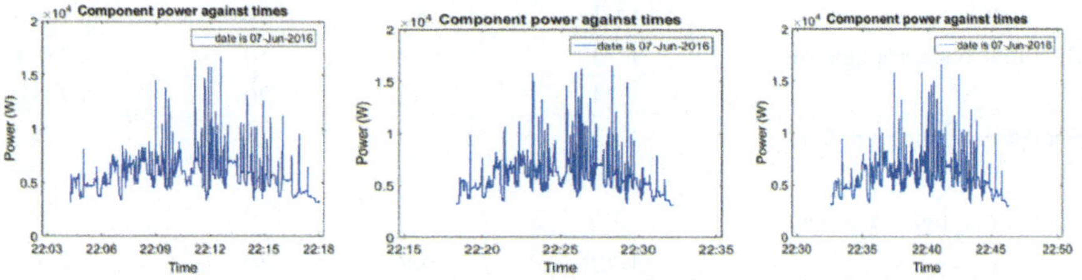

(b) 3 samples for Component 2

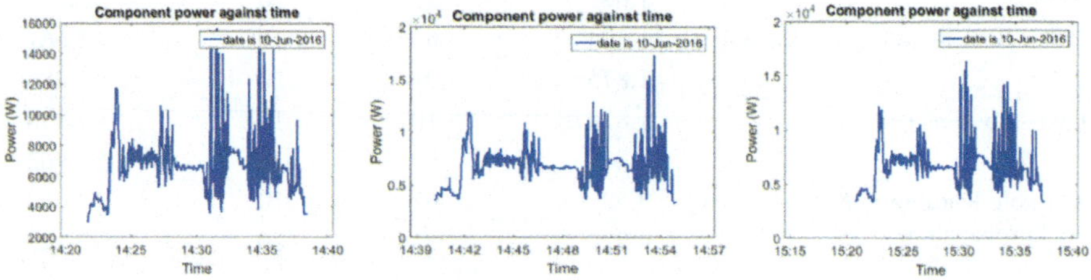

(c) 3 samples for Component 3

Fig. 6.18 Training samples in MX520, MAZAK and HAAS, respectively

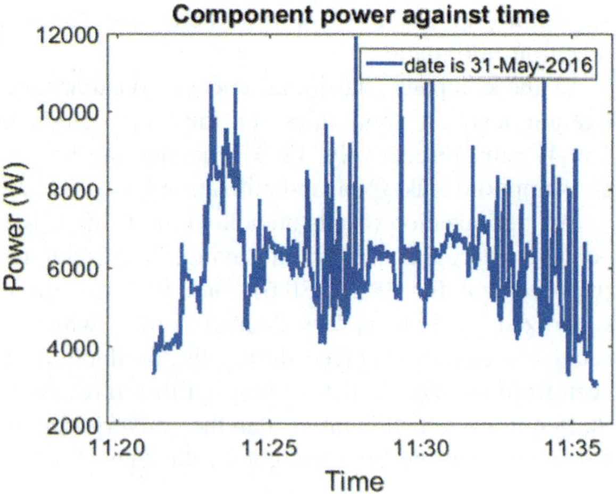

Fig. 6.19 New pattern for Component 1 (with an energy deviation of 19.24%, which indicates a possible tool wear or machine degrading)

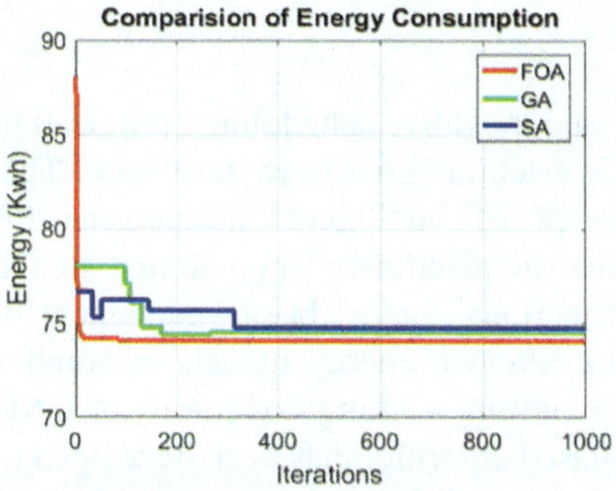

(a) Optimization of scheduling in terms of energy consumption.

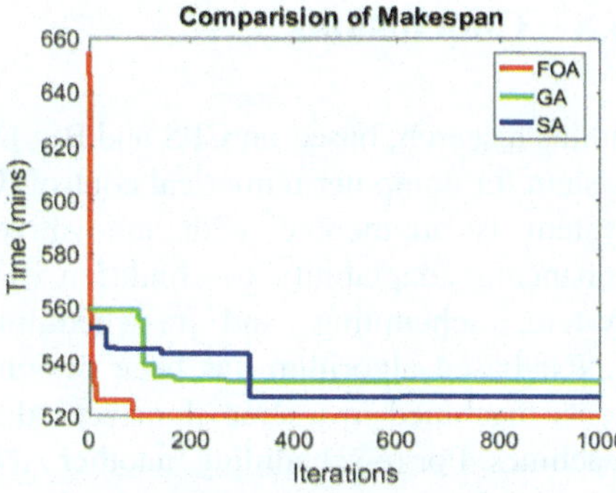

(b) Optimization of scheduling in terms of makespan.

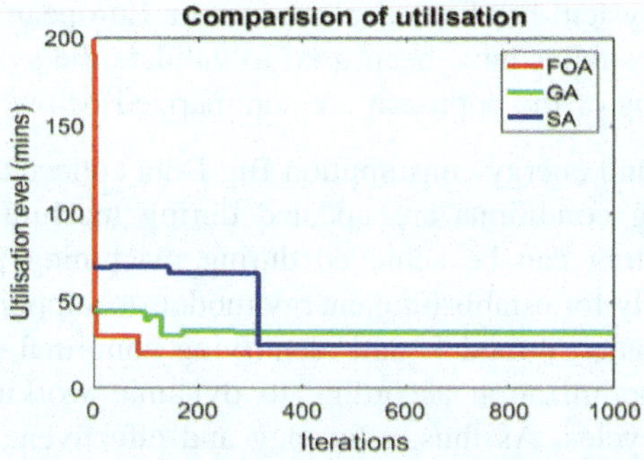

(c) Optimization of scheduling in terms of utilization level

Fig. 6.20 Optimization results of scheduling

Table 6.10 Optimization comparisons

Algorithms	FFO	GA	SA
Iterations to achieve optimal result	85	268	314
Initial energy consumption (KWh)	120.00	120.00	120.00
Energy consumption (KWh)	74.09	74.57	74.74
Optimization percentage (%)	38.26	37.86	37.72
Initial makespan (min)	750.00	750.00	750.00
Optimized makespan (min)	520.43	533.54	527.23
Optimization percentage (%)	30.61	28.86	29.70
Initial utilization level (min)	250.47	250.47	250.47
Optimized utilization level (min)	20.47	32.80	25.22
Optimization percentage (%)	91.83	86.90	89.93

6.8 Conclusions

In this research, based on CPS and Big Data, an innovative scheduling optimization system for computer numerical control (CNC) machining has been developed. This system is augmented with innovative design of intelligent mechanisms for enhancing adaptability to condition dynamics in machining shop floors. In the system, scheduling and re-scheduling functions have been implemented. ANNs-based algorithm has been designed to establish energy models of components machined in a shop floor according to current working conditions of CNC machines. For re-scheduling, another ANNs-based algorithm has been developed to monitor the energy consumption of machining components during production in real time. An enhanced FFO algorithm has been developed and applied to generate a multi-objective optimized schedule or re-schedule if working conditions vary significantly. The system has been deployed into a European machining company, and industrial case studies have been used to validate the system's applicability.

The contributions of the approach are summarized below:

- Based on CPS and energy consumption Big Data collected and analyzed in real time, machining conditions are updated during machining processes so that optimal scheduling can be achieved during machining processes. ANNs are designed not only for establishing energy models to support scheduling, but also for fine-tuning energy models and identifying abnormal conditions to support re-scheduling optimization according to dynamic working conditions during machining lifecycles. As thus, efficiency and effectiveness of scheduling and re-scheduling are significantly improved;
- Innovative design of CPS, Big Data, ANNs, and optimization algorithm are integrated for systematic and effective implementation of manufacturing intelligence. The deployment of the system in real-world shop floor and validation of the system by using various industrial case studies in a European machining company have been achieved. The significant potentials of applicability in practice have been demonstrated.

Acknowledgements This research was carried out as a part of the Smarter and Cloudflow projects which are supported by the European Commission 7th Framework Programme under the grant agreement PEOPLE-2013-IAPP-610675 and FP7-2013-NMP-ICT-FOF- 609100.

References

1. O'Driscoll, E., & O'Donnell, G. (2013). Industrial power and energy metering—a state-of-the-art review. *Journal of Cleaner Production, 41,* 53–64.
2. Bunse, K., Vodicka, M., Schönsleben, P., Brülhart, M., & Ernst, F. (2011). Integrating energy efficiency performance in production management—gap analysis between industrial needs and scientific literature. *Journal of Cleaner Production, 19*(6–7), 667–679.

3. Li, W. D., & McMahon, C. A. (2007). A simulated annealing—based optimization approach for integrated process planning and scheduling. *International Journal of Computer Integrated Manufacturing, 20*(1), 80–95.
4. Wang, S., Lu, X., Li, X., & Li, W. D. (2015). A systematic approach of process planning and scheduling optimization for sustainable machining. *Journal of Cleaner Production, 87,* 914–929.
5. Gahm, C., Denz, F., Dirr, M., & Tuma, A. (2016). Energy-efficient scheduling in manufacturing companies: A review and research framework. *European Journal of Operational Research, 248*(3), 744–757.
6. Babiceanu, R., & Seker, R. (2016). Big Data and virtualization for manufacturing cyber-physical systems: A survey of the current status and future outlook. *Computers in Industry, 81,* 128–137.
7. Fang, K., Uhan, N., Zhao, F., & Sutherland, J. (2011). A new approach to scheduling in manufacturing for power consumption and carbon footprint reduction. *Journal of Manufacturing Systems, 30*(4), 234–240.
8. He, Y., Liu, F., Wu, T., Zhong, F., & Peng, B. (2011). Analysis and estimation of energy consumption for numerical control machining. *Proceedings of the Institution of Mechanical Engineers, Part B: Journal of Engineering Manufacture, 226*(2), 255–266.
9. Yan, J., & Li, L. (2013). Multi-objective optimization of milling parameters—the trade-offs between energy, production rate and cutting quality. *Journal of Cleaner Production, 52,* 462–471.
10. Winter, M., Li, W., Kara, S., & Herrmann, C. (2014). Determining optimal process parameters to increase the eco-efficiency of grinding processes. *Journal of Cleaner Production, 66,* 644–654.
11. Yan, J., Li, L., Zhao, F., Zhang, F., & Zhao, Q. (2016). A multi-level optimization approach for energy-efficient flexible flow shop scheduling. *Journal of Cleaner Production, 137,* 1543–1552.
12. Tang, D., Dai, M., Salido, M., & Giret, A. (2016). Energy-efficient dynamic scheduling for a flexible flow shop using an improved particle swarm optimization. *Computers in Industry, 81,* 82–95.
13. Liu, Y., Dong, H., Lohse, N., & Petrovic, S. (2016). A multi-objective genetic algorithm for optimisation of energy consumption and shop floor production performance. *International Journal of Production Economics, 179,* 259–272.
14. Xu, W., Shao, L., Yao, B., Zhou, Z., & Pham, D. (2016). Perception data-driven optimization of manufacturing equipment service scheduling in sustainable manufacturing. *Journal of Manufacturing Systems, 41,* 86–101.
15. Salido, M., Escamilla, J., Barber, F., & Giret, A. (2016). Rescheduling in job-shop problems for sustainable manufacturing systems. *Journal of Cleaner Production* (online).
16. Dai, J., Huang, J., Huang, S., Huang, B., & Liu, Y. (2011). HiTune: Dataflow-based performance analysis for big data cloud. In *Proceedings of the 2011 USENIX ATC* (pp. 87–100). Berkeley.
17. Nagorny, K., Colombo, A., & Schmidtmann, U. (2012). A service- and multi-agent-oriented manufacturing automation architecture. *Computers in Industry, 63*(8), 813–823.
18. Loshin, D. (2014). *Big data analytics*. Burlington, NJ: Morgan Kaufmann Pub.
19. Noor, A. (2013). Putting big data to work. *ASME Mechanical Engineering*, 32–37.
20. Prabhu, N. (2013). *Design and construction of an RFID-enabled infrastructure*. Florida: CRC Press.
21. Chaplin, J., Bakker, O., de Silva, L., Sanderson, D., Kelly, E., Logan, B., et al. (2015). Evolvable assembly systems: A distributed architecture for intelligent manufacturing. *IFAC-PapersOnLine, 48*(3), 2065–2070.
22. Liu, C., & Jiang, P. Y. (2016). A cyber-physical system architecture in shop floor for intelligent manufacturing. *Procedia CIRP, 56,* 372–377.

23. Zhou, R., Nee, A. Y. C., & Lee, H. (2009). Performance of an ant colony optimisation algorithm in dynamic job shop scheduling problems. *International Journal of Production Research, 47*(11), 2903–2920.
24. Adibi, M., Zandieh, M., & Amiri, M. (2010). Multi-objective scheduling of dynamic job shop using variable neighborhood search. *Expert Systems with Applications, 37*(1), 282–287.
25. Li, W. D., Ong, S. K., & Nee, A. Y. C. (2006). *Integrated and collaborative product development environment—Technologies and implementation*. Singapore: World Scientific Publisher.
26. Mausser, H. (2006). Normalization and other topics in multi-objective optimization. In *Proceedings of the Fields–MITACS Industrial Problems Workshop* (pp. 59–101).
27. Pan, W. (2012). A new fruit fly optimization algorithm: Taking the financial distress model as an example. *Knowledge-Based Systems, 26,* 69–74.
28. Mayer-Schonberger, V., & Cukier, K. (2013). *Big data: A Revolution that will transform how we live*. Boston: Houghton Miffin Harcourt Publishing Company.
29. Zheng, X., Wang, L., & Wang, S. (2014). A novel fruit fly optimization algorithm for the semiconductor final testing scheduling problem. *Knowledge-Based Systems, 57,* 95–103.

Chapter 7
Sustainable Machining Process: Qualitative Analysis and Energy Efficiency Optimization

L. C. Moreira, W. D. Li, X. Lu and M. E. Fitzpatrick

Abstract Computer numerical control (CNC) machining is one of the major manufacturing activities. It is imperative to develop energy-efficient CNC machining processes to achieve the overall goal of sustainable manufacturing. Due to the complexity of machining parameters, it is challenging to develop effective energy consumption modelling and optimization approaches to implement energy-efficient CNC machining. In this chapter, via experiments and qualitative analysis, the impact that key cutting parameters generates on energy consumption of milling processes on BS EN24T alloy (AISI 4340) has been conducted in detail. This facilitates machining process planners to choose suitable scopes of machining parameters, (e.g. cutting speed, feed per tooth, engagement depth) to improve energy efficiency. Based on the above, optimization has been carried out to formulate a multi-objective optimization model, and a novel-improved multi-swarm fruit fly optimization algorithm (iMFOA) has been developed to identify optimal solutions. Case studies and algorithm benchmarking have been conducted to validate the effectiveness of the optimization approach. The characteristics and novelty of the research include: (1) the relationships between energy consumption and key machining parameters have been analysed to support process planners in implementing energy saving measures efficiently; and (2) the optimization approach is effective in fine-tuning the key parameters for enhancing energy efficiency while meeting the dynamic production requirements.

Keywords CNC machining · Energy saving · Optimization · Sustainable manufacturing

L. C. Moreira · W. D. Li (✉) · X. Lu · M. E. Fitzpatrick
Faculty of Engineering, Environment and Computing,
Coventry University, Coventry, UK
e-mail: weidong.li@coventry.ac.uk

W. Li and S. Wang (eds.), *Sustainable Manufacturing and Remanufacturing Management*, https://doi.org/10.1007/978-3-319-73488-0_7

7.1 Introduction

Ambitious goals to achieve significant energy savings have been set by major economies such as Europe, China and USA. The manufacturing sector is a major consumer of energy and critical raw material. Therefore, it is imperative to develop effective sustainable manufacturing approaches so as to achieve the targets of energy savings for societies. Among the manufacturing sector, computer numerical control (CNC) machining is one of the major processes. For CNC machining processes, process planning is a significant decision-making stage to determine the quality and productivity of machining. According to [1], process planning for CNC machining is increasingly concerned with reducing energy consumption during machining processes. The exponential growth in research publications related to process planning for energy-efficient CNC machining processes, which has been recently summarized by Moreira et al. [2], demonstrates the importance of this topic worldwide.

Energy information from machining process is important to assist process planning or lifecycle analysis and keep energy efficiently used [3]. Further, it is crucial to develop effective energy consumption modelling and optimization methodologies of process planning to implement energy-efficient CNC machining. To date, some energy consumption optimization approaches for process planning for CNC have been developed [2, 4]. On the other hand, CNC machining processes are complex in terms of various machining parameters, machining strategies and operations. For this reason, to disclose how key cutting parameters generate impacts on energy consumption of machining processes is essential. Based on that, it is important to develop an effective optimization solution for sustainable CNC machining processes. To address the current research gap, this chapter presents qualitative analysis and optimization considering key CNC machining parameters to achieve energy efficiency during machining. Research characteristics and contributions are as follows:

- Qualitative investigation on the relationship between key cutting parameters and energy consumption has been conducted by experimental design and the application of the main effect analysis approach. This facilitates machining process planners to choose suitable machining parameters to minimize energy consumption during machining;
- Optimization has been carried out to formulate a multi-objective optimization model considering the energy efficiency, productivity and cutting tool life. An improved multi-swarm fruit fly optimization algorithm (iMFOA) has been developed for solving the optimization problem. Case studies and algorithm benchmarking have been conducted to validate the effectiveness of the algorithm.

This research work is organized as follows: Sect. 7.2 presents the literature survey. In Sect. 7.3, the experimental set-up and results are given. In Sect. 7.4, a qualitative analysis on the experimental results is described. The optimization algorithm, case studies and benchmarking analysis are presented in detail in Sect. 7.5. The conclusions are drawn in Sect. 7.6.

7.2 Literature Survey

7.2.1 Energy Consumption (EC) Modelling

Energy consumption is a topic of concern not only in board rooms but also in the practices of manufacturing management in factories. Rising energy costs and proposed environmental taxes have driven industrial enterprises to improve their energy efficiency [5]. The exponential growth in research publications in the last two decades clearly shows the academic and practitioners' strong interests on this topic. The electricity demanded by computer numerical control (CNC)-enabled machine tools' servomotors on a factory shop floor produces energy consumption data that, when well-processed, is a treasured information source. Based on the data, energy consumption (EC) predictive models can be developed as promising means to enhance the sustainability of machining. EC models can be used to assess and improve the overall efficiency of shop floors, aid production engineers in scheduling optimization, and support machining systems to be self-controlled and self-optimized through embedded optimal control algorithms. To develop effective EC models, research work must be carried out for both qualitative understanding and quantitative understanding.

Recently, methods such as ANOVA, response surface methodology (RSM), Taguchi signal-to-noise ratio and artificial neural network (ANN) have been employed to model the relationships between cutting parameters and energy consumption to establish EC models [6–12]. Also, [13] carried out an experimental investigation on different machine tools using nonlinear regression. The results show that the motion of CNC machine tool is the root cause of energy consumption.

Many other researchers have used several approaches and techniques for understanding the EC of CNC machining processes. A common way of energy and productivity assessment is through the material removal rate (MRR) [8, 9]. That is because the MRR is estimated based on key cutting parameters spindle speed (*S*), feed rate (*f*) and engagement depth ($a_p \cdot a_e$). This simplifies the modelling process, and by doing this, it is assumed that each cutting parameter has the same effect on the energy consumption. However, Sealy et al. [14] has observed low predictive accuracy when employed to estimate the net-specific energy, which is strict to the process level of machining process.

To date, only few work carried investigations focused on the net-specific energy [15]. Further, no efforts have been made towards the implementation of machining net power and time estimation models to obtain optimum cutting parameters that maximize the energy efficiency of milling operations. Considerations of other factors involved in the machining process, such as tool wear, mode of milling, cutter tool holder type and workpiece holding systems are still lacking analysis and should be involved in the empirical modelling to develop more robust predictive models.

Based on that, this chapter develops an EC model considering the machining cutting variables S, f and $a_p \cdot a_e$, individually. Also, the machining net power (power load) is introduced for the first time to assess the cutting tool life.

7.2.2 *Optimization Approaches for Machining*

The use of optimization algorithms for performance optimality is a key step towards increasing machining efficiency, cost reduction and manufacturing sustainability. Significant effort has been taken by the research community to address complex manufacturing scenarios, involving environmental, legal, economic, and quality requirements.

Table 7.1 shows related work and summarizes the optimization methods and objectives that have been used in the past years.

The EC modelling and optimization method developed in this chapter follow the required steps highlighted by [16, 17], respectively, which are as follows:

- Knowledge of the machining processes under analysis.
- Empirical equations of the objective(s) and constraint(s) to define the optimization problem.
- Specifications for the CNC machine capabilities.
- Draw optimization criteria and the problem formulation.
- Knowledge of mathematical and numerical optimization techniques.

Table 7.1 Related work on the use of optimization methods for machining processes

Related work	Methods	Objectives	Cutting parameters
Wang et al. [18]	Pattern search (PS), genetic algorithm (GA) and simulated annealing (SA)	Energy consumption and productivity	Cutting speed (v_c), a_p and a_e
Sonmez et al. [17]	Dynamic programming and geometric programming	Production rate	v_c and feed per tooth (s_z)
Ozcelik et al. [19]	GA	Surface roughness	v_c, f, a_p and a_e
Sreeram et al. [20]	GA	Tool life	a_p
Li et al. [22]	GA	SEC and machining time	S, f, a_p and a_e
Baskar et al. [21]	GA, Hill climbing algorithm and memetic algorithm	Maximum profit	S and f

As shown in Table 7.1, GA is among the most used algorithm for solving machining optimization problems. Also, a considerable number of optimization objectives have been observed. However, an efficient and reconfigurable optimization strategy considering the specific energy and the manufacturing requirements for cutting tool life and productivity altogether has not been accomplished yet. The trade-offs involved among these criteria are the core motivating challenges of this work.

7.3 Experiment Design

This section provides the details of the experimental trials and machining conditions for data collection. These are crucial for the qualitative and optimization presented in this chapter.

7.3.1 Experiments Set-up

The experimental trials were carried out on a 3-axis vertical milling machine Haas VF-3, which comprises of a 30HP (22.4 kW) vector drive, has the maximum spindle speed of 8100 rpm and voltage 415 V (Fig. 7.1).

The BS EN24T alloy steel (AISI 4340) was selected as the workpiece material for experimental trials. There are two reasons for choosing this material: (1) the material is suitable for several engineering applications, aircraft and automotive such as gear shafts, propeller and so on [23]; (2) the BS EN24T alloy steel is a hard material; the EC for machining hard materials is more significant than that of mild and soft materials owing to the greater torque required during the cutting process.

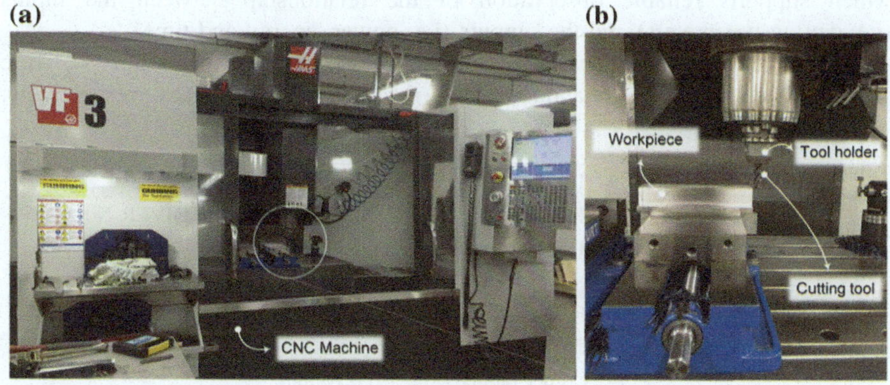

Fig. 7.1 Experiments set-up. **a** Haas VF-3 vertical milling machine; **b** machined workpiece and cutting tool

Table 7.2 Material properties for the workpiece

BS EN24T Alloy Steel (AISI 4340)		
Composition: C 0.36–0.44/Si 0.10–0.35/Mn 0.45–0.70/ S < 0.040/P < 0.035/Cr 1.00–1.40/Mo 0.20–0.35/Ni 1.30–1.70		
Property	Value	Unit
Density	7.85e3	kg/m^3
Young's modulus	210	GPa
Hardness-Brinell	248–302	HB

Therefore, it is more opportune to disclose the impact of key machining parameters on energy consumption to conduct energy saving research. The material's properties are displayed in Table 7.2.

The cutter tool used is a solid carbide (Table 7.3), held by a side lock tool holder. The machining processes were carried out under dry conditions and up milling mode.

The part selected is a jaw-type geometry with slotting features on both sides (Fig. 7.2). The toolpath strategy is a unidirectional route with constant tool engagement (linear motion). A safe clearance distance of 8 mm is set in the *X* direction for the cutter tool on the start and end of the cutting process, and 1 mm clearance in *Z*. The cutting tool travels for 8 mm in G01 on a motion with the supplied experimental parameters to engage onto the workpiece.

The power consumption is monitored by a cyber-physical system (CPS) mounted on CNC machines with the measuring frequency of 10 Hz, further specifications provided in Lu et al. [24].

Experiments were designed to analyse the significance and the interaction effects of spindle speed (*S*), feed rate (*f*) and width of cut (a_e) on the energy consumption for the roughing stage of milling. Different levels of S, f and a_e were selected: recommended (Re), low (Lo), middle-low (M-L), middle-high (M-H) and high (Hi). The selection of the several levels provides a good coverage for each parameter, which supports reliable observation of the relationship between the inputs (i.e. cutting parameters) and the outputs (i.e. power, energy and time).

The Re, Lo and Hi values of cutting speed (v_c) and feed per tooth (s_z) were selected from machining handbook and machinists' experiences, respectively.

Table 7.3 Cutting tool specifications

Tool property	Specifications
Tool ID	End mills RF 100 DIVER No. 6736
Tool diameter (*D*)	16 mm
No. of teeth	4
Feed per tooth (S_z)	0.025–0.1 mm/tooth
Cutting speed (v_c)	150–250 mm/min
Corner radius	0.16 mm
Cutter material	Solid carbide

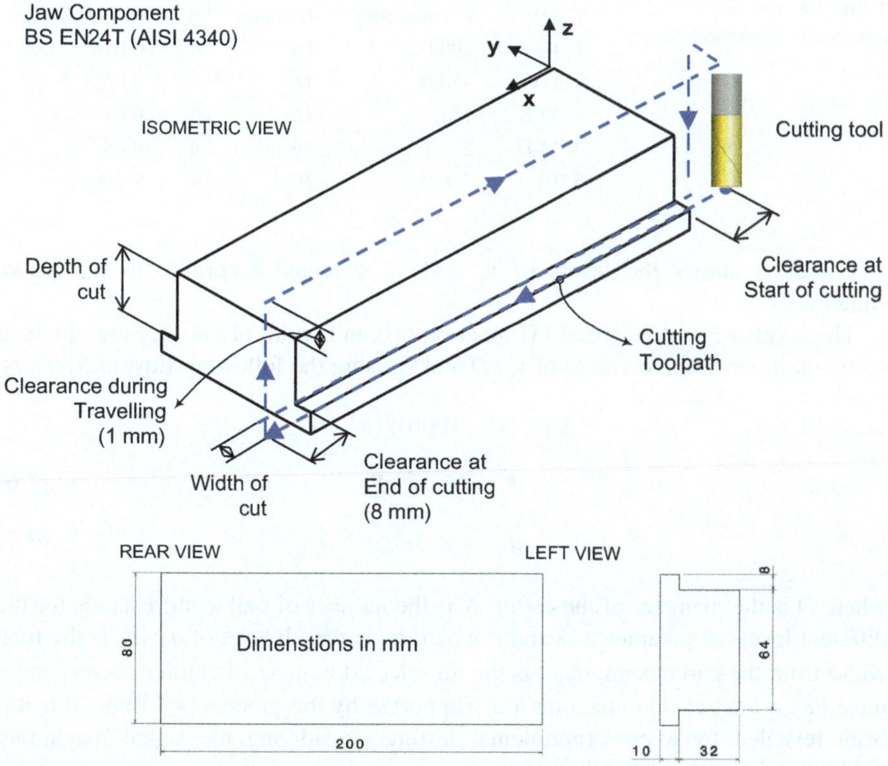

Fig. 7.2 CAD design of the machined metal component and dimensions

While the M-L and M-H values of (v_c) (i.e. v_{c_i}, $v_{c_{i-1}}$) and s_z (i.e. s_{z_i}, $s_{z_{i-1}}$) can be calculated based on the following Eqs. (7.1)–(7.4):

$$I_{v_c} = (v_{c\mathrm{Hi}} - v_{c\mathrm{Lo}})/(n_{\mathrm{level}} - 1) \tag{7.1}$$

$$v_{c_i} = v_{c_{i-1}} + I_{v_c} \tag{7.2}$$

$$I_{s_z} = (s_{z\mathrm{Hi}} - s_{z\mathrm{Lo}})/(n_{\mathrm{level}} - 1) \tag{7.3}$$

$$s_{z_i} = s_{z_{i-1}} + I_{s_z} \tag{7.4}$$

where I is the interval between the levels; i stands for the different levels of v_c and s_z; $v_{c\mathrm{Hi}}$ and $v_{c\mathrm{Lo}}$ are the highest and lowest boundaries for v_c, and $s_{z\mathrm{Hi}}$ and $s_{z\mathrm{Lo}}$ are the highest and lowest boundaries of s_z, which ranges can be decided by machinists according to their experiences; n_{level} is the number of levels desired, which 5 and 4 levels are chosen in this study (this impacts on the number of experimental trials and resources available).

Table 7.4 Cutting parameters boundaries

Levels	v_c (mm/min)	D (mm)	N	S_z (mm/tooth)
1. Re	200.0	16	4	0.070
2. Lo	150.0	16	4	0.025
3. M-L	184.5	16	4	0.059
4. M-H	217.7	16	4	0.082
5. Hi	250.0	16	4	0.100

Table 7.4 shows the levels of v_c and s_z, obtained according to the above equations.

The levels of spindle speed (*S*), feed rate (*f*) and width of cut (a_e) are obtained based on the respective values of v_c, *D* and s_z using the following Eqs. (7.5)–(7.7).

$$S_i = (v_{C_i} \cdot 1000)/(\pi \cdot D) \tag{7.5}$$

$$f_i = N \cdot s_{z_i} \cdot S_i \tag{7.6}$$

$$a_{e_i} = a_{e_f}/n_{\text{pass}_i} \tag{7.7}$$

where *D* is the diameter of the cutter; *N* is the number of tool teeth; *i* stands for the different levels of parameters *S* and *f*; where a_{e_i} is the *i*th level of a_e; a_{e_f} is the final width from the part design; n_{pass_i} is the *i*th selected number of cutting passes, and it must be an integer. The maximum a_e supported by the process is 4 mm—this has been revealed by a pre-experimental testing considering the actual machining holding and fixtures capabilities.

As a result, the calculated values of levels for *S*, *f* and a_e are presented in Table 7.5. These values are used to design the experimental trials.

Taguchi fractional factorial was used to define the design of experiments, and a total of 24 experiments were carried out based on the orthogonal principle (Table 7.6). Moreover, MRR is a significant evaluation factor on the energy consumption [15]. Thus, to evaluate the results considering this factor, the MRR of each trial is calculated using Eq. (7.8).

$$\text{MRR} = f \cdot a_e \cdot a_p = (v_c \cdot 1000 \cdot N \cdot S_z/\pi \cdot D) \cdot a_e \cdot a_p \tag{7.8}$$

where a_p is the depth of the cut (in this research, it was chosen 32 mm as the full depth of the designed part), and MRR is the material removal rate in mm^3/min.

Table 7.5 Machining cutting parameters and levels

Levels	*S* (rpm)	*f* (mm/min)	a_e (mm)
1. Re	4000	1115	4.00
2. Lo	3000	300	1.60
3. M-L	3670	870	2.00
4. M-H	4350	1430	2.67
5. Hi	5000	2000	4.00

Table 7.6 Experimental design based on orthogonal principle

Trial	*S* (rpm)	*f* (mm/min)	a_e (mm)	a_p (mm)	MRR (mm^3/min)	Productivity level
1	3000	1115	4.00	32	142,720	M
2	3670	1115	4.00	32	142,720	M
3	4350	1115	4.00	32	142,720	M
4	5000	1115	4.00	32	142,720	M
5	4000	300	4.00	32	38,400	Lo
6	4000	870	4.00	32	111,360	M-L
7	4000	1430	4.00	32	183,040	M-H
8	4000	2000	4.00	32	256,000	Hi
9	3000	870	4.00	32	111,360	M-L
10	3000	1430	4.00	32	183,040	M-H
11	3000	2000	4.00	32	256,000	Hi
12	3670	870	4.00	32	111,360	M-L
13	3670	1430	4.00	32	183,040	M-H
14	3670	2000	4.00	32	256,000	Hi
15	4350	870	4.00	32	111,360	M-L
16	4350	1430	4.00	32	183,040	M-H
17	4350	2000	4.00	32	256,000	Hi
18	5000	870	4.00	32	111,360	M-L
19	5000	1430	4.00	32	183,040	M-H
20	5000	2000	4.00	32	256,000	Hi
21	4000	1115	1.60	32	57,088	M-Lo
22	4000	1115	2.00	32	71,360	M-L
23	4000	1115	2.67	32	95,265.6	M-L
24	4000	1115	4.00	32	142,720	M

7.3.2 Experiment Results

During the 24 experimental trials, the power data monitored in the time domain show that different sets of milling parameters generated different power profiles. Figure 7.3 shows the power profiles of trials, which demonstrate the impacts of parameter sets on machining time and power loads.

The data obtained from CPS and sensors were analysed considering two distinct machining states: state of engagement (SoE) and state of non-engagement travelling (SoT). The former represents the process of material removal (actual cutting), while the latter represents non-cutting movements (air cutting). $\overline{P}_{\mathrm{SoE}}$, which is the average of the power of SoE (i.e. P_{SoE}), is introduced to assess the electricity consumption performance during a machining process. Similarly, $\overline{P}_{\mathrm{SoT}}$ is the average of the power of SoT (i.e. P_{SoT}). Energy consumption $\mathrm{EC}_{\mathrm{SoE}}$ and $\mathrm{EC}_{\mathrm{SoT}}$ are the total energy consumption for SoE and SoT, respectively. Specific energy consumption

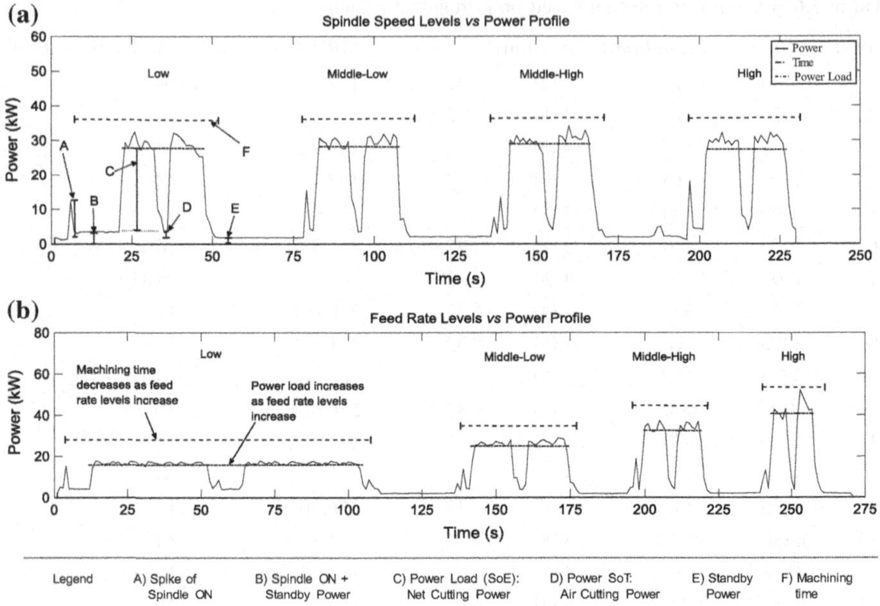

Fig. 7.3 Power profile of machining experiments on BS EN24T alloy workpiece. **a** Spindle speed analysis; **b** feed rate analysis

(SEC) during the SoE is used to indicate the machine energy efficiency when removing materials [25]. The relevant computations are in the following Eqs. (7.9)–(7.13):

$$\overline{P}_{\mathrm{SoE}} = \int_{t_1}^{t_{\mathrm{SoE}}} P_{\mathrm{SoE}} \cdot \mathrm{d}t / t_{\mathrm{SoE}} \tag{7.9}$$

$$\overline{P}_{\mathrm{SoT}} = \int_{t_1}^{t_{\mathrm{SoT}}} P_{\mathrm{SoT}} \cdot \mathrm{d}t / t_{\mathrm{SoT}} \tag{7.10}$$

$$\mathrm{EC}_{\mathrm{SoE}} = \int_{t_1}^{t_{\mathrm{SoE}}} P_{\mathrm{SoE}} \cdot \mathrm{d}t \tag{7.11}$$

$$\mathrm{EC}_{\mathrm{SoT}} = \int_{t_1}^{t_{\mathrm{SoT}}} P_{\mathrm{SoT}} \cdot \mathrm{d}t \tag{7.12}$$

Table 7.7 Experimental results for milling on BS EN24T alloy steel

Trial	EC (kJ)		% SoT	*t* (s)		$\overline{P}_{SoE}$ (kW)	Cutting tool life level	SEC_{SoE} (kJ/cm^3)	Energy efficiency level
	SoE	SoT		SoE	SoT				
1	580.02	91.51	14	20	10	29.00	M	11.33	M-H
2	595.42	93.88	14	20	10	29.77	M	11.63	M-H
3	607.02	97.35	14	20	10	30.4	M	11.88	M-H
4	603.06	99.75	14	20	10	30.15	M	11.78	M-H
5	1297.26	127.25	9	78	21	16.63	Hi	25.34	Lo
6	694.14	94.08	12	26	11	26.7	M-H	13.56	M-H
7	543.72	79.35	13	16	8	33.98	M-Lo	10.62	M-H
8	497.26	63.05	11	12	6	41.44	M-Lo	9.71	Hi
9	1517.79	184.5206	21	80	23	19.3	Hi	19.78	M
10	1222.33	145.57	19	63	17	20.73	M-H	15.92	M-H
11	1197.56	102.63	8	48	13	24.97	M-H	15.61	M-H
12	1010.82	61.05	6	32	9	31.59	M-L	13.16	M-H
13	1065.61	85.62	7	40	10	26.64	M-H	13.88	M-H
14	877.3	87.4	9	24	8	36.6	M-Lo	11.44	M-H
15	649.78	142.82	18	16	9	40.61	Lo	7.46	Hi
16	1104.79	97.24	8	40	10	27.62	M-H	14.39	M-H
17	827.36	106.85	11	24	8	34.51	M-L	10.79	M-H
18	675.18	142.07	17	16	9	42.2	Lo	7.79	Hi
19	1117.84	97.16	8	40	10	27.95	M-H	14.56	M-H
20	847.17	106.08	11	24	8	35.34	M-L	11.04	M-H
21	685.91	145.54	18	16	9	42.87	Lo	7.93	Hi
22	1157.24	106.66	8	40	10	27.96	M	15.08	M-H
23	902.2	112.88	11	24	8	37.59	Lo	11.75	M-H
24	687.21	157.82	19	16	9	43.01	Lo	7.96	Hi

$$SEC = EC_{SoE}/V \tag{7.13}$$

where V is the volume removed during machining, t_{SoE} is the machining time during SoE for each cutting pass n.

The data collected using the smart sensor network for power consumption and time of all experimental trials are summarized in Table 7.7.

7.4 Qualitative Analysis on Experiments

Qualitative analysis is an efficient means for obtaining knowledge from a complex environment, and thus, this method is used in this section to understand the relationships of key cutting parameters in machining processes and the energy consumption to produce BS EN24T (AISI 4340) parts.

The analysis of the significance of the key parameters on the energy consumption reveals the important order of relationships between each input and this response, and this way, it supports the selection of the correct mathematical model for the optimization.

The results of $\overline{P}_{\mathrm{SoE}}$ and SEC in Table 7.7 show that the machining performance (analysed through the power, energy and time) is highly affected by the selection of machining parameters and key trade-offs have been identified. For instance, Trial 24 requires the highest power load, 43.01 kW, while Trial 5 presents the lowest, 16.53 kW. Nevertheless, the energy efficiency of Trial 24 (SEC = 7.96 kJ/cm^3) is lower than that of Trial 5 (SEC = 25.34 kJ/cm^3), that is due to the greater machining time spent for Trial 5.

In addition, there are two main observations based on the results for the energy consumed during the SoE and SoT:

- The energy required for air travelling (SoT) is between 6 and 21% of the overall EC ($\mathrm{EC_{SoE}} + \mathrm{EC_{SoT}}$) for all trials. The results for the experiments reveal that the amount of energy consumed during the SoE is the most representative over the SoT. Moreover, SoE is varied from 79 to 94% of the overall energy consumed. Consequently, the investigation finds that the machining parameters play an even more critical role on the energy efficiency of the production.
- Based on the energy results for SoT, it is observed that the amount of energy varies significantly between the experimental trials. It was caused by the different safe clearance distance set in the NC code, in which the feed table moves along with the commanded feed rate (G01 movement) to approach the workpiece, as well as different spindle speed levels. These observations are machine dependent (e.g. vector drive horsepower and drives technology).

The effects of machining parameters, spindle speed (S), feed rate (f) and width of cut (a_e) on the power, energy and time required during SoE are investigated below.

7.4.1 *Spindle Speed Effects*

The main effects of spindle speed on the power load and energy are analysed. The results of the experiments are presented in Fig. 7.4.

The main results from the experimental trials show that:

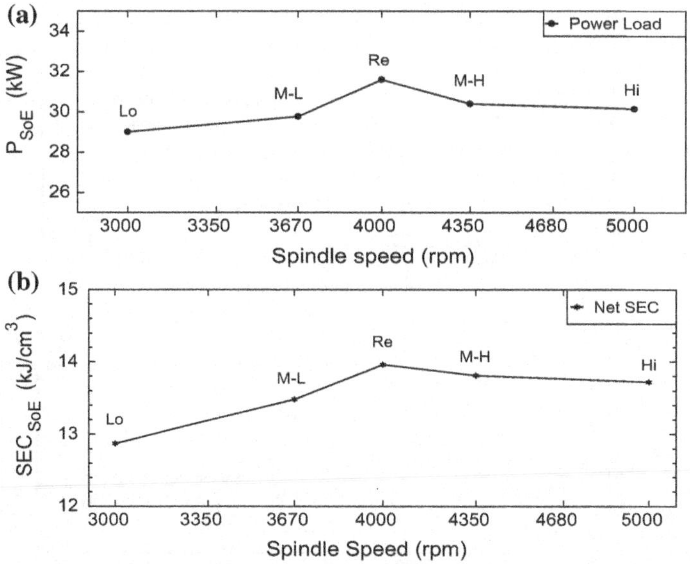

Fig. 7.4 Experimental results on BS EN24T alloy. **a** Relationship between *S* and $\overline{P}_{SoE}$; **b** relationship between *S* and SEC

- Changes on *S* will not generate substantial effects on the $\overline{P}_{SoE}$. This is revealed by the low standard deviation of $\overline{P}_{SoE}$ to the changes on *S*: 0.8473 kW. *S* does not affect the machining time as prior known.
- During the travelling time, more energy is wasted at higher levels of spindle speed, since the spindle motor requires more power at higher speeds. An increasing energy demand of approximately 3% between each level of *S* is revealed.
- The power load $\overline{P}_{SoE}$ increases from the Lo until the Re level of *S*. Beyond this level, a drop of $\overline{P}_{SoE}$ is identified (shown in Fig. 7.4a). This way, the M-H level is the point at which increasing *S*, when all other parameters are kept constant, the amount of material removed per cutting tool revolution has a positive effect on the energy consumption. Consequently, the cutting load per unit time is smaller. Higher *S* promotes a decrease on the power load ($\overline{P}_{SoE}$).

The results show that a selection of Lo or Hi levels of *S* is more appropriate to achieve energy efficiency in machining processes (Fig. 7.4b), although higher cutting speeds are known to decrease the cutting tool life [26].

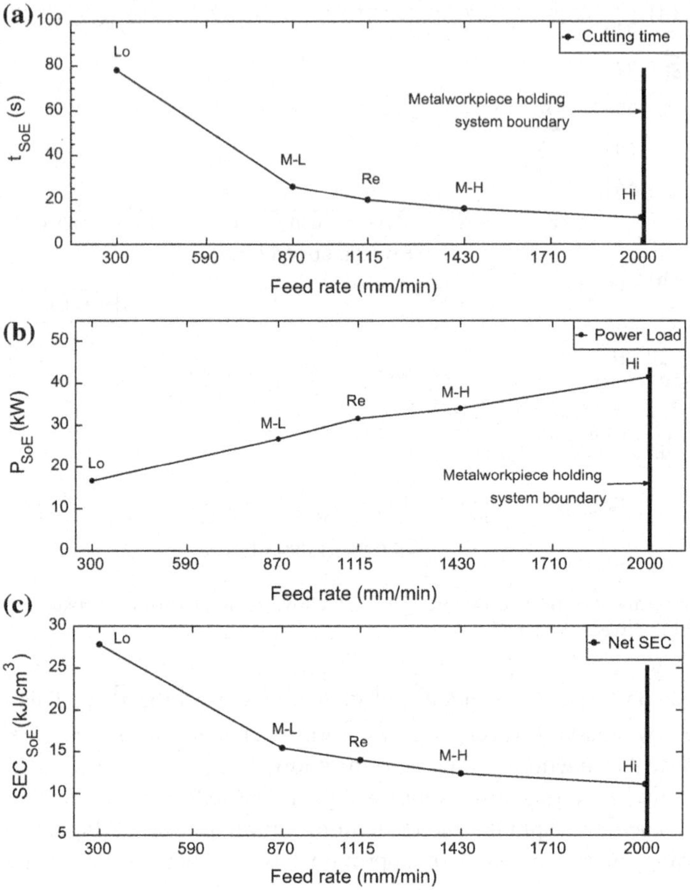

Fig. 7.5 Experimental results on BS EN24T alloy. **a** Relationship between t_{SoE} and f; **b** relationship between $\overline{P}_{SoE}$ and f; **c** Relationship between SEC and f

7.4.2 Feed Rate Effects

Feed rate (f) is one of the major factors that determine the MRR, as shown in Eq. (7.8). That is, the increase in f and maintaining other parameters unchanged will lead to a greater MRR. Figure 7.5 shows the results for the experimental trials for the feed rate analysis.

The main findings of this experimental investigation are as follows:

- Substantial effects of the feed rate f on the power load $\overline{P}_{SoE}$ and machining time t_{SoE} are observed—through the standard deviations of the power load and machining time: 7.2318 kW and 26.4764s, respectively. It shows that f generates a greater impact on the machining time compared to the power load. It could be conflicting when considering a sustainable process, since the increase in the feed rate would increase the productivity rate but, at the same time, increase the power load.

- Increasing the feed rate reduces the machining time, as shown in Fig. 7.5a. The machining time is approximately 7 times less at the maximum rate of *f* when compared to the lowest rate of *f*.
- Increasing the feed rate increases the machining power load, as shown in Fig. 7.5b. The power load *f* at the Hi level is approximately 3 times greater than at the Lo level *f*.
- A high feed rate promotes the better energy efficiency owing to savings in machining time. The process at the Lo level required 2.6 times more energy compared to the Hi level, per unit volume of material removed, shown in Fig. 7.5c. However, the drawback is that it produces higher cutting forces and higher temperatures at the cutting tool, consequently, shortening the tool life.

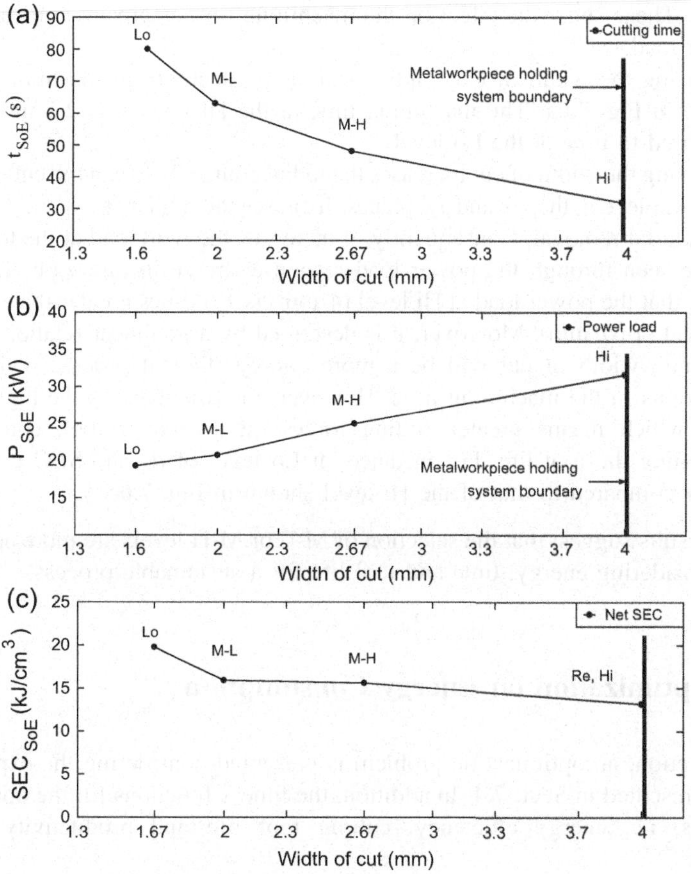

Fig. 7.6 Experimental results on BS EN24T alloy. **a** Relationship between the t_{SoE} and a_e; **b** relationship between the P_{SoE} and a_e; **c** Relationship between the SEC and a_e

- The results suggest that the selection of M-L or M-H cutting feed levels are more appropriate to address the trade-offs between energy, time and cutting tool life.

7.4.3 Width of Cut Effects

Width of cut influences MRR in a machining process, as shown in Eq. (7.8). The experimental results of a_e on machining processes are presented in Fig. 7.6.

Significant effects of a_e on the power load, machining time and energy efficiency are revealed. A summary of the observations is provided below:

- The substantial effects on the power load and machining time can be measured through the standard deviations, which are 6.56014 kW and 24.3721s, respectively. These show the effect on the machining time is greater compared to the power load.
- Increasing the width of cut implies significant decrease in machining time, as shown in Fig. 7.6a. The machining time at the Hi level was 2.5 times shorter compared to time at the Lo level.
- Increasing the width of cut increases the radial contact between the cutter tool and the workpiece in the *XY* and *YZ* planes. It causes the higher stress and power load for material removal. Consequently, it increases the workload at the tool, which can be seen through the power load response shown in Fig. 7.6b. The results reveal that the power load at Hi level (4 mm) is 1.6 times greater than that at the Lo level (1.67 mm). Moreover, it is described by a nonlinear relationship.
- The high width of cut will be a more energy-efficient process owing to the reductions in the machining time. However, the drawback is the higher power load, which means greater cutting forces on the cutter tool, consequently, shortening the tool life. For instance, at Lo level of a_e, the SEC is 1.5 times higher compared to that of the Hi level shown in Fig. 7.6c.

The results suggest that the selection of M-L or M-H levels are more appropriate when considering energy, time and tool life for a sustainable process.

7.5 Optimization on Energy Consumption

In this section, an optimization problem is presented considering the experimental results, presented in Sect. 7.4. In addition, the fitness functions for the optimization objectives, i.e. energy efficiency, cutting tool life and productivity are also provided.

7.5.1 Optimization Modelling

The energy required during the state of engagement (SoE) for the milling on BS EN24T alloy (AISI 4340) is highly representative, and it accounted for 79–94% of the overall energy. Therefore, it will be significant to save energy of the machining process if the energy during SoE (i.e. EC_{SoE}) will be minimized. The following formulas represent EC_{SoE} and the related parameters:

$$t_{\mathrm{SoE}} = V/\mathrm{MRR} \tag{7.14}$$

$$EC_{\mathrm{SoE}} = \overline{P}_{\mathrm{SoE}} \cdot t_{\mathrm{SoE}} = \overline{P}_{\mathrm{SoE}} \cdot (V/\mathrm{MRR}) \tag{7.15}$$

$$\overline{P}_{\mathrm{SoE}} = f_1(S, f, a_e \cdot a_p) \tag{7.16}$$

$$\mathrm{MRR} = f_2(f, a_e, a_p) = f \cdot a_e \cdot a_p \tag{7.17}$$

where $\overline{P}_{\mathrm{SoE}}$ is the average power used during SoE, V is the removed volume of material, MRR is the material removal rate, S, f, a_e, and a_p are the cutting parameters spindle speed, feed rate, width of cut and depth of cut, respectively.

In order to establish the function of $\overline{P}_{\mathrm{SoE}}$, a responsive surface regression model is developed. The model structure is presented below:

$$y = \beta_0 + \beta_1 \cdot x_1 + \beta_2 \cdot x_2 + \beta_3 \cdot x_3 + \beta_{11} \cdot x_1^2 + \beta_{22} \cdot x_2^2 + \beta_{12} \cdot x_1 \cdot x_2 \tag{7.18}$$

where y will be $\overline{P}_{\mathrm{SoE}}$, x_1, x_2 and x_3 will be S, f, $a_e \cdot a_p$, respectively, $\beta_{0,1,2}$ and $\beta_{11,12,22}$ are coefficients to be determined.

Apart from experimental trials 4, 5, 9, 10, 14 and 23, other trials are used to generate the coefficients. The output data are filtered using a single exponential smoothing technique. This is an additional step prior to the coefficient estimation process. Such step reduces the random fluctuations in the time series for the collected data, thus providing a more accurate pattern of the power load of each experimental trial. By taking this step, the accuracy of the final predictive model is increased by 2.92%. Subsequently, nonlinear regression and least squares methods are employed

Table 7.8 Power load model coefficients

Coefficient	Value	Significance (P value: $\alpha < 0.05$)*
β_0	−16.1700	0.000
β_1	0.00577	0.036
β_2	0.01225	0.000
β_3	0.1751	0.000
β_{11}	−0.0000010	0.001
β_{22}	−0.0000020	0.000
β_{12}	0.0000020	0.005

*Interval of confidence is 95%, i.e. $\alpha = 0.05$

Table 7.9 Production weighting strategy

Description	Weighting selection[a]
Cutting tools are the major constraint	$0.8 \leq \rho_1 \leq 0.9$
Cutting tools are more constrained than lead time.	$0.5 < \rho_1 < 0.8$
Both resources are constrained	$\rho_1 = \rho_2 = 0.5$
Lead time is more constrained than cutting tools	$0.5 < \rho_2 < 0.8$
Lead time is the major constraint	$0.8 \leq \rho_2 \leq 0.9$

[a]Weights law: $\rho_1 + \rho_2 = 1$

to estimate the model's coefficients. The estimated coefficients $\beta_{0,1,2}$ and $\beta_{11,12,22}$ are given in Table 7.7. The accuracy of the smoothed model is R^2-adjusted equal to 0.9406, which shows the achievement of satisfactory accuracy (Table 7.8).

This model is validated using data collected from experimental trials 4, 5, 9, 10, 14–23. The results of the estimated $\overline{P}_{\mathrm{SoE}}$ present a predictive accuracy R^2 of 0.9789, which shows good performance.

From the above Formula (7.15), it can be observed that if it is going to minimize $\mathrm{EC}_{\mathrm{SoE}}$, $\overline{P}_{\mathrm{SoE}}$ should be minimized and MRR should be increased. Based on this, an optimization objective (fitness) to minimize the $\mathrm{SEC}_{\mathrm{SoE}}$, the indicator for the energy efficiency, is set up below:

$$\begin{cases} \text{Minimize} & \mathrm{SEC}_{\mathrm{SoE}} = \rho_1 \cdot \overline{P}_{\mathrm{SoE}} + \rho_2 \cdot \frac{1}{\mathrm{MRR}} \\ \text{Subject to :} & \\ & 200 \leq x_1 \leq 250 \\ & 0.07 \leq x_2 \leq 0.1 \\ & 51.2 \leq x_3 \leq 128 \end{cases} \tag{7.19}$$

where x_1, x_2 and x_3 are cutting speed, feed per tooth and engagement depth, respectively; and, ρ_1 and ρ_2 are weights, where $\rho_1 + \rho_2 = 1$.

$\overline{P}_{\mathrm{SoE}}$ is also related to the cutting tool's life. Increases of $\overline{P}_{\mathrm{SoE}}$ will generate the increases in cutting forces and temperature on the cutting tool so that the life of the tool will be reduced. MRR represents the process productivity. Regarding the setting of the two weights, industrial surveys were made and a strategy for the setting can be defined as presented in Table 7.9.

The appropriate strategy is chosen by the engineer or process planer based on the immediate availability of the resources cutting tools and lead time—or which has the greatest priority—in the factory. After that, the appropriate weights, ρ_1 and ρ_2, are selected from the weighting strategy table and combined with the objective function for energy saving. Consequently, the importance of the objective within the optimization process is reconfigured to align these with the factory's immediate requirements. As a result, the optimal solution for the operation is also the best solution for the factory.

7.5.2 Optimization Algorithm: Improved Multi-swarm Fruit Fly Optimization Algorithm (IMFOA)

An improved optimization algorithm, fruit fly optimization algorithm (FFOA), is employed to solve the optimization problem formulated in Sect. 7.5.1.

FFOA is a recent nature-inspired algorithm for solving optimization problems by mimicking the highly advanced sense of smell of insects to detect foods' locations [27]. This recent algorithm has presented outstanding performance on solving optimization problems, especially in business and finance areas which requires highly reliable predictions [28–31]. However, its ability to solve machining trade-offs has not been investigated yet.

To address this gap, a multi-swarm fruit fly optimization algorithm (MFOA) developed by [32] is improved to cope with the machining optimization of this study. The problem formulated in Sect. 7.5.1 comprises of three input variables (i.e. cutting parameters) which are constrained by the safe boundaries. However, the MFOA algorithm is designed to solve problems with non-constrained two input variables. Thus, improvements have been made to the original algorithm, in which major changes to achieve the improved MFOA (iMFOA) can be found below:

- A third axis has been included to specify the fruit fly coordinates (i.e. positions), so the algorithm can cope with the three input variables.
- A sphere function is embedded to define the search space, i.e. the fruit flies flying space, this way, ensuring the cutting parameters selected are within the safe boundaries.
- A penalty function is included to constrain the power load fitness function, which cannot be above a certain level to guarantee energy sustainability.

Figure 7.7 shows the algorithm schematic and illustration of the iMFOA.

Firstly, the engineer or process planner defines the production weights (i.e. ρ_1 and ρ_2), to align the optimization engine with the production constraints so the algorithm can be initialized (STEP I). Then, based on the process safe boundaries (calculated in STEP II) the fruit flies' populations (i.e. sub-swarms) are generated in STEP III. Each fruit fly position, i.e. $(x, y, z)_i$, represents a combination of the cutting parameters S, f and $a_p \cdot a_e$. This process can be represented as below:

$$X_{\text{new}}(i,j) = x_{\text{initial}}(i,j) + \text{randi}(\text{boundaries}_{\text{SpindleSpeed}})$$
$$Y_{new}(i,j) = y_{\text{initial}}(i,j) + \text{randi}(\text{boundaries}_{\text{FeedRate}})$$
$$Z_{new}(i,j) = z_{\text{initial}}(i,j) + \text{randi}(\text{boundaries}_{\text{EngagementDepth}})$$

where X, Y and Z_{new} are the fruit flies' positions of the new populations; i is the fruit fly and j is the sub-swarm; x, y and z_{initial} are the initial positions which are set to be zero at the start; randi is a computational function to select the respective values within the cutting parameters minimum and maximum boundaries;

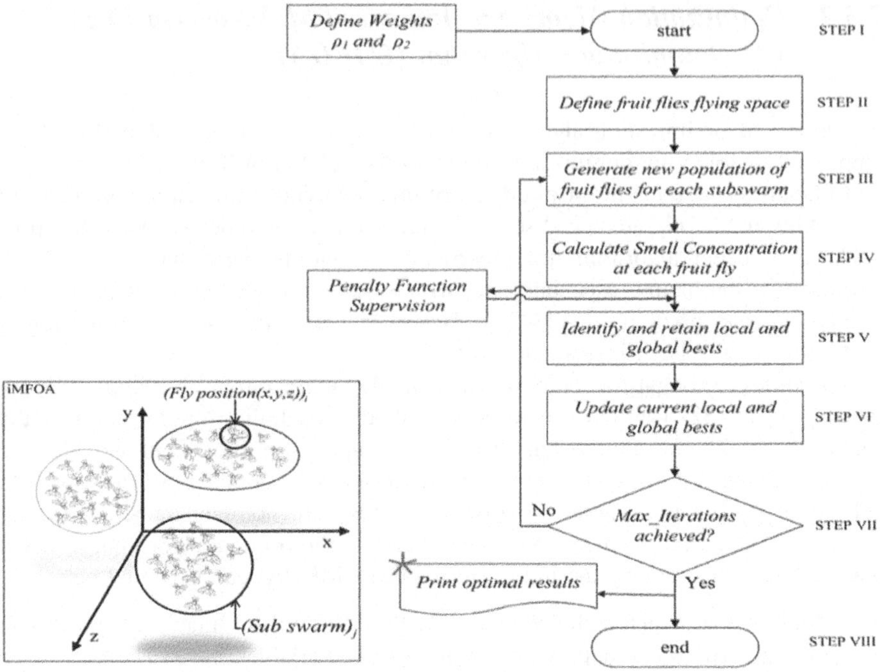

Fig. 7.7 Flow chart of the improved MFOA (iMFOA) algorithm

To calculate the smell concentration (fitness) of each fruit fly, in STEP IV, the new populations for fruit flies are called into each of the fitness functions, i.e. SEC, $\overline{P}_{\mathrm{SoE}}$ and MRR. In the optimization problem, these fitness functions are combined to save computational time as follows:

$$\mathrm{Smell\_SEC}(i,j) = \rho_1 \cdot P_{\mathrm{SoE}}(i,j) + \rho_2 \cdot \frac{1}{\mathrm{MRR}}(i,j) \tag{7.20}$$

The output values of $\overline{P}_{\mathrm{SoE}}$ and smell concentration are evaluated by a penalty function which judges the energy efficiency and cutting tool life based on the knowledge embedded into the system and act accordingly: if the power load is above the thresholds defined empirically, it reduces the smell concentration considerably. This supervisory loop ensures that inefficient cutting conditions are not identified as local or global best, in STEP V and, consequently, not retained in STEP VI.

Fruit flies (i) with the highest smell concentration within a sub-swarm (j) are identified as local bests, while the global best is represented by the fruit fly with highest smell concentration among all sub–swarms. Further, the local bests are used to substitute the initial positions and generate the new populations in the next iteration. This process occurs recursively until the maximum number of iterations is reached, so the global best fruit fly, which holds the optimal cutting parameters, and smell concentration path are achieved.

7.6 Case Study for Validation of Optimization Approach

A case study including three real-case manufacturing scenarios is presented in this section. This way, the proposed optimization problem and iMFOA algorithm can be assessed. This will be done by evaluating the optimization outputs considering some key rules of sustainable machining.

The details of the manufacturing scenarios are given in Table 7.10.

SEC, power load (P_{SoE}) and material removal rate (MRR) are used as key sustainable indicators (KSI) for the energy efficiency, cutting tool life and productivity, respectively. Furthermore, the optimal performances are analysed considering the rules for sustainable machining, as below:

(i) The smaller the SEC, the better the energy efficiency will be.
(ii) The greater the MRR, the better the productivity will be.
(iii) The smaller the $P_{SoE,}$ the better the cutting tool life will be.

Accordingly, the optimization results for each manufacturing scenario will be discussed based on the above rules. This further supports the selection of the best result among the three optimization algorithms employed for benchmarking analysis: GA [33], FFOA [27] and the iMFOA, presented in Sect. 7.5.2.

The details for the algorithm initialisation are as follows: the production constraints' weights were defined heuristically based on each of the scenario characteristics. Then, the initial set-up for the algorithm engine is defined as: number of sub-swarms equal to 10, size of population of fruit flies per sub-swarm equal to 25, and maximum number of iterations equal to 1000.

The optimization algorithm was run under the initial set-up. Figure 7.8 shows the smell concentration path containing the global best values during the convergence to the optimal solution from the iMFOA algorithm.

Table 7.10 Manufacturing scenarios for the optimization problem

Real-case scenarios of factory immediate requirements	Production constraints
(a) The production batch requires highly expensive cutting tools; meanwhile, the lead time of production is also an important aspect to be optimised	Both resources are constrained $\rho_1 = \rho_2 = 0.5$
(b) The deadline for delivering the production order has been extended; the manager asks to reconfigure the machining operations to a smoother strategy to prolong cutting tool life	Cutting tools become the Constraint $\rho_1 = 0.8,\ \rho_2 = 0.2$
(c) The deadline for delivering the production order has been shortened; the manager asks to reconfigure the machining operations to an aggressive strategy to boost the productivity	Lead time becomes the constraint $\rho_2 = 0.8,\ \rho_1 = 0.2$

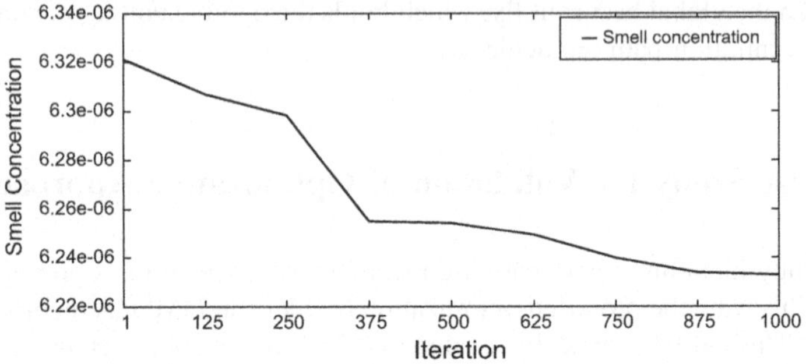

Fig. 7.8 Smell concentration path during optimization using the iMFOA algorithm

Table 7.11 Optimization results and KSI

Scenario constraint	Optimization algorithm	Optimal cutting parameters			Key sustainable indicators		
		Cutting speed (mm/min)	Feed (feed/tooth)	Engagement depth (mm^2)	SEC (kJ/cm^3)	MRR (cm^3/min)	Power Load (kW)
Lead time and cutting tools	iMFOA	250.3	0.0336	80.10	17.63[a]	53.67	15.77
	FOA	167.8	0.0444	103.30	20.13	61.26	20.56
	GA	250.4	0.0338	77.59	17.68	52.22	15.38
Cutting tools	iMFOA	151.1	0.0188	55.00	15.84	20.57	5.43[a]
	FOA	175.2	0.0259	57.80	24.78	21.23	7.77
	GA	237.8	0.0212	52.00	17.97	20.90	6.26
Lead time	iMFOA	250.2	0.1236	105.70	12.74	157.18[a]	33.38
	FOA	152.5	0.0611	90.60	17.30	67.13	20.47
	GA	163.4	0.1096	107.12	13.14	152.75	33.45

[a]Optimal value based on rules and manufacturing requirements

Table 7.11 shows the optimization results, i.e. optimal cutting parameters and estimated SEC, MRR and P_{SoE}, obtained from the algorithms used to solve the three manufacturing scenarios.

Thus, based on the rules for sustainable machining, the optimization results from the iMFOA algorithm showed better performance, especially when compared to the FOA algorithm. This validates the adaptation of the FOA and the feasibility of using this swarm algorithm for machining optimization.

Moreover, this case study through real-case manufacturing requirements validates the optimization approach proposed in this research. Furthermore, it proves that the novel-weighting strategy is an easy and effective method to align the manufacturing requirements with the key sustainable indicators to enhance the impact of the machining optimization objectives. That is, the approach ensures the optimal cutting parameters at the operational level are aligned with the dynamic production requirements at the factory level.

7.7 Conclusions

To achieve energy-efficient CNC machining processes, it is essential to develop effective analysis and optimization approaches: to evaluate the impact of machining parameters on energy consumption and identify optimal parameters. In this chapter, (via experiments and qualitative analysis) key cutting parameters affecting energy efficiency have been analysed in detail. The findings facilitate machining process planners in choosing the suitable scopes of machining parameters to minimize energy consumption during machining. Based on the analysis, an improved multi-swarm fruit fly optimization algorithm (iMFOA) has been developed to optimize machining parameters. Case studies and benchmarking have been conducted to justify the algorithm. The research innovations are from the following two aspects:

(1) Detailed analyses on experiments have been made to disclose the relationships between energy consumption and key machining parameters. It facilitates process planners in implementing energy saving measures efficiently. The main conclusion is: the feed per tooth has the most significant effect on the machining time, specific energy and power load. For the energy-efficient CNC machining, high feed rates are suggested due to the savings in machining time; however, if cutting tools are limiting production, the optimal cutting conditions should be reconfigured to low levels of feed per tooth and cutting speed, while the engagement depth should be the recommended.

(2) The developed optimization approach (iMFOA) is an effective tool to fine-tune the key machining parameters to guarantee energy-efficient machining, and furthermore meet the requirements for shorter lead time and longer cutting tool life. iMFOA provides a better performance compared to traditional FFOA and GA algorithms.

To finalize, the further research includes generalizing optimization approach to facilitate energy-efficient CNC machining for other types of operations such as turning, boring, WEDM; and enhancing the robustness of the developed approach for online decision and optimization.

Acknowledgements The authors would acknowledge Mr G. Booth for the support and knowledge transferred during the machining experiments.

References

1. Shin, S. J., Woo, J., & Rachuri, S. (2017). Energy efficiency of milling machining: Component modeling and online optimization of cutting parameters. *Journal of Cleaner Production, 161,* 12–29.
2. Moreira, L. C., Li, W. D., Fitzpatrick, M. E., & Lu, X. (2017). Research on energy consumption and energy efficiency of machine tools : a comprehensive survey. *International Journal of Nanomanufacturing* (Forthcoming issue). http://www.inderscience.com/info/ingeneral/forthcoming.php?jcode=IJNM.

3. Yingjie, Z. (2014). Energy efficiency techniques in machining process: A review. *International Journal of Advanced Manufacturing Technology, 71,* 1123–1132.
4. Peng, T., & Xu, X. (2012). The state of the art in energy consumption model—The key to sustainable machining. *Applied Mechanics and Materials, 232,* 592–599.
5. O'Driscoll, E., & O'Donnell, G. E. (2013). Industrial power and energy metering e a state-of-the-art review. *Journal of Cleaner Production, 41,* 53–64.
6. Newman, S. T., Nassehi, A., Imani-Asrai, R., & Dhokia, V. (2012). Energy efficient process planning for CNC machining. *CIRP Journal of Manufacturing Science and Technology, 5,* 127–136.
7. Mori, M., Fujishima, M., Inamasu, Y., & Oda, Y. (2011). A study on energy efficiency improvement for machine tools. *CIRP Annals-Manufacturing Technology, 60,* 145–147.
8. Diaz, N., & Redelsheimer, E. D. D. (2011). Energy consumption characterization and reduction strategies for milling machine tool use. *Manufacturing Chemist, 82,* 22–24.
9. Kara, S., & Li, W. (2011). Unit process energy consumption models for material removal processes. *CIRP Annals-Manufacturing Technology, 60,* 37–40.
10. Draganescu, F., Gheorghe, M., & Doicin, C. V. (2003). Models of machine tool efficiency and specific consumed energy. *Journal of Materials Processing Technology, 141,* 9–15.
11. Camposeco-Negrete, C., de Dios, C. N. J., & Miranda-Valenzuela, J. C. (2016). Optimization of cutting parameters to minimize energy consumption during turning of AISI 1018 steel at constant material removal rate using robust design. *International Journal of Advanced Manufacturing Technology, 83,* 1341–1347.
12. Peng, T., & Xu, X. (2014). Energy-efficient machining systems: A critical review. *International Journal of Advanced Manufacturing Technology, 72,* 1389–1406.
13. Lv, J., Tang, R., Jia, S., & Liu, Y. (2016). Experimental study on energy consumption of computer numerical control machine tools. *Journal of Cleaner Production, 112,* 3864–3874.
14. Liu, Z. Y., Guo, Y. B., Sealy, M. P., & Liu, Z. Q. (2016). Energy consumption and process sustainability of hard milling with tool wear progression. *Journal of Materials Processing Technology, 229,* 305–312.
15. Sealy, M. P., Liu, Z. Y., Zhang, D., Guo, Y. B., & Liu, Z. Q. (2016). Energy consumption and modeling in precision hard milling. *Journal of Cleaner Production, 135,* 591–1601.
16. Moreira L. C. (2016). *An innovative method for process planning for sustainable manufacturing*. Dissertation, Coventry University.
17. Sonomez, A. I., Baykasoglu, A., Dereli, T., & Filiz, I. H. (1999). Dynamic optimization of multipass milling operations via geometric programming. *International Journal of Machine Tools and Manufacture, 39,* 297–320.
18. Wang, S., Lu, X., Li, X. X., & Li, W. D. (2015). A systematic approach of process planning and scheduling optimization for sustainable machining. *Journal of Cleaner Production, 87,* 914–929.
19. Ozcelik, B., Oktem, H., & Kurtaran, H. (2005). Optimum surface roughness in end milling Inconel 718 by coupling neural network model and genetic algorithm. *International Journal of Advanced Manufacturing Technology, 27,* 234–241.
20. Sreeram, S., Kumar, A. S., Rahman, M., & Zaman, M. T. (2006). Optimization of cutting parameters in micro end milling operations in dry cutting condition using genetic algorithms. *International Journal of Advanced Manufacturing Technology, 30,* 1030–1039.
21. Li, J. G., Lu, Y., Zhao, H., Li, P., & Yao, Y. X. (2014). Optimization of cutting parameters for energy saving. *International Journal of Advanced Manufacturing Technology, 70,* 117–124.
22. Nee, A. Y. C., Song, B., Ong, S.-K., Peng, T., & Xu, X. (2013). A universal hybrid energy consumption model for CNC machining systems. *Re-engineering Manufacturing for Sustainability* 251–256. https://doi.org/10.1007/978-981-4451-48-2_41.
23. Steel Express. EN24T Steel Properties. Available at https://www.steelexpress.co.uk/engineeringsteel/EN24T-properties.html. Accessed January 30, 2018.
24. Lu, X., Wang, S., Li, W. D., Jiang, P., & Zhang, C. (2015). Development of a WSN based real time energy monitoring platform for industrial applications. In *IEEE 19th International Conference on Computer Support Cooperative Work in Design* (pp. 337–342).

25. Zhou, L., et al. (2016). Energy consumption model and energy efficiency of machine tools: A comprehensive literature review. *Journal of Cleaner Production, 112,* 3721–3734.
26. Davim, J. P. (2015). *Traditional machining processes.* London: Springer.
27. Xing, B., & Gao, W. (2014). Fruit fly optimization algorithm. In Springer International Publishing (Ed.), *Innovative computational intelligence: A rough guide to 134 clever algorithms* (pp 167–170). London: Springer. https://doi.org/10.1007/978-3-319-03404-1_11.
28. Chen, P.-W., Lin, W., Huang, T., & Pan, W. T. (2013). Using fruit fly optimization algorithm optimized grey model neural network to perform satisfaction analysis for e-business service. *Applied Mathematics & Information Sciences, 7,* 459–465.
29. Pan, W. T. (2012). A new Fruit Fly Optimization Algorithm: Taking the financial distress model as an example. *Knowledge-Based System, 26,* 69–74.
30. Li, H., Guo, S., Li, C., & Sun, J. (2013). A hybrid annual power load forecasting model based on generalized regression neural network with fruit fly optimization algorithm. *Knowledge-Based System, 37,* 378–387.
31. Kanarachos, S., Griffin, J., & Fitzpatrick, M. E. (2017). Efficient truss optimization using the contrast-based fruit fly optimization algorithm. *Computers & Structures, 182,* 137–147.
32. Yuan, X., Dai, X., Zhao, J., & He, Q. (2014). On a novel multi-swarm fruit fly optimization algorithm and its application. *Applied Mathematics and Computation, 233,* 260–271.
33. Yang, X. S. (2014). *Nature-inspired optimization algorithms.* Amsterdam: Elsevier.

Chapter 8
A Multi-granularity NC Program Optimization Approach for Energy Efficient Machining

X. X. Li, W. D. Li and F. Z. He

Abstract NC programs are widely developed and applied to various machining processes. However, the lack of effective NC program optimization strategy for the machining energy efficiency has been crippling the implementation of sustainability in companies. To address this issue, a multi-granularity NC program optimization approach for energy efficient machining has been developed and presented in this paper. This approach consists of two levels of granularities: the granularity of a group of NC programs for a setup where the features are machined on a single CNC machine with the same fixture and the granularity of a NC program. On the former level of granularity, the execution sequence of the NC programs for the setup of a part is optimized to reduce the energy consumed by the cutting tool change among the NC programs. On the latter level of granularity, the execution sequence of the features in the same NC program is optimized to reduce the energy consumed by the cutting tool's traveling among the machining features. Experiments on the practical cases show that the optimization results from this approach are promising and the approach has significant potential of applicability in practice.

Keywords Multi-granularity optimization · Energy efficient machining
NC program · Sustainable manufacturing

X. X. Li
College of Informatics, Huazhong Agricultural University, Wuhan, China

W. D. Li (✉)
Faculty of Engineering, Environment and Computing, Coventry University, Coventry, UK
e-mail: weidong.li@coventry.ac.uk

F. Z. He
School of Computer Science and Technology, Wuhan University, Wuhan, China

W. Li and S. Wang (eds.), *Sustainable Manufacturing and Remanufacturing Management*, https://doi.org/10.1007/978-3-319-73488-0_8

8.1 Introduction

In recent years, economic, environmental, and legislative drivers have raised the energy-saving awareness of both manufacturers and customers. The soaring electricity price has brought the manufacturers the increasing energy cost which has increased by almost 70% since the late 1990s [1]. The rapidly growing production demands have incurred more manufacturing activities [2–7] where plenty of greenhouse gas has been emitted from the usage of energy sources such as electricity, coal, and oil. Statistics has shown that the greenhouse gas from manufacturing accounts for more than 37% even 50% of the world's total greenhouse gas emissions [8]. In order to balance the multifaceted dimensions of economic growth and environmental protection, a series of regulations and guidelines [9] have been developed. Additionally, the rising energy-saving awareness of customers always drives them to choose a product with lower life-cycle energy consumption. Hence, it is imperative for the manufacturing companies to take energy-saving measures to enhance their competitiveness.

As the most widely used machine tool in manufacturing companies, computer numerical control (CNC) machines highly contribute to energy consumption in the manufacturing sector. However, statics has shown that the energy efficiency of machines tools is generally less than 30% [10]. Hence, the optimization of numerical control (NC) machining process is of great importance for energy saving.

For any NC machining process, NC programs which are composed of a series of coded instructions are required to control it. The NC programs can be generated by not only the hand programming but also the commercial CAD/CAM packages. However, no optimal techniques have been used to optimize the NC programs to save energy. Thus, the optimization of the NC programs used to control the CNC machining processes is imperative to achieve energy efficient machining.

In order to realize energy efficient machining by optimizing NC programs, a multi-granularity optimization approach has been developed. The approach optimizes the NC programs used to control the NC machining processes from the following two granularities: (1) the execution sequence of the NC programs used to execute all the features in the setup is optimized to reduce the energy consumed by the cutting tool change and (2) the optimization of the tool path connecting all the machining features in the same NC program is achieved by considering the criteria of energy consumption. The practical cases are used to demonstrate the feasibility and effectiveness of the proposed method. The energy efficiency improvement after the multi-granularity optimization can be at least 10% which is more than 5% expected by the empirical value.

The rest of the paper is organized as follows. In Sect. 8.2, related work is reviewed. In Sect. 8.3, the problem description and formulation are presented. In Sect. 8.4, the multi-granularity optimization approach is presented. In Sect. 8.5, the prototype system and the corresponding implementation of our solution are presented. Finally, the research is concluded in Sect. 8.6.

8.2 Related Work

In the last decade, the optimization for energy efficient machining, as a challenge in sustainable manufacturing [9], has attracted many researchers' attention. The related work focused on two different levels, i.e., manufacturing system and machining process.

8.2.1 *The Optimization on Manufacturing System Level*

The research on manufacturing system level concentrates on scheduling optimization to improve the energy efficiency of the whole manufacturing system.

Mouzon et al. developed a multiple objective mathematical programming model and several algorithms to optimize the scheduling on a single CNC machine and reduce the energy consumption and total completion time. In order to minimize the total energy consumption and the total tardiness on a machine, a greedy randomized adaptive search algorithm was further developed to achieve the multi-objective schedule optimization [11]. In Fang et al.'s work [12], a new mixed integer linear programming model was built to schedule a classical flow shop that combined the peak total power consumption and associated carbon footprint with the makespan. Bruzzone et al. [13] developed an energy-aware scheduling algorithm based on a mixed integer programming formulation to realize energy savings for a given flexible flow shop which was required to keep fixed original jobs' assignment and sequencing. Dai et al. [14] presented an energy-efficient model for flexible flow shop scheduling and an improved genetic-simulated annealing algorithm to evaluate the total energy consumption in the flexible flow shop and optimize the total energy consumption and the makespan, respectively. Li et al. [9] developed a multi-objective optimization model for scheduling to improve material removal rate (i.e., MRR) and energy efficiency.

8.2.2 *The Optimization on Machining Process Level*

Different from the research on the manufacturing system level, the research on the machining process level focuses on modeling for decision-making [15, 16] and the optimization of the aspects involved in the NC machining processes, which mainly include the optimization of NC machining parameters [17–27] and the optimization of tool path [28–38].

In order to support decision-making for energy efficient machining, some research work focuses on developing specific models of unit process energy consumption. Gutowski et al. [39] built a theoretical power consumption equation based on thermal equilibrium approach to generally describe unit process energy for machining processes. In this model, the process rate was identified as the main factor for the unit process energy consumption. However, other factors in this model, such as the fixed power P_0 and the constant k, were lack of clear definition and quantification. Thus, the model cannot be used to predict the energy consumption yet. Taking up the missing specification in the model of Gutowski et al., empirical modeling approaches were adopted by Li and Kara [40] to build an empirical model to characterize the relationship between the specific energy consumption (SEC, the energy consumption of the machine tool for removing 1 cm^3 material) and the material removal rate (MRR). This model can be used to predict the energy consumption of manufacturing processes. However, it is difficult to precisely assign the factors for each coefficient of the model. In order to explore an energy consumption model with high accuracy and well-defined coefficients, a hybrid modeling approach of thermal equilibrium and empirical modeling was used to build an improved model [41]. This model was tested on a CNC micromachining center, and a reliable prediction of energy consumption for given process parameters with a higher accuracy was obtained. However, the further extension of the model is still needed to achieve a more generic energy consumption model.

Simultaneously, other research work was carried out to model the energy consumption to characterize the relationship between the process parameters and the energy consumption for cutting processes and optimize the process parameters. Newman et al. [8] developed an empirical model to establish the relationship between the power consumption and the process parameters such as spindle speed, federate, cutting depth, and cutting width. An experimental design method was adopted by Lin et al. [42] to establish a machining parameter optimization model of multi-pass turning operations in dry and wet cut environments. Design expert was used to optimize the cutting parameters of the turning operation by Anand et al. [43], and the optimized values were further checked and compared by those being generally used. A response surface method was used by Campatelli et al. [44] to optimize the process parameters to minimize the power consumption in the milling of carbon steel. Kant and Sangwan [45] considered power consumption and surface roughness by optimizing the machining parameters. On the basis of the experimental data obtained by the sensors mounted on the cutter, another empirical model involving spindle speed, federate, and cutting depth was presented by Hu et al. [46]. In Camposeco-Negrete's work [47], the Taguchi method was used to analyze the relationship among cutting parameters, energy consumption, and surface roughness to optimize cutting parameters and achieve the minimum energy consumption and the best surface roughness. Based on weighted gray relational analysis and response surface methodology, a multi-objective method was developed by Yan and Li [48] to evaluate trade-offs between sustainability, production rate, and cutting quality. In Sheng's work [49], artificial neural networks were used to establish the complex nonlinear relationships between the process parameters including spindle speed, federate, cutting depth and cutting width, and energy consumption. Intelligent algorithms were then applied to identify the optimal process parameters.

In addition, the influence of tool path on energy consumption was also studied. Several tool path generation schemes were tested to explore the influence of tool path generation schemes on the amount of energy required to machine the same part [50]. The non-proportional relation between energy consumption and machining time was verified by [51]. The tool path was considered in [52] to develop a model for the energy evaluation. All these studies have indicated that tool path has a significant impact on energy saving in machining process. However, the existing research work on tool path optimization mainly aims at high-productivity machining. Much research has been done on minimizing the cutting time and the airtime (i.e., the time to move spindle in the air) by optimizing tool path from tool path generation and the connection among the tool path and so on. A comprehensive survey can be found from [53].

Based on the above-detailed literature survey, it can be observed:

- Energy consumption is influenced by scheduling, process parameters, and tool path significantly. However, compared with the research work on the optimization of scheduling and process parameters, less attention has been paid to optimizing another aspect in the NC programs, i.e., tool path for energy saving;
- Previous work in the tool path optimization has mostly focused on exploring the influence on the machining energy consumption and minimizing the machining time. However, the research on the tool path optimization for energy saving is still preliminary.
- In order to bridge the research gap, it is imperative to understand the characteristics of energy consumption influenced by the tool path and develop systematic and efficient approaches to achieve energy saving by optimizing the tool path involved in the NC programs.

8.3 Problem Description and Formulation

8.3.1 Representations for NC Program Optimization

For any part to be produced using CNC machines, NC programs are needed to control the machining process and complete all the machining features in the part. The machining features in the part can be divided into one or more setups. Each setup consists of a group of features that are machined on a single CNC machine with the same fixture [54]. Figure 8.1 shows a part with a single setup which contains nine machining features (i.e., F_1–F_9). Hence, the purpose of reducing the energy consumed by producing a part can be achieved by optimizing the NC programs for each setup in the part.

In order to obtain the NC programs for each setup, two kinds of programming methods are always employed. One is hand programming which is suitable for simple cutting process. The other is automatic programming by the CAD/CAM packages that provide automatic NC programming. Compared with the hand programming, the automatic programming is more suitable for the complex part that

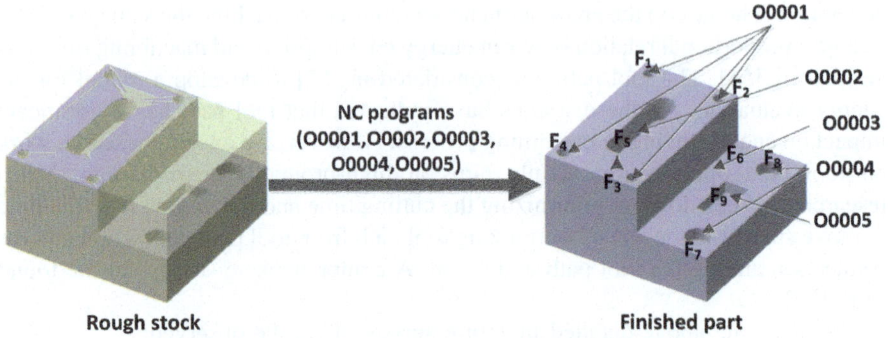

(a) The part to be produced and the NC program files

```
......
(Tool=DR16-70-64-2-16_BT40-32DS_70;
D=16.00; R=0.80)
......
G73 X0.0 Y60.0 Z-36. R3. F300. Q8.
X-60.0
Y-60.0
X0.0
G80
G00 Z10.
......
%
```

O0001

(b) The G-codes in the NC program O0001

Fig. 8.1 Examples for tool path optimization based on NC programs

cannot be programmed easily. However, with the hand programming and the automatic NC programming, no techniques are used to optimize the NC programs to reduce the energy consumed during machining the features in a setup.

Aiming at reducing the energy consumption of a setup, the tool path influenced by the execution sequence of the features in the setup should be considered because it influences not only the energy consumed by the tool change but also the energy consumed by the tools' traveling among the features. Take the part shown in Fig. 8.1 as an example, the execution sequence of the NC programs used to machine the features is O0001-O0002-O0003-O0004-O0005. The NC program O0001 is used to execute the four hole features (i.e., F_1, F_2, F_3, and F_4) on the top surface. The NC program O0002, O0003, and O0005 are used to execute the pocket features F_5, F_6, and F_9, respectively. File O0004 is used to complete the hole features F_7 and F_8. It can be seen that the cutting tool is changed between O0001 and O0002, O0003 and O0004, and O0005. Another example is that the code "G73" included in the NC program O0001 (c.f. Fig. 8.1b) is used to execute the four hole features on the top surface. During the execution of this kind of NC programs, energy is consumed in not only the cutting process but also the traveling among the features to be machined. Therefore, the energy consumption of a setup can be reduced by optimizing the execution sequence of the features in the setup, which consists of two parts: the sequence of the features in the different NC programs and the sequence of the features in the same NC program. That is, the NC program optimization for energy efficient machining can be achieved from two granularities: the granularity of a group of NC programs for a setup and the granularity of a NC program.

8.3.2 Energy Consumption Modeling for the NC Program Optimization

As one of the granularities to be optimized, the NC programs for a setup influence the energy consumed by the tool change which is decided by the execution sequence of the NC programs. If the adjacent NC programs do not share the same cutting tool, the cutting tool will be changed by the tool change system to support the features in the next NC program. Thus, the energy consumed by the tool change is influenced by the execution sequence of the NC programs. According to the execution sequence of the NC programs, the energy consumed by changing the tool among the NC programs can be evaluated using the following equation:

$$\mathrm{EC}_{\mathrm{tc}_{\mathrm{nc}}} = \sum_{i \in V, j \in V} x_{ij} \times y_{ij} \times P_{\mathrm{tc}} \times T_{\mathrm{tc}} \tag{8.1}$$

where $EC_{\mathrm{tc\_nc}}$ is the energy required for changing the cutting tool among the NC programs, x_{ij} can be 1 (if the NC program O_j is the next one to be executed after NC program O_i), y_{ij} can be 1 (if tool change is needed between O_i and O_j) or 0, V is the NC program set, P_{tc} is the power demand for the tool change motor, T_{tc} is the time required for tool change.

For the granularity of a NC program, the energy consumption mainly refers to the energy consumption of the traveling among the features in the same NC program. The energy consumed by traveling among the features consists of the energy consumed by spindle motor and the energy required to move the cutting tool in *X*-, *Y*-, and *Z*-direction at the given feedrate. Therefore, the energy consumption of traveling among all the features can be evaluated using the following model:

$$EC_{travelling} = EC_{spindle} + EC_{feedrate} \tag{8.2}$$

$$EC_{spindle} = \sum_{i \in V, j \in V} x_{ij} \times P_{spindle} \times (T_{ijx} + T_{ijy} + T_{ijz}) \tag{8.3}$$

$$EC_{feedrate} = \sum_{i \in V, j \in V} x_{ij} \times (P_x T_{ijx} + P_y T_{ijy} + P_z T_{ijz}) \tag{8.4}$$

where $EC_{travelling}$, $EC_{spindle}$, and $EC_{feedrate}$ are the total, spindle, and feedrate energy consumption, respectively, x_{ij} can be 1 (if the feature Fj is the next one to be visited after feature F_i), $P_{spindle}$ is the spindle power that is an unloaded power of the spindle motor, P_x, P_y, and P_z are the power of the feed axis in *x*-, *y*-, and *z*-direction, respec*tively,* T_{ijx}, T_{ijy}, *and* T_{ijz} *are the time to move the axis from* F_i *to* F_j *in x*-, *y*- *and z*-direction, respectively. The time can be obtained using distance divided by feedrate, and the distance between two machining features can be calculated by the coordinates of the approaching location of the features.

8.3.3 Modeling for the NC Program Optimization

As mentioned above, the NC program is optimized from two granularities. On the level of the setup granularity, the NC programs' execution sequence is optimized to minimize the energy consumed by cutting tool change. On the level of a NC program granularity, the sequence of the features in the same NC program is optimized to minimize the energy consumed by the traveling among the features in the NC program. If the NC programs and the features in them are used to represent the cities (c.f. Fig. 8.2) and the cutting tool is used to represent the salesman, both the optimization of the NC programs' execution sequence and the optimization of the sequence of the features in the same NC program can be modeled as a special case of traveling salesman problem (TSP) where there are a number of nodes (i.e., cities to be visited) and a salesman whose goal is to visit each node exactly once with the minimum tour. TSP can be described mathematically as follows:

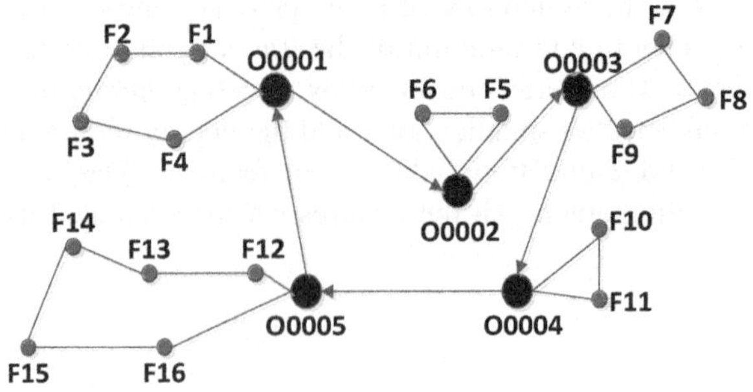

Fig. 8.2 Examples for the optimization of the tool path influenced by the features' execution sequence

$$\text{minimize} \quad \sum_{i \in V, j \in V} d_{ij} x_{ij} \tag{8.5}$$

$$\text{subject to} \quad x_{ij} \in \{0, 1\} \tag{8.6}$$

$$\sum_{i \in V} x_{ij} = 1 \quad j \in V \tag{8.7}$$

$$\sum_{j \in V} x_{ij} = 1 \quad j \in V \tag{8.8}$$

where d_{ij} is the distance between cities i and j, V is the node set, x_{ij} can be 1 (if the city j is the next city to be visited after city i is visited) or 0, constraints 10, 11, and 12 ensure that each city is visited exactly once.

There are four main differences between the NC program optimization for energy efficient machining and TSP. One is that the NC program optimization for energy efficient machining consists of two TSPs: the optimization of the NC programs' execution sequence to minimize the energy consumed by tool change and the optimization of the sequence of the features in the same NC program to minimize the energy consumed by the traveling tool path among the features. The second difference is that the salesman in the TSP will return to the starting city, where the cutting tool in the optimization of the tool path influenced by the features' execution sequence completes its tour when it reaches the last NC program or the last feature in the same NC program. The third difference is that the goal is no longer limited to the minimum length tour, where energy consumption during the tour is also considered. The fourth difference is that the travel between the cities in the NC program optimization for energy efficient machining has some precedence constraints to keep the dependent relationship among the machining features.

8.4 The Solution

8.4.1 The General Scheme

Based on the above problem description and the established models, our method is presented. The NC programs for a setup are chosen as the input and the output of our method. As is shown in Fig. 8.3, there are three main steps in our method, which are listed as follows:

- The NC programs for a setup are imported and the information of the cutting tool and the machining features is obtained.
- The imported NC programs are optimized from the following two granularities:

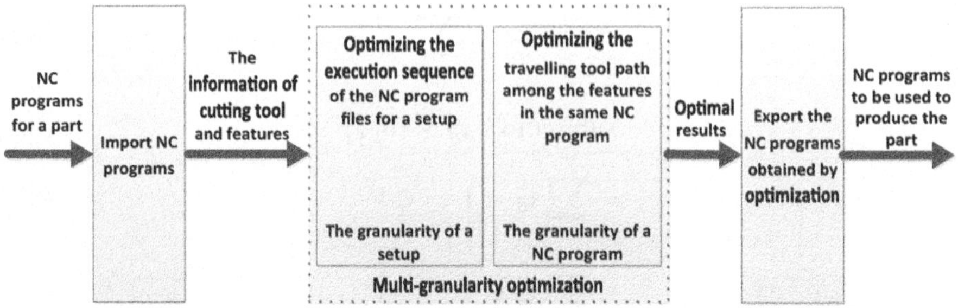

Fig. 8.3 General scheme

- The granularity of a setup: optimize the execution sequence of all the NC programs for the setup.
- The granularity of a NC program: optimize the traveling tool path among the features in the same NC program.

For both granularities, the exhaustive algorithm and an improved honeybee mating optimization (HBMO)-simulated annealing (SA) algorithm are employed to search for the optimal solution. The exhaustive algorithm can be used to search for the best one from all the solutions. However, the efficiency of the algorithm is influenced by the number of the solutions. Thus, in order to find the best solution and take the efficiency of the algorithm into account, the exhaustive algorithm is used to deal with the optimization where the number of NC programs or the number of features in the same NC program is less than some value (supposing it be m). For the granularities which involve more than m NC programs or features, an improved HBMO-SA algorithm is proposed to achieve the optimization.

- The NC programs obtained by the above multi-granularity optimization are exported to be used to produce the part.

8.4.2 *The Improved HBMO-SA Algorithm for the Multi-Granularity Optimization*

There are many heuristic algorithms that have been implemented in combinatorial optimization problems, such as the genetic algorithm [55–57], the simulated annealing algorithm (SA) [58, 59], the ant colony optimization [60–62], particle swarm optimization [63, 64], the honeybee mating optimization algorithm (HBMO) [65, 66]. It has been proved that the HBMO can reach the optimization solution quickly. However, it is liable to converge to local optima. Fortunately, SA has the ability to jump out of the local optimization because it can accept some probability.

Therefore, in this paper, the HBMO and the SA are incorporated to rapidly search for an optimal or near-optimal solution and then a better one.

8.4.2.1 The Fitness Function

The goal of the optimization on the level of the setup granularity is to minimize the energy consumption of the tool change by reducing the number of tool change among the NC programs for a setup. With the decrease of the number of tool change, the time to change the cutting tools is reduced. That is, minimal energy consumption will bring improved machining productivity. Thus, only the energy consumed by the tool change is chosen as the goal of the optimization on this level and the reciprocal of the energy consumption function for the tool change is used as the fitness function to follow the rule that the individual with a greater fitness has a higher chance to be chosen. The fitness function for a solution is as follows:

$$\mathrm{Fitness}_{\mathrm{nc}} = \frac{1}{\mathrm{EC}_{\mathrm{tc\_nc}}} \tag{8.9}$$

Different from the optimization on the level of the setup granularity, the optimization on the NC program level aims at minimizing both the energy consumption of the traveling tool path and the corresponding traveling time because there is a trade-off between them [see the Eq. (8.8)]. Thus, the optimization on this level of granularity is a multi-objective optimization problem.

Based on the established models, the weighted additive utility function is adopted to solve the multi-objective optimization problem. The total weighted performance criteria (TWPC) can be described as:

$$\mathrm{TWPC} = W_1 \mathrm{EC}_{\mathrm{travelling}} + W_2 T_{\mathrm{travelling}} \tag{8.10}$$

where w_1 and w_2 are the weights. The value of each weight is between 0 and 1, and the sum of them equals to one.

Following the rule that the individual with a greater fitness has a higher chance to be chosen, and the reciprocal of the objective function is used as the fitness function [c.f. Eq. (8.2)].

$$\mathrm{Fitness}_{\mathrm{travelling}} = \frac{1}{\mathrm{TWPC}} \tag{8.11}$$

8.4.2.2 The HBMO Operation Phase

The HBMO algorithm is developed by simulating the honeybee mating process. That is, the HBMO algorithm should contain a number of different procedures which correspond to the different phase of the honeybee mating process.

The honeybee mating process mainly consists of four stages: mating flight, generating broods, feeding the broods, and the selection of the new queen. During the mating flight, the strong drones catch up with the queen and mate with her. After the queen's spermatheca is full of sperms, she will fly back to the nest and lay eggs. Each time the queen lays eggs, she randomly retrieves a sperm from her spermatheca to fertilize the eggs and a set of broods is generated [65]. Then, the workers will take care of and improve the broods (e.g., feeding them with royal jelly). If a brood is better than the queen, it will be the new queen and starts its mating flight.

As a consequence of the above, the HBMO algorithm can be described as follows:

(1) Initially, the population of the honeybees (i.e., initial solution) is created to configure the initial hive. In the proposed algorithm, the initial population where each individual is composed of the NC program sequence and feature sequence and is generated randomly. Then, the solution with the maximum fitness value is selected as the queen. All the other members of the population are used as drones.

(2) The probabilistic rule in Eq. 8.2 is used to select the strong drones to mate with the queen. The selected drones' sperms are stored in the queen's spermatheca. This procedure will not stop until the spermatheca is full. The queen's size of spermatheca is defined before the mating process begins. A drone mates probabilistically by using an annealing function as follows [21]:

$$\frac{-|\Delta f|}{e^{\text{Speed}(t)}} > r \tag{8.12}$$

where Δf is the difference between the fitness of the drone and queen, speed(t) is the queen's flight speed at the t-th mating, r is randomly generated, the queen's flight speed is also generated at random before each mating flight and decays with the mating. It decays according to the following equation:

$$\text{Speed}(t+1) = \alpha \times \text{Speed}(t) \tag{8.13}$$

where α is between 0 and 1.

(3) A brood is generated by crossover the queen's genotype with the sperm selected from the queen's spermatheca. First, both the queen and the selected sperm are separated into two parts from the crossover point chosen randomly. Then, the queen's left part and the sperm's right part are copied to generate a brood. The redundant genes in the brood are replaced with the lost ones. Finally, the brood is adjusted according to the precedence constraints among the features. The new brood is obtained.

(4) For each brood, a worker (i.e., a local search heuristics) is chosen randomly to improve it. If the improved brood (i.e., the new solution) is better than the current queen, it will replace the queen. All the other brood will take part in the next mating flight as drones. Two operators are used to generate the workers. One is the adjacent swapping which is realized by exchanging two adjacent and

interchangeable NC programs or features. The other is the shift operator which is done by removing a NC program or a feature from its present position to insert it at another position.

(5) If the number of mating flight is still not more than the maximum which has been defined before the mating process, a new mating flight will begin. That is, step 2–4 will be repeated until the number of mating flight is large enough.

8.4.2.3 The SA Operation Phase

In the SA operation phase, the good individuals generated by the HBMO are sent to the SA for improvement. The simulated annealing algorithm can be described as follows:

(1) Decide an initial solution: Use the individual selected by the HBMO as the initial solution S_0.
(2) Choose the initial solution as the current solution S.
(3) Determine the start and end temperatures T_{start} and T_{end} and Use T_{start} as the current temperature T.
(4) Generate a new temporary solution S': The mutation strategy used to generate S' is described as follows:
    (a) Two NC programs or features are chosen randomly.
    (b) If the precedence constraint among the features can be satisfied, the chosen NC programs or features are exchanged. If not, the iteration of the random choice will be judged to decide whether (a) will be repeated or not.
(5) The fitness function defined above is used to compute the difference between the performance criteria of S' and S. The following steps are executed to judge whether keep S' as the original solution for the next iteration.
    (a) Compute the difference δ between the performance criteria between S'
    (b) and S: The criterion function defined above is used to compute the performance criteria of S' and S. The difference $\delta = \mathrm{Fitness}_{tc}(S') - \mathrm{Fitness}_{tc}(S)$.
    (c) Judge the difference δ: if δ is less than 0 or the following probability function can be met, S' will be adopted as the original solution for the next iteration.

$$e^{\frac{-|\delta|}{T_{start}}} \geq \mathrm{random} \tag{8.14}$$

where random is between 0 and 1 and randomly generated.

(6) Reduce the temperature: the temperature T will be reduced at the ratio α which is between 0 and 1.
(7) Judge the iteration: (4)–(6) will be repeated until the current temperature $T <= T_{end}$.

8.5 Case Studies and Discussions

The presented solution has been used to establish a pre-optimization prototype system to optimize the NC programs before they are used to control the CNC machine to produce the part. In this section, the implementations of our solution are presented.

Two groups of NC programs from practical manufacturing are presented to validate the effectiveness and feasibility of the presented approach. In the case studies, the machines tested by Aramcharoen and Mativenga were used as the machining resource [51]. The experiments were performed on the Windows 8 operating system with Intel Dual-Core CPU at 2.40 GHz and 8.00 GB of main memory. The approach was carried out using the Java programming language.

In order to select the appropriate parameters, a number of different alternative values of the proposed algorithm's parameters were tested. For the exhaustive algorithm, the parameter m equals to 8. Parameters of HBMO-SA are set as: the size of queen's spermathecal equals to 25, the number of drones is 50, the number of mating flights is 200, the initial speed(t) is 1000, the ration α in Eq. (8.2) is 0.85, the start and end temperature T_{start} and T_{end} equal to 1000 and 2.4775e−034, respectively, and α in the SA equals to 0.9.

8.5.1 Case Study 1

The first case study used the NC programs for the part shown in Fig. 8.4a to validate the presented approach. The part has a single setup which consists of 13 machining features (i.e., F_1–F_{13}). NC program O0001 is used to machine the surface feature F_1. NC program O0002, O0004, and O0006 illustrated in Fig. 8.4b are used to machine the corresponding hole features. O0003 and O0005 are used to machine the pocket features F_8 and F_{11}, respectively.

First, the NC programs O0001-O0006 are imported and the name of the cutting tool in each NC program and the cutting tool coordinates for each feature in the NC programs is obtained.

Then, the multi-granularity NC program optimization is done. For the granularity of the setup, the exhaustive algorithm is chosen to find the optimal execution sequence of the NC programs because the number of the NC programs is less than 8. The optimal result is O0001-O0003-O0005-O0002-O0004-O0006. For the granularity of the NC program, the sequence of the features in each NC program is optimized. Take the NC program O0002 (c.f. Fig. 8.4b) as an example, the exhaustive algorithm is used again to optimize the execution sequence of the six hole features (i.e., F_2-F_3-F_4-F_5-F_6-F_7). The optimal result is F_2-F_7-F_6-F_5-F_4-F_3. Based on the multi-granularity tool path optimization, the energy efficiency of the machining process is improved by 10%.

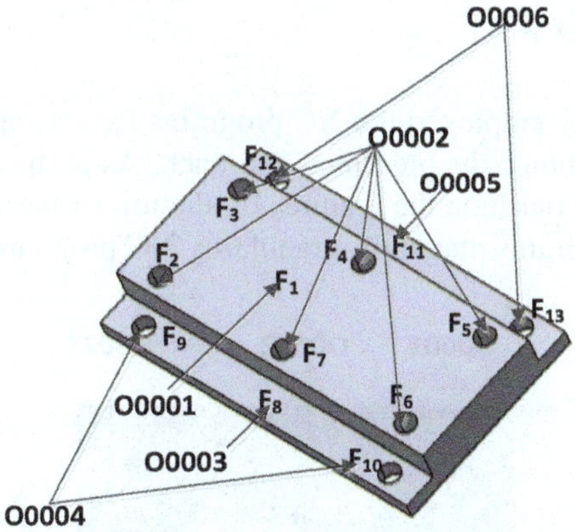

(a) The part to be machined

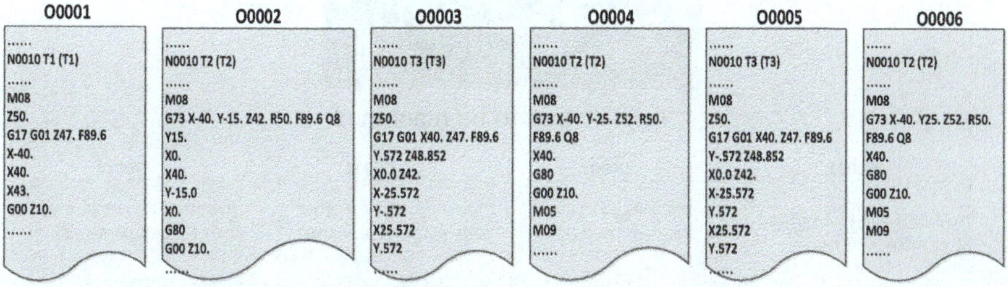

(b) The original NC programs to be imported

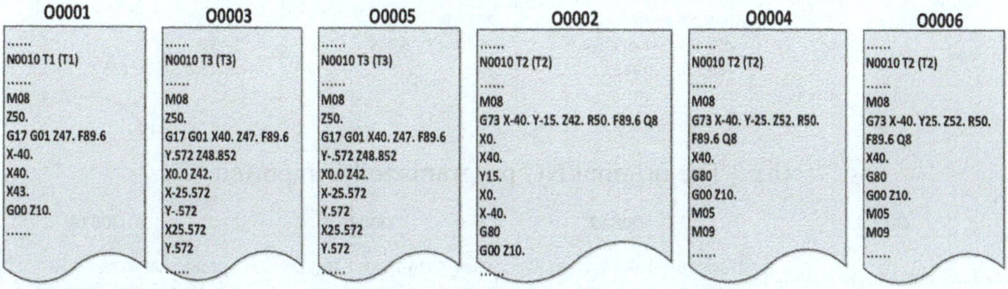

(c) The optimal NC programs to be used to produce the part

Fig. 8.4 First part and the NC programs

Finally, according to the optimal sequence, the cutting tool coordinates of the corresponding features in each NC program are adjusted to obtain the optimal NC programs and then the optimal NC programs (c.f. Fig. 8.4c) are exported to be used to operate the part.

8.5.2 *Case Study 2*

The second case study employed the NC programs for a setup of the part shown in Fig. 8.5 to further validate the presented approach. As is shown in Fig. 8.5, 23 NC programs are used to machine the features on the top surface of the part. Based on the imported NC programs, the multi-granularity NC program optimization is done.

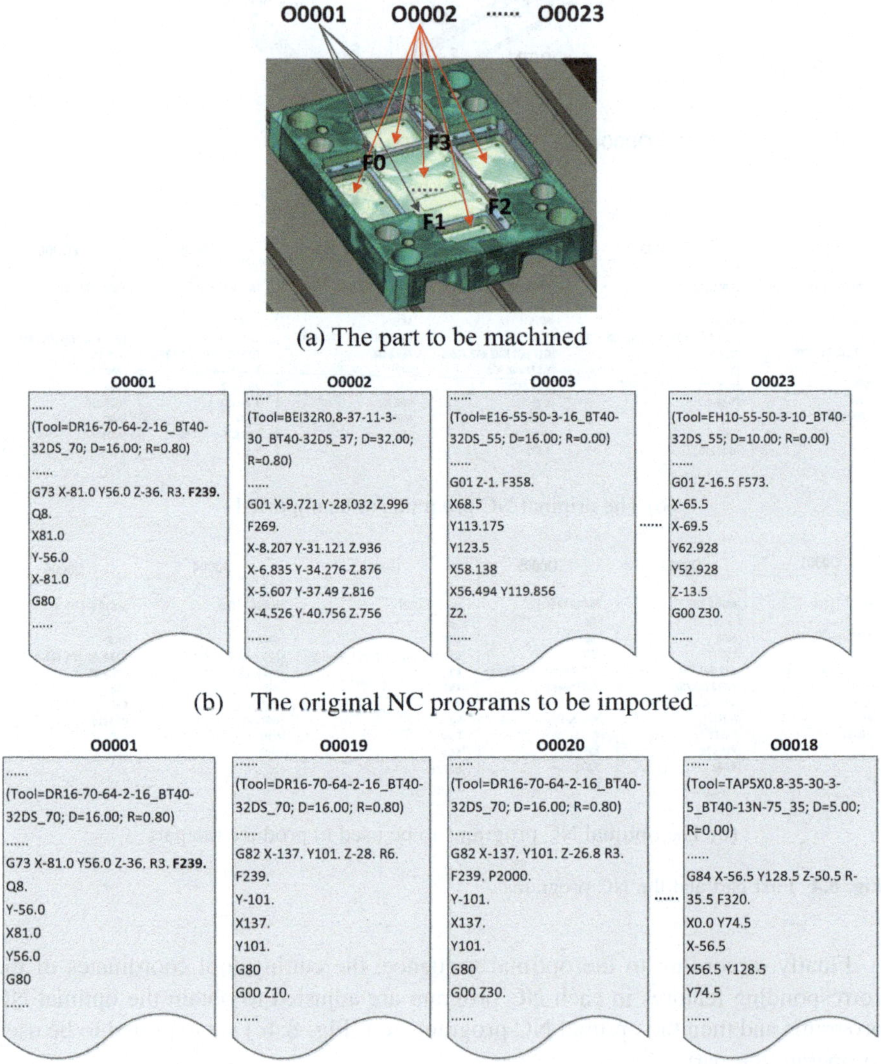

(a) The part to be machined

(b) The original NC programs to be imported

(c) The optimal NC programs to be used to produce the part

Fig. 8.5 Second part and the NC programs

On the level of the setup granularity, the energy consumed by the tool change is used as the performance criterion and the execution sequence of the NC programs is optimized using HBMO-SA. As is shown in Fig. 8.5c, the optimal execution sequence of the NC programs (O0001-O0019-O0020-O0002-O0003-O0004-O0005-O0006-O0021-O0022-O0023-O0007-O0008-O0009-O0010-O0011-O0012-O0013-O0016-O0014-O0015-O0017-O0018) is obtained.

On the level of the NC program granularity, both traveling time and energy consumed by the traveling among the features are taken into account and the traveling tool path among the features in the same NC program is optimized. Take NC program O0001 as an example, four hole features (c.f. Fig. 8.5b) are included in this NC program. The original machining sequence is (F_0-F_1-F_2-F_3). The exhaustive algorithm was used to optimize the traveling tool path among these features. The optimal result is (F_0-F_3-F_2-F_1). Another example is O0002 (c.f. Fig. 8.5b). According to the retracting coordinates, 20 machining features are obtained. The sequence of these features is optimized using HBMO-SA. The convergence curves of HBMO-SA under the condition of minimizing machining energy consumption are illustrated in Fig. 8.6. It can be observed that the HBMO drops fast in the first phase of the proposed algorithm and converges to an intermediate solution. Based on this intermediate solution, a further optimization is obtained by the SA in the second phase. The energy efficiency of the machining process is improved by 15.9%.

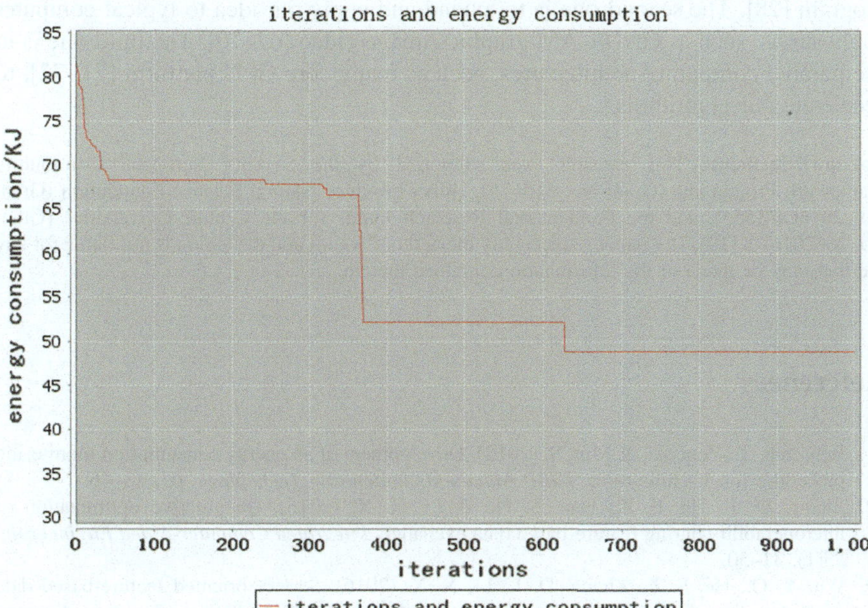

Fig. 8.6 Convergence curve of HBMO-SA

8.6 Conclusions

NC program optimization is critical for energy efficient NC machining. In this paper, the NC program optimization for energy efficient machining is explored and a multi-granularity optimization approach is presented. In summary, the contributions of the presented approach are from the following aspects:

- Multi-granularity optimization strategy is presented. On the basis of the strategy, the energy consumption of NC machining process is effectively considered to optimize the cutting tool change and the tool path connection to achieve energy efficient machining process.
- The energy consumption models for the setup granularity and the NC program granularity are built to evaluate the energy consumed by the tool change among the different NC programs for the setup and the tool's traveling among the features in the same NC program, respectively.
- Systematic models for the NC program optimization in the terms of energy consumption and machining time are established. Based on the established models, exhaustive algorithm and HBMO-SA are adopted to take into account both optimal result and efficiency in the further optimization of the NC programs.

The further research is intended to be focused on following directions but not limited. The first one is to extend our optimization to the features' tool path in a NC program [28]. The second one is to extend and apply our idea to typical computer applications, such as CAD/CAM/graphics/image/video [67–70]. The third one is to use parallel computing architectures, such as many-core GPU platform [71, 72], to accelerate our algorithms.

Acknowledgements This research was supported by the Seventh European Community Framework Programme (Grant No. 610675), Hubei Province Natural Science Foundation (Grant No. 2016CFB555), and the Fundamental Research Funds for the Central Universities (Grant No. 2662016PY119). The paper reflects only the authors' views and the Union is not liable for any use that may be made of the information contained therein.

References

1. Behrendt, T., Zein, A., & Min, S. (2012). Development of an energy consumption monitoring procedure for machine tools. *CIRP Annals-Manufacturing Technology, 61,* 43–46.
2. Zhang, D. J., He, F. Z., Han, S. H., & Li, X. X. (2016). Quantitative optimization of interoperability during feature-based data exchange. *Integrated Computer-Aided Engineering, 23*(1), 31–50.
3. Wu, Y. Q., He, F. Z., Zhang, D. J., Li, X. X. (2016). Service-oriented feature-based data exchange for cloud-based design and manufacturing. *IEEE Transactions on Services Computing, 11*, 1939–1974.

4. Cai, X. T., Li, W. D., He, F. Z., & Li, X. X. (2015). Customized encryption of computer aided design models for collaboration in cloud manufacturing environment. *Journal of Manufacturing Science and Engineering, 137*(4), 040905.
5. Kim, B. C., Mun, D., & Han, S. (2010). Retrieval of CAD model data based on Web Services for collaborative product development in a distributed environment. *International Journal of Advanced Manufacturing Technology, 50*(9–12), 1085–1099.
6. Yuan, W. Q., Liu, Y. S., Zhao, J. J., & Wang, H. W. (2016). Pattern-based Integration of System Optimization in Mechatronic System Design. *Advances in Engineering Software, 98,* 23–37.
7. Buckholtz, B., Ragai, I., & Wang, L. H. (2015). Cloud manufacturing: Current trends and future implementations. *Journal of Manufacturing Science and Engineering, 137*(4), 040902.
8. Newman, S. T., Nassehi, A., Imani-Asrai, R., & Dhokia, V. (2012). Energy efficient process planning for CNC machining. *CIRP Journal of Manufacturing Science and Technology, 5,* 127–136.
9. Li, X. X., Li, W. D., Cai, X. T., & He, F. Z. (2015). A hybrid optimization approach for sustainable process planning and scheduling. *Integrated Computer-Aided Engineering, 22*(4), 311–326.
10. Hu, S. H., Liu, F., He, Y., & Hu, T. (2012). An on-line approach for energy efficiency monitoring of machine tools. *Journal of Cleaner Production, 27,* 133–140.
11. Mouzon, G., & Yildirim, M. B. (2008). A framework to minimise total energy consumption and total tardiness on a single machine. *International Journal of Sustainable Engineering, 1* (2), 105–116.
12. Fang, K., Uhan, N., Zhao, F., & Sutherland, J. W. (2011). A new approach to scheduling in manufacturing for power consumption and carbon footprint reduction. *Journal of Manufacturing Systems, 30*(4), 234–240.
13. Bruzzone, A. A. G., Anghinolfi, D., Paolucci, M., & Tonelli, F. (2012). Energy-aware scheduling for improving manufacturing process sustainability: A mathematical model for flexible flow shops. *CIRP Annals-Manufacturing Technology, 61*(1), 459–462.
14. Dai, M., Tang, D. B., Giret, A., Salido, M. A., & Li, W. D. (2013). Energy-efficient scheduling for a flexible flow shop using an improved genetic-simulated annealing algorithm. *Robotics and Computer-Integrated Manufacturing, 29,* 418–429.
15. Wang, L. (2013). Machine availability monitoring and machining process planning towards cloud manufacturing. *CIRP Journal of Manufacturing Science and Technology, 6*(4), 263–273.
16. Zhu, H. W., Shao, Y. L., Liu, Y. S., & Zhao, J. J. (2016). Automatic hierarchical mid-surface abstraction of thin-walled model based on rib decomposition. *Advances in Engineering Software, 97,* 60–71.
17. Kumar, H., Abbas, M., Mohammad, A., & Jafri, H. Z. (2013). Optimization of cutting parameters in CNC turning. *Optimization, 3,* 331–334.
18. Ojolo, S. J., & Ogunkomaiya, O. (2014). A study of effects of machining parameters on tool life. *International Journal of Materials Science and Applications, 3,* 183–199.
19. Sarıkaya, M., & Güllü, A. (2014). Taguchi design and response surface methodology based analysis of machining parameters in CNC turning under MQL. *Journal of Cleaner Production, 65,* 604–616.
20. Sood, P. K., Sehgal, R., Dwivedi, D. K. (2011). Optimization of turning parameters for surface roughness in CNC turning. *Journal of Mechanical Engineering, 1,* 26–31.
21. Saini, S. K., & Pradhan, S. K. (2014). Soft computing techniques for the optimization of machining parameter in CNC turning operation. *International Journal of Emerging Technology and Advanced Engineering, 4*(3), 117–124.
22. Das, S. R., Nayak, R. P., & Dhupal, D. (2012). Optimization of cutting parameters on tool wear and workpiece surface temperature in turning of AISI D2 steel. *International Journal of Lean Thinking, 3,* 140–156.

23. Islam, M., Buijk, A., Rais-Rohani, M., & Motoyama, K. (2015). Process parameter optimization of lap joint fillet weld based on FEM–RSM–GA integration technique. *Advances in Engineering Software, 79,* 127–136.
24. Jiang, P., Wang, C. C., Zhou, Q., Shao, X. Y., Shu, L. S., & Li, X. B. (2016). Optimization of laser welding process parameters of stainless steel 316L using FEM, Kriging and NSGA-II. *Advances in Engineering Software, 99,* 147–160.
25. Zhao, L., Liang, J., Zhong, Q. P., Yang, C., Sun, B., & Du, J. F. (2014). Numerical simulation on the effect of welding parameters on welding residual stresses in T92/S30432 dissimilar welded pipe. *Advances in Engineering Software, 68,* 70–79.
26. Huang, J. D., Gao, L., & Li, X. Y. (2015). An effective teaching-learning-based cuckoo search algorithm for parameter optimization problems in structure designing and machining processes. *Applied Soft Computing, 36,* 349–356.
27. Zhang, G. X., Cheng, J. X., Gheorghe, M., & Meng, Q. (2013). A hybrid approach based on differential evolution and tissue membrane systems for solving constrained manufacturing parameter optimization problems. *Applied Soft Computing, 13*(3), 1528–1542.
28. Jin, Y. A., He, Y., Fu, J. Z., Gan, W. F., & Lin, Z. W. (2014). Optimization of tool-path generation for material extrusion-based additive manufacturing technology. *Additive Manufacturing, 1–4,* 32–47.
29. Barclaya, J., Dhokiaa, V., Nassehi, A. (2015). Generating milling tool paths for prismatic parts using genetic programming. In *9th CIRP Conference on Intelligent Computation in Manufacturing Engineering—CIRP ICME '14*, Vol. 33, pp. 490–495.
30. Manav, C., Bank, H. S., & Lazoglu, I. (2013). Intelligent toolpath selection via multi-criteria optimization in complex sculptured surface milling. *Journal of Intelligent Manufacturing, 24*(2), 349–355.
31. Medina-Rodriguez, N., Montiel-Ross, O., Sepulveda, R., & Castillo, O. (2012). Tool path optimization for computer numerical control machines based on parallel ACO. *Engineering Letters, 20*(1), 101.
32. Petunin, A. (2011). Tools path optimization for cnc cutting machines. *Vestnik UGATU Systems Engineering and Information Technologies, 15*(44), 179–182.
33. Pan, R., Zhang, Y. J., Ding, J. B., Cao, C., Wang, Z. Z., Jiang, T., et al. (2016). Rationality optimization of tool path spacing based on dwell time calculation algorithm. *International Journal of Advanced Manufacturing Technology, 84*(9), 2055–2065.
34. Prajapati, R., Rajurkar, A., & Chaudhary, V. (2013). Tool path optimization of contouring operation and machining strategies for turbo machinery blades. *International Journal of Engineering Trends and Technology, 4*(5), 1731–1736.
35. Hsieh, H. T., & Chu, C. H. (2012). Optimization of tool path planning in 5-axis flank milling of ruled surfaces with improved PSO. *International Journal of Precision Engineering and Manufacturing, 13*(1), 77–84.
36. Lu, Y. A., Ding, Y., & Zhu, L. M. (2016). Smooth tool path optimization for flank milling based on the gradient-based differential evolution method. Transactions of the ASME. *Journal of Manufacturing Science and Engineering, 138*(8), 081009.
37. Kersting, P., & Zabel, A. (2009). Optimizing NC-tool paths for simultaneous five-axis milling based on multi-population multi-objective evolutionary algorithms. *Advances in Engineering Software, 40,* 452–463.
38. Kuo, C. L., Chu, C. H., Li, Y., Li, X. Y., & Gao, L. (2015). Electromagnetism-like algorithms for optimized tool path planning in 5-axis flank machining. *Computers & Industrial Engineering, 84,* 70–78.
39. Gutowski, T., Dahmus, J., Thiriez, A. (2006). Electrical energy requirements for manufacturing processes. In *Proceedings of 13th CIRP International Conference on LCE.*
40. Kara, S., & Li, W. (2011). Unit process energy consumption models for material removal processes. *CIRP Annals-Manufacturing Technology, 60*(1), 37–40.
41. Li, L., Yan, J., & Xing, Z. (2013). Energy requirements evaluation of milling machines based on thermal equilibrium and empirical modeling. *Journal of Cleaner Production, 52,* 113–121.

42. Lin, W. W., Yu, D. Y., Zhang, C. Y., Tian, Y. H., Liu, S. Q., Luo, M. (2016). Multi-objective optimization of machining parameters in multi-pass turning operations for low-carbon manufacturing. *Journal of Engineering Manufacturing*. https://doi.org/10.1177/0954405416629098.
43. Anand, Y., Gupta, A., Abrol, A., Ayush, Gupta, Kumar, V., Tyagi, S. K., et al. (2016). Optimization of machining parameters for green manufacturing. *Cogent Engineering, 3*(1), 1–16.
44. Campatelli, G., Lorenzini, L., & Scippa, A. (2014). Optimization of process parameters using a response surface method for minimizing power consumption in the milling of carbon steel. *Journal of Cleaner Production, 66,* 309–316.
45. Kant, G., & Sangwan, K. S. (2014). Prediction and optimization of machining parameters for minimizing power consumption and surface roughness in machining. *Journal of Cleaner Production, 83,* 151–164.
46. Hu, S. H., Liu, F., He, Y., & Hu, T. (2012). An on-line approach for energy efficiency monitoring of machine tools. *Journal of Cleaner Production, 27,* 133–140.
47. Camposeco-Negrete, C. (2013). Optimizing of cutting parameters for minimizing energy consumption in turning of AISI 6061 T6 using Taguchi methodology and ANOVA. *Journal of Cleaner Production, 53,* 195–203.
48. Yan, J., & Li, L. (2013). Multi-objective optimization of milling parameters—The trade-offs between energy, production rate and cutting quality. *Journal of Cleaner Production, 52,* 462–471.
49. Wang, S., Lu, X., Li, X. X., & Li, W. D. (2015). A systematic approach of process planning and scheduling optimization for sustainable machining. *Journal of Cleaner Production, 87,* 914–929.
50. Kong, D., Choi, S., Yasui, Y., Pavanaskar, S., Dornfeld, D., & Wright, P. (2011). Software-based tool path evaluation for environmental sustainability. *Journal of Manufacturing Systems, 30,* 241–247.
51. Aramcharoen, A., Mativenga, P. T. (2014). Critical factors in energy demand modelling for CNC milling and impact of toolpath strategy. *Journal of Cleaner Production*, *78,* 1–12.
52. Avram, O. I., & Xirouchakis, P. (2011). Evaluating the use phase energy requirements of a machine tool system. *Journal of Cleaner Production, 19,* 699–711.
53. Narooei, K. D., & Ramli, R. (2014). Application of artificial intelligence methods of tool path optimization in CNC machine: A review. *Research Journal of Applied Sciences, Engineering and Technology, 8*(6), 746–754.
54. Li, W. D., & McMahon, C. A. (2007). A simulated annealing-based optimization approach for integrated process planning and scheduling. *International Journal of Computer Integrated Manufacturing, 20*(1), 80–95.
55. Tairidis, G., Foutsitzi, G., Koutsianitis, P., & Stavroulakis, G. E. (2016). Fine tuning of a fuzzy controller for vibration suppression of smart plates using. *Advances in Engineering Software, 101,* 123–135.
56. Marthinus, Ras N., Daniel, Wilke N., Albert, Groenwold A., & Schalk, Kok. (2014). On rotationally invariant continuous-parameter genetic algorithms. *Advances in Engineering Software, 78,* 52–59.
57. Buiga, O., & Tudose, L. (2014). Optimal mass minimization design of a two-stage coaxial helical speed reducer with Genetic Algorithm. *Advances in Engineering Software, 68,* 25–32.
58. Bank, M., Ghomi Fatemi, S. M. T., Jolai, F., & Behnamian, J. (2012). Application of particle swarm optimization and simulated annealing algorithms in flow shop scheduling problem under linear deterioration. *Advances in Engineering Software, 47*(1), 1–6.
59. Musharavati, F., & Hamouda, A. S. M. (2012). Enhanced simulated-annealing-based algorithms and their applications to process planning in reconfigurable manufacturing systems. *Advances in Engineering Software, 45*(1), 80–90.
60. Rezaei, G., Afshar, H. M., & Rohani, M. (2014). Layout optimization of looped networks by constrained ant colony optimisation algorithm. *Advances in Engineering Software, 70,* 123–133.

61. He, J. J., & Hou, Z. E. (2012). Ant colony algorithm for traffic signal timing optimization. *Advances in Engineering Software, 43*(1), 14–18.
62. Uğur, A., & Aydin, D. (2009). An interactive simulation and analysis software for solving TSP using Ant Colony Optimization algorithms. *Advances in Engineering Software, 40*(5), 341–349.
63. Trivedi, R. R., Pawaskar, D. N., & Shimpi, R. P. (2016). Optimization of static and dynamic travel range of electrostatically driven microbeams using particle swarm optimization. *Advances in Engineering Software, 97,* 1–16.
64. Yan, X. H., He, F. Z., Chen, Y. L., & Yuan, Z. Y. (2015). An efficient improved particle swarm optimization based on prey behavior of fish schooling. *Journal of Advanced Mechanical Design, Systems, and Manufacturing, 9*(4), 1–10.
65. Abbass, H. A. (2001). Marriage in honey-bee optimization (MBO): A haplometrosis polygynous swarming approach. In *The Congress on Evolutionary Computation (CEC2001)* (pp. 207–214). Seoul, Korea.
66. Wen, X. Y., Li, X. Y., Gao, L., & Sang, H. Y. (2014). Honey bees mating optimization algorithm for process planning problem. *Journal of Intelligent Manufacturing, 25*(3), 459–472.
67. Cheng, Y., He, F., Wu, Y., & Zhang, D. (2016). Meta-operation conflict resolution for human-human interaction in collaborative feature-based CAD systems. *Cluster Computing, 19*(1), 237–253.
68. Li, K., He, F. Z., & Chen, X. (2016). Real-time object tracking via compressive feature selection. *Frontiers of Computer Science, 10*(4), 689–701.
69. Sun, J., He, F. Z., Chen, Y. L., & Chen, X. (2016). A multiple template approach for robust tracking of fast motion target. *Applied Mathematics-A Journal of Chinese Universities, 31*(2), 177–197.
70. Ni, B., He, F. Z., Pan, Y., & Yuan, Z. Y. (2016). Using shapes correlation for active contour segmentation of uterine fibroid ultrasound images in computer-aided therapy. *Applied Mathematics-A Journal of Chinese Universities, 31*(1), 37–52.
71. Zhou, Y., He, F. Z., & Qiu, Y. M. (2016). Optimization of parallel iterated local search algorithms on graphics processing unit. *Journal of Supercomputing, 72*(6), 2394–2416.
72. Zhou, Y., He, F. Z., & Qiu, Y. M. (2017). Dynamic strategy based parallel ant colony optimization on GPUs for TSPs. *Science China Information Sciences, 60*(6), 06812. https://doi.org/10.1007/s11432-015-0594-2.

Chapter 9
Energy Efficiency, Robustness, and Makespan Optimality in Job-Shop Scheduling Problems

M. A. Salido, J. Escamilla, F. Barber, A. Giret, D. B. Tang and M. Dai

Abstract Many real-world problems are known as planning and scheduling problems, where resources must be allocated so as to optimize overall performance objectives. The traditional scheduling models consider performance indicators such as processing time, cost, and quality as optimization objectives. However, most of them do not take into account energy consumption and robustness. We focus our attention in a job-shop scheduling problem where machines can work at different speeds. It represents an extension of the classical job-shop scheduling problem, where each operation has to be executed by one machine, and this machine can work at different speeds. The main goal of the paper is focused on the analysis of three important objectives: energy efficiency, robustness, and makespan, and the relationship among them. We present some analytical formulas to estimate the ratio/relationship between these parameters. It can be observed that there exist a clear relationship between robustness and energy efficiency and a clear trade-off between robustness/energy efficiency and makespan. It represents an advance in the state of the art of production scheduling, so obtaining energy-efficient solutions also supposes obtaining robust solutions, and vice versa.

Keywords Job-shop scheduling · Energy efficiency · Robustness
Makespan

M. A. Salido · J. Escamilla · F. Barber
Instituto de Atuomática e Informática Industrial,
Universitat Politècnica de Valencia, Valencia, Spain

A. Giret (✉)
Departamento de Sistemas Informáticos, Computación,
Universitat Politècnica de Valencia, Valencia, Spain
e-mail: adgibog@upvnet.upv.es

D. B. Tang · M. Dai
College of Mechanical and Electrical Engineering,
Nanjing University of Aeronautics and Astronautics, Nanjing, China

W. Li and S. Wang (eds.), *Sustainable Manufacturing and Remanufacturing Management*, https://doi.org/10.1007/978-3-319-73488-0_9

9.1 Introduction

Nowadays, the main objective of many companies and organizations is to improve profitability and competitiveness. These improvements can be obtained with a good optimization of resources allocation. But in the last years, many companies are not only facing complex and diverse economic trends of shorter product life cycles, quick changing science and technology, increasing customer demand diversity, and production activities globalization, but also enormous and heavy environmental challenges of global climate change (e.g., greenhouse effect), rapid exhaustion of various non-renewable resources (e.g., gas, oil, coal), and decreasing biodiversity. Scheduling problems are widely discussed in the literature, and two main approaches can be distinguished [2]:

- Classical deterministic methods consider that the data are deterministic and that the machine environment is relatively simple. Some traditional constraints are taken into account (precedence constraints, release dates, due dates, preemption, etc.). The criterion to optimize is often standard (makespan). A number of methods have been proposed (exact methods, greedy algorithms, approximate methods, etc.), depending on the difficulty of a particular problem. These kinds of studies are the most common in the literature devoted to scheduling problems.
- On-line methods; When the algorithm does not have access to all the data from the outset, the data become available step by step, or "on-line." Different models may be considered here. In some studies, the tasks that we have to schedule are listed and appear one by one. The aim is to assign them to a resource and to specify a start time for them. In other studies, the duration of the tasks is not known in advance.

Flexibility occurs at the boundary between these two approaches: some information is available concerning the nature of the problem to be solved and concerning the data. Although this information is imperfect and not wholly reliable, it cannot be totally ignored. It is well-known that there will be discrepancies, for a number of reasons, between the initial plan and what is actually realized. Given a set of disruptions that can occur in unforeseen circumstances, the aim is to propose one or more solutions that adapt well to disruptions and then produce reactive decisions in order to ensure a smooth implementation [2].

The job-shop scheduling problem (JSP) represents a problem where there are some specific resources or machines which have to be used to carry out some tasks. Many real-life problems can be modeled as a job-shop scheduling problem and can be applied in some variety of areas, such as production scheduling in the industry, departure and arrival times of logistic problems, the delivery times of orders in a company. Most of the solving techniques try to find the optimality of the problem for minimizing the makespan, minimizing tardiness, minimizing flow time, etc.

Recently, some works have focused on minimizing the energy consumption in scheduling problems [8, 18, 19], mainly from the Operations Research Community [5, 20]. Furthermore, some works have been carried out to obtain robust schedules to absorb incidences in dynamic scheduling [6]. However, no works have related energy efficiency with robustness although there exists a clear relationship between them in job-shop scheduling.

We focus our attention in a job-shop scheduling problem with different speed machine (JSSM). It represents an extension of the classical job-shop scheduling problem, where each operation must be executed on a machine at a determined speed (by a classical deterministic method). If the speed of a machine is high, the energy consumption increases, but the processing time of the task decreases, meanwhile if the speed is low, the energy consumption decreases and the processing time increases. Thus, in on-line scheduling, if disruptions occur, reactive decisions are needed, so machines can accelerate the speed to absorb these disruptions and recover the original schedule, obtained by the classical deterministic method.

To this end, we analyze the relationship between some important parameters in order to obtain a multi-objective solution. We show that there is a clear trade-off between makespan and energy consumption; and between makespan and robustness. Therefore, there is a close relationship between energy consumption and robustness. However, this close relationship has not been analyzed in the literature and new techniques can be developed to achieve these objectives jointly. Thus, our main goal is to find a solution that minimizes the energy consumption and the makespan. Furthermore, we extend this goal to determine the saved time by energy efficiency as a robustness measure in order to be used if incidences appear. Thus, if a task is delayed, the lost time can be recovered by increasing the speed of the machine to recover the original solution.

9.2 Problem Description

Formally, the job-shop scheduling problem with different speed machine (JSSM) can be defined as follows. We are given a set of n jobs $\{J_1, ..., J_n\}$, a set of m resources or machines $\{R_1, ..., R_m\}$. Each job J_i consists of a sequence of v_i tasks $(\theta_{i1}, \ldots, \theta_{iv_i})$. Each task θ_{il} has a single machine requirement $R_{\theta_{il}}$ and a start time $st_{\theta_{il}}$ to be determined. Each machine can work with different speeds, so each task is linked up to an integer duration $p_{\theta_{il}}$ and an integer energy $e_{\theta_{il}}$ used by the corresponding machine.

A feasible schedule is a complete assignment of starting times to tasks that satisfy the following constraints: (i) The tasks of each job are sequentially scheduled, (ii) each machine can process at most one task at any time, (iii) no preemption is allowed. The objective is finding a feasible schedule that minimizes the completion time of all the tasks and the energy used. The problem is a standard job-shop problem denoted as $J||C_{\max}$ according to classification scheme proposed in [3]. But the association between duration and energy has been created so the problem JSSM can be denoted as J(Speed)//Cmax, Energy. For each task, three different speeds have been defined. Each speed has a duration and an energy consumption. When the working speed increases, the energy also increases but the duration decreases.

9.3 Energy Efficiency

Nowadays, manufacturing enterprisers are not only facing complex and diverse economic trends of shorter product life cycles, quick changing science and technology, increasing customer demand diversity, and production activities globalization, but also enormous and heavy environmental challenges of global climate change (e.g., greenhouse effect), rapid exhaustion of various non-renewable resources (e.g., gas, oil, coal), and decreasing biodiversity. Statistical data in 2009 shows the Germany industrial sector was responsible for approximately 47% of the total national electricity consumption, and the corresponding amount of CO_2 emissions generated by this electricity summed up to 18–20% [4]. Thus, manufacturing companies are responsible for the environmental outcome and are forced to have manufacturing systems that demonstrates major potential to reduce environmental impacts [9].

Recently, there has been growing interest in the development of energy savings due to a sequence of serious environmental impacts and rising energy costs. Research on minimizing the energy consumption of manufacturing systems has focused on three perspectives: the machine level, the product level, and the manufacturing system level. From the machine-level perspective, developing and designing more energy-efficient machines and equipment to reduce the power and energy demands of machine components is an important strategic target for manufacturing companies [18, 21]. Unfortunately, previous studies show that the share of energy demand for removal of metal material compared to the share of energy needed to support various functions of manufacturing systems is quite small (less than 30% of total energy consumption) [7, 12].

From the product-level perspective, modeling embodied product energy framework based on a product design viewpoint for energy reduction approach is beneficial to support the improvements of product design and operational decisions [23, 26]. It requires strong commercial simulation software to facilitate the analysis and evaluation of the embodied product energy. The results cannot be applied easily in most manufacturing companies, especially in small- and medium-sized enterprises due to the enormous financial investments required. From the manufacturing system-level perspective, thanks to decision models that support energy savings, it is feasible to achieve a significant reduction in energy consumption in manufacturing applications. In the specialized literature about production scheduling, the key production objectives for production decision models, such as cost, time, and quality, have been widely discussed. However, decreasing energy consumption in manufacturing systems through production scheduling has been rather limited. One of the most well-known research works is the work of Mouzon et al. [20], who developed several algorithms and a multi-objective mathematical programming model to investigate the problem of scheduling jobs on a single CNC machine in order to reduce energy consumption and total completion time. They pointed out that there was a significant amount of energy savings when non-bottleneck machines were turned off until needed; the relevant share of savings in total energy consumption could add up to 80%. They also reported that the inter-arrivals would be forecasted, and therefore, more energy-efficient dispatching rules could be adopted for scheduling.

In further research, Mouzon and Yildirim [19] proposed a greedy randomized adaptive search algorithm to solve a multi-objective optimization schedule that minimized the total energy consumption and the total tardiness on a machine. Fang et al. [10] provided a new mixed-integer linear programming model for scheduling a classical flow shop that combined the peak total power consumption and the associated carbon footprint with the makespan. Bruzzone et al. [5] presented an energy-aware scheduling algorithm based on a mixed-integer programming formulation to realize energy savings for a given flexible flow shop that was required to keep fixed original job assignment and sequencing.

Although the majority of the research on production scheduling has not considered energy-saving strategies completely, the efforts mentioned above provide a starting point for exploring an energy-aware schedule optimization from the viewpoint of energy consumption. However, no work has been carried out to consider a multi-objective optimization schedule to minimize the total energy consumption, the makespan, and to maximize the robustness of the schedule.

9.4 Robustness

Robustness is a common feature in real-life problems. Biological life, functional systems, physical objects, etc., persist if they remain running and maintain their main features despite continuous perturbations, changes, incidences, or aggressions [24]. Thus, robustness is a concept related to the persistence of the system, of its structure, of its functionality, etc., against external interferences: *A system is robust, if it persists*.

It is really difficult to give a unique definition for robustness, as this concept is differently defined in several domains. Furthermore, the definition often remains implicit in the literature or is determined by the specific target application. Finally, most authors prefer to use the concept of robust solution (and here, of robust schedule).

The data associated with a scheduling problem are the processing times, occurrence dates of some events, some structural features and the costs. None of this data is free from factors of uncertainty. The duration of tasks depends on the conditions of their execution, in particular, on the necessary human and material resources. They are thus inherently uncertain, regardless of contingent factors that may impair their execution. For instance, transportation times for components between separate operations in a manufacturing system will depend on the characteristics of the transportation resources available.

Finally, in a production scheduling, some resources such as versatile machines require a reconfiguration time between operations. This time depends on the type of tools needed and the location of these tools in the shop, not to mention the operator carrying out the reconfiguration [2].

Let us first propose some consensus definition: A schedule is robust if its performance is rather insensitive to the data uncertainties. Performance must be understood here in the broad sense of solution quality for the person in charge; this naturally encompasses this solution value relatively to a given criterion, but also the structure itself of the proposed solution. The robustness of a schedule is a way to characterize its performance.

In the literature, it is sometimes difficult to separate sensitivity analysis and robustness. In fact, the sensitivity analysis tries to answer the "what if…" questions. It deals with disturbances more than with general uncertainty: Data are fixed but might be disturbed [2].

In scheduling problems, robustness can be defined as:

Definition 1 Robustness: is the ability of a solution to maintain its feasibility when incidences appear during execution in the scheduling problem.

In this paper, the robustness of a schedule will be used to answer what if questions, mainly related to small disruptions that daily occur in real-life scheduling problems. In this way, the robustness of a schedule can be used to obtain energy-aware schedules that do not modify the start time of tasks. To this end, the slack between tasks that makes the schedule robust (to absorb incidences) can be profitable by machines to work at lower speed and therefore saving energy consumption. However, if this slack is needed, due to a disruption, the involved machine can increase its speed in order to recover the disrupted time and finalize the task on time. In this way, there exists a relationship between robustness and energy saving that can be applied to many scheduling contexts.

9.5 Modeling and Solving a JSSM

The more natural way to solve the job-shop scheduling problem involves all variables and constraints related to jobs, tasks, and machines [11, 13, 22]. However, the solution obtained is an optimal solution that minimizes the makespan but it does not guarantee a certain level of robustness. Generally, this solution is not able to absorb incidences and a delay in a task is propagated along the rest of the schedule.

Several reactive/proactive techniques have been developed in the literature to manage incidences in scheduling problems [2]. Thus, computing a new solution from scratch after each problem change is possible (reactive technique), but it has two important drawbacks: inefficiency and instability of the successive solutions [25]. While reactive methods merely deal with the consequences of an unexpected change, taking a more proactive approach may guarantee a certain level of robustness. We are interested in this proactive approach so that our goal is searching for a equitable trade-off between robustness and optimality of a solution.

Robustness (as in Sect. 9.4 in job-shop scheduling) can be obtained through allocating buffer times between tasks in order to absorb small disruptions (task delays, etc.) that can occur stochastically along the schedule.

In an optimized solution of a JSSM, some natural buffers appear to satisfy the involved constraints (non-overlapping constraints). These buffers give the schedule some robustness degree.

However, if more buffers must be included to make the final solution more robust, the involved tasks must be moved and the effect must be propagated to the rest of the schedule.

To add robustness to JSSM solutions, we use the extra speed that machines can work only in cases where machines are not working at top speed. Several solutions are obtained with different weights to minimize makespan or energy used. When the main objective is to minimize the energy used, the solutions are composed of several tasks that are processed by machines in a low speed. If some incidences appear, this speed can be increased and the solutions remain valid. Following this idea, the energy roominess can be considered as robustness.

IBM ILOG CPLEX CP Optimizer

The problem is modeled and solved with IBM ILOG CPLEX CP Optimizer tool (CP Optimizer). CP Optimizer uses constraint programming technology to solve detailed scheduling problems and other hard combinatorial optimization problems [15].

CP Optimizer is a commercial solver embedding powerful constraint propagation techniques and a self-adapting large neighborhood search method dedicated to scheduling [17]. This solver is expected to be very efficient for a variety of scheduling problems as it is pointed in [14], in particular, when the cumulative demand for resources exceeds their availability as it happens, for example, in the Satellite Control Network Scheduling Problem confronted in [16].

The problem has been modeled as a typical job-shop scheduling problem. The extension with different machine speeds has been implemented considering that each task is executed by a machine and this machine has different optional modes where each mode represents the duration of the task and an associated consumption energy.

The objective is to find a solution that minimizes the multi-objective makespan and energy consumption. The weight of each objective can be changed by λ parameter. Following is the expression (9.1).

$$\lambda * \text{Makespan} + (1 - \lambda) * \text{Energy Consumption} \tag{9.1}$$

Since the values of energy consumption and makespan are not proportional, it is necessary to normalize both measures (NormEnergy) (NormMakespan). NormEnergy value is calculated by summing the energy used in the execution of all the tasks, divided the maximum energy (maxEnergy). maxEnergy is the sum of the energy needed to execute all task at top speed. The NormMakespan is the makespan divided the sum of the task durations when the machines are working at the lowest speed (maxMakespan). The objective function is the expression (9.2).

$$\lambda * \text{Makespan} + (1 - \lambda) * \text{Energy Consumption} \tag{9.2}$$

Algorithm 1 shows a pseudo-code of the model to solve the problem.

Algorithm 1: Model in CP Optimizer

Data: tasks: Set of tasks; modes: Set of 3 modes for each task;
Result: A solution minimizing the objective function depending of λ
Interval itvs : = Define interval, one for each task;
Interval modes : = Define mode, three for each task;
Sequence mchs : = Each itvs with the same machine is linked up;

$Minimize((1 - \lambda) * NormEnergy + \lambda * NormMakespan)$

Subject to

- noOverlap (*mchs*)
- endBeforeStart (*itvs*[*j*][*o*], *itvs*[*j*][*o* + 1])
- alternative (*itvs*[*j*][*o*], all(md in *modes*: if(md.id == itvs.id)))

Figure 9.1 shows two different schedules obtained by CP Optimizer for a given instance of the JSSM proposed in [1]. This instance represents a scheduling

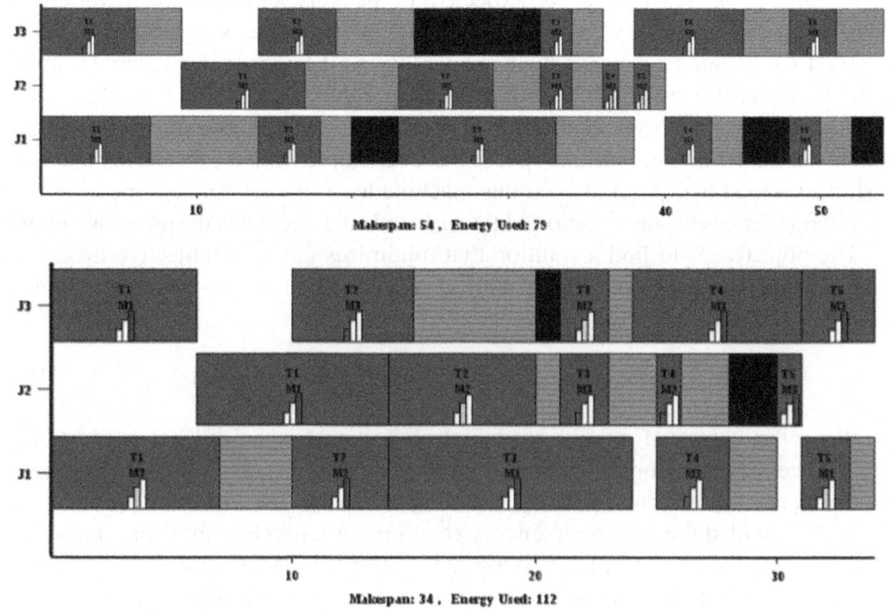

Fig. 9.1 Schedules with different λ values

problem with 3 machines, 3 jobs, each with 5 tasks, and each task has a processing time between 1 and 10 time units when the machine works at full speed.

Each task is represented by a gray rectangle which can be divided into two regions: A solid black gray color represents the processing time when the machine is working at full speed (mandatory), and a light gray color with horizontal lines represents the extra processing time if the machine does not work at full speed (optional). This region represents the used time to save energy. However, this time can also be used to absorb incidences if a disruption occurs (EEBuffer). Each task is labeled with the number of tasks, the machine used, and the speed used by the corresponding machine (green: low speed, yellow: medium speed, red: full speed). Finally, the black rectangles represent natural buffer times generated by the solution. They can also be used to absorb incidences.

Two solutions (schedules) have been obtained for the same instance with different lambda values (λ) between 0.1 and 0.9.

For $\lambda = 0.1$, the solution X_1 give more importance (0.9) to energy efficiency and less importance (0.1) to minimize makespan. It can be observed that the makespan was 54, no tasks were carried out by machines at full speed (red), 2 tasks at medium speed (yellow) and 13 tasks at low speed (green). It generated an energy consumption of 79 units, and it can be observed that all tasks maintain slack to absorb incidences so the robustness of the schedule is considered high.

For $\lambda = 0.9$, the solution X_2 gives more importance (0.9) to minimize makespan and less importance (0.1) to energy efficiency. It can be observed that the makespan was 34, and 7 tasks were carried out by machines at full speed (red), 4 tasks at medium speed (yellow) and 4 tasks at low speed (green). It generated an energy consumption of 112 units, and it can be observed that only 8 tasks maintain slack to absorb incidences so the robustness of the schedule is considered low.

By modifying the value of λ, an approximate Pareto front for the bicriteria optimization schedule is generated. It must be taken into account that no single solution between the X_1 and X_2 can be said, a priori, to be the best one. Indeed, they are non-comparable, so choosing a solution from an approximate Pareto front can only be done by the user, depending on the requirements. This is why we advocate producing, for a given problem instance, the Pareto front rather than a single solution.

9.6 Definition of the Benchmark Set

To analyze the relationship among makespan, energy consumption, and robustness, we have evaluated the behavior on the benchmarks proposed in [1]. According to the benchmark results (small and large instances), several analytical formulas have been developed to estimate these parameters. All analyzed instances are characterized by the number of machines (m), the maximum number of tasks by job (v_{max}), and the range of processing times (p). The number of jobs (j) is set to 3. A set of instances was generated by combining values of each parameter: m = 3, 5,

7; $v_{\max}$ = 5, 7, 10, 20, 25, 30 and p = [1, 10], [1, 50], [1, 100], [1, 200]. In these benchmarks, the number of operators was not considered so that we fixed it to the number of machines. We model the instances to be solved by the optimizer.

We have also extended the original instances of Agnetis [1] to add different energy consumptions (e_1, e_2, e_3) to each task according to three processing times $(\mathrm{pt}_1, \mathrm{pt}_2, \mathrm{pt}_3)$, where pt_1 is equal to the original value of processing time in the Agnetis instances. pt_2 and pt_3 were calculated following the expressions (9.3) and (9.4), respectively. These instances can be found in our webpage[1].

$$\mathrm{pt}_2 = \mathrm{Max}(\max \mathrm{dur} * 0.1 + \mathrm{pt}_1, \mathrm{Rand}(1, 25 * \mathrm{pt}_1, 2.25 * \mathrm{pt}_1) \tag{9.3}$$

$$\mathrm{pt}_3 = \mathrm{Max}(\max \mathrm{dur} * 0.1 + \mathrm{pt}_2,\ \mathrm{Rand}(1, 25 * \mathrm{pt}_2,\ 2.25 * \mathrm{pt}_2) \tag{9.4}$$

The value max dur represents the maximum duration of a task for the corresponding instance, and the expression rand represents a random value between both expressions. Similar expressions were developed to calculate the energy consumption (9.5, 9.6, 9.7).

$$e_1 = \mathrm{Rand}(\mathrm{pt}_1, 3 * \mathrm{pt}_1) \tag{9.5}$$

$$e_2 = \mathrm{Max}(1, \mathrm{Min}(e_1 - \max \mathrm{dur} * 0.1, \mathrm{Rand}(0.25 * e_1, 0.75 * e_1) \tag{9.6}$$

$$e_3 = \mathrm{Max}(1, \mathrm{Min}(e_2 - \max \mathrm{dur} * 0.1, \mathrm{Rand}(0.25 * e_2, 0.75 * e_2) \tag{9.7}$$

Following these expressions, the processing times of pt_1, pt_2, pt_3 increase as the energy consumption of e_1, e_2, e_3 decrease.

For example, given an instance with 5 tasks per job, three triplets are represented for each task: the id of the task, the energy used, and the processing time (< id, e, pt >):

$\langle \mathrm{id}, \mathrm{e}_3, \mathrm{pt}_3\rangle, \langle \mathrm{id}, \mathrm{e}_2, \mathrm{pt}_2\rangle, \langle \mathrm{id}, \mathrm{e}_1, \mathrm{pt}_1\rangle$

$\$\langle 1,\ 14,\ 14\rangle, \langle 1,\ 16,\ 10\rangle, \langle 1,\ 19,\ 7\rangle,$

$\ldots$

$\langle 15,\ 3,\ 6\rangle, \langle 15,\ 5,\ 4\rangle, \langle 15,\ 6,\ 3\rangle$

9.7 Makespan Versus Energy Consumption

In this section, we analyze the trade-off between makespan and energy consumption in job-shop scheduling problems with different machine speeds. Figure 9.2 shows an approximate Pareto front for a set of 10 instances with 5 machines, 10 tasks per

[1]http://gps.webs.upv.es/jobshop/

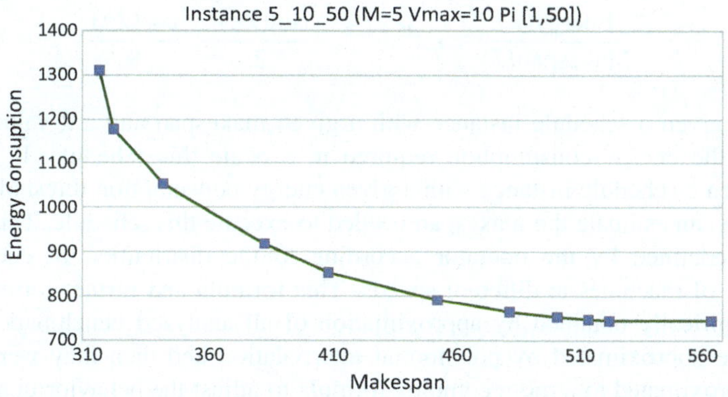

Fig. 9.2 Approximate Pareto front for the bicriteria makespan–energy consumption

Table 9.1 Makespan and energy consumption in instances $\langle m, v_{max}, p \rangle$

	5_10_50		7_10_100		3_20_50		3_25_100		3_30_200	
λ	Mk	Energ.	Mk	Energ.	Mk	Energ.	Mk	Energ.	Mk	Energ.
0	565.4	745	1088.4	1571.4	1296	1507.4	3160	3827.1	7289.6	9162.7
0.1	524.4	745	1004.3	1571.6	1168.7	1507.4	2768.4	3827.5	6600.8	9163.5
0.2	515.8	747.1	992.1	1574.5	1145.7	1513.7	2734.2	3835.7	6513.2	9184
0.3	502.6	752.3	970.2	1584.2	1112.6	1527.3	2667.1	3866.7	6364.8	9258.8
0.4	483.6	764.2	918.3	1616.3	1079	1559.9	2553	3946.1	6158	9386.5
0.5	454.3	792.1	884.5	1650.3	1011.7	1628.6	2421.8	4077	5825	9771.5
0.6	410.1	854.1	843.7	1709.8	946.8	1722.5	2280.6	4300.6	5452	10,239.4
0.7	384.7	917.4	768.2	1879.5	854.6	1933.7	2056.1	4742.9	4953.3	11,370
0.8	343.1	1053	690.1	2147.3	777.8	2216.3	1854.4	5466.1	4530.8	12,935.4
0.9	322.8	1179.1	635.7	2466.1	728.3	2498	1738.4	6164.9	4182.9	14,696.1
1	317	1311.5	625.9	2664.1	707.8	2764.3	1656.1	6667.1	4048.9	16,235.5

job and a maximum processing time of 50 time units. For $\lambda = 1$, it can be observed that the average energy consumption was 1311 and the average makespan was minimized (317). However, for $\lambda = 0$, the average energy consumption was minimized (745) and the average makespan was maximized (564.4). As we pointed out above, depending on the user requirements, a value of λ must be selected to obtain the desired level of makespan/energy consumption. Table 9.1 shows the makespan and energy consumption for each value of λ in different instances. It must be taken into account the relationship/ratio between makespan and energy consumption is similar in all instances, so that this trade-off is not dependent on the number of machines, number of tasks per job, neither the range of processing times.

According to the analyzed instances, the ratio between energy consumption and makespan can be estimated by using the formula (9.8):

$$\frac{\text{Energy}(\lambda)}{\text{Makespan}(\lambda)} \approx e^{0.25+1.2\lambda} - \frac{\sin(\lambda\pi)}{2} + \frac{\cos(\lambda\pi)}{8} \tag{9.8}$$

Thus, given a schedule instance with a given makespan and a λ value, we can estimate the energy consumption required to execute this schedule. In the same way, given a schedule instance with a given energy consumption threshold and a λ value, we can estimate the makespan needed to execute this schedule. This formula can be redefined by the operator according to the distribution of energy consumption of machines at different speeds. This formula and further formulas have been empirically obtained by approximation of all analyzed benchmarks. Firstly, they were approximated by polynomial interpolation and then they were empirically approximated to a more complex formula to adjust the behavior in all desired points. Thus, they show that there is a clear relationship between the involved parameters.

9.8 Robustness Versus Energy Consumption

The main goal of this paper is to show the trade-off between robustness and energy consumption. In this way, the advantage could be twofold. By developing new techniques for searching energy-efficient schedules also mean searching for robust schedules. Thus, these techniques will generate energy-aware and robust solutions in production scheduling, so small disruptions can be repaired by accelerating the needed machines to recover the original schedule. In this way, no rescheduling is needed and the user can adjust the parameters to obtain the optimal solution based on the problem preference.

To carry out this study, we have simulated 100 incidences to each instance in order to analyze the amount of incidences that can be absorbed by the resultant schedule. An incidence is a delay to a random task of the schedule. The duration of the incidence (%incid) was bounded by a 20% of total duration of the involved task. Figure 9.3 shows an approximate Pareto front for a set of 10 instances with 7 machines, 10 tasks per job and a maximum processing time of 100 time units. It can be observed that as the robustness increased the energy consumption decreased. This is due to the fact that more robust solutions allow machines to work at minimum speed, so the energy consumption decreased; i.e., if all machines work at minimum speed, all tasks have a slack (time between solving the task at minimum speed minus solving the task at minimum speed). Thus, if a disruption occurs in a machine m_i at speed (s_{i1}) during the task t_i, this machine can accelerate its speed to s_{i2} in this task t_i in order to finish on time (before the next task t_{i+1} starts). In this case, we consider the schedule is robust. If the delay of task t_i affects to the following task t_{i+1}, the machine m_j that works in this task accelerates its speed in order to finish on time. Finally, the disruption is absorbed in some steps. In this

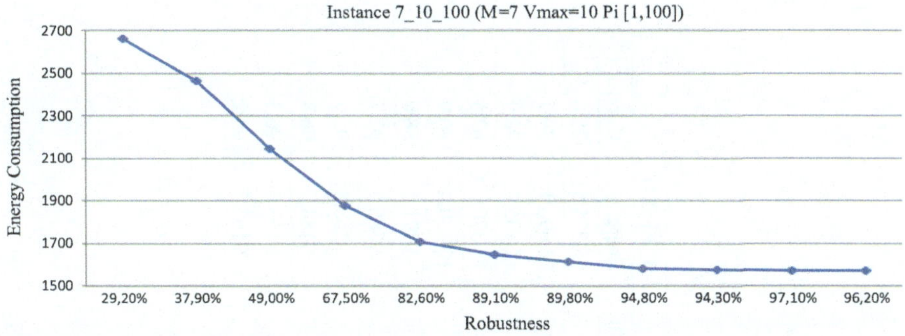

Fig. 9.3 Approximate Pareto front for the bicriteria energy consumption–robustness

case, we consider the schedule is stable due to the fact that the disruption has been propagated to some other tasks before the original solution is recovered.

Table 9.2 shows the energy consumption and robustness in different instances. It must be taken into account that the robustness maintained the same behavior in all instances, so that the robustness is not directly dependent on the number of machines, number of tasks per job, neither the range of processing times. However in most instances, for $\lambda = 0,1$ and $\lambda = 0$, the energy needed is similar, but the robustness is different (see last two rows in instances 3-5-10, 3-7-10, and 5-10-50). Thus, given an energy consumption threshold, we can obtain different solutions with different robustness and makespan level.

The relationship between energy consumption and robustness can be estimated by using the formula (9.9):

$$\frac{\text{Energy}(\lambda)}{\text{Robustness}(\lambda, \%\text{incid})} \approx \frac{\text{ttasks}.p/2}{(6,3 - 6\frac{\%\text{incid}}{100}) * (8 * \cos(\lambda^2 \pi) - \sin(\lambda \pi) + 11)} \quad (9.9)$$

This formula is more accurate for λ values close to 0 (from 0,6 to 0), due to the fact that the energy consumption is more considered for these values in the objective function.

Thus given a percentage of robustness for a given incidence duration (%incid) and a λ value of a schedule, we can estimate the energy needed to carry out this schedule. In the same way, a schedule with a given energy consumption, a λ value, and a threshold of the duration of the incidences (%incid), we can estimate the robustness of this schedule. This formula can be refined by the operator according to the distribution of energy consumption of machines at different speeds.

Table 9.2 Energy consumption and robustness in instances $\langle m, v_{\max}, p\rangle$

	5_10_50		7_10_100		3_20_50		3_25_100		3_30_200	
λ	Energ.	Robust. (%)	Energ.	Robust. (%)	Energ.	Robust. (%)	Energ.	Robust. (%)	Energ.	Robust. (%)
1	11,311	26.7	2664.1	29.2	2764.3	26.8	6667.1	27.0	16,235.5	25.6
0.9	1179.1	37.2	2466.1	37.9	2498	36.3	6164.9	36.8	14,696.1	39.7
0.8	1053	48.2	2147.3	49.0	2216.3	50.6	5466.1	49.5	12,935.4	48.9
0.7	917.4	68.1	1879.5	67.5	1933.7	66.0	4742.9	67.9	11,370	62.5
0.6	854.1	79.8	1709.8	82.6	1722.5	79.6	4300.6	78.9	10,239.4	80.9
0.5	792.1	86.6	1650.3	89.1	1628.6	88.0	4077	86.3	9771.5	88.0
0.4	764.2	91.8	1616.3	89.8	1559.9	93.4	3946.1	94.1	9386.5	92.7
0.3	752.3	94.2	1584.2	94.8	1527.3	95.3	3866.7	94.9	9258.8	95.9
0.2	747.1	97.0	1574.5	94.3	1513.7	96.6	3835.7	96.0	9184	96.8
0.1	745	93.7	1571.6	97.1	1507.4	97.3	3827.5	97.6	9163.5	96.3
0	745	97.2	1571.4	96.2	1507.4	99.0	3827.1	98.9	9162.7	99.2

9.9 Makespan Versus Robustness

There is a direct relationship between Makespan and Robustness because as makespan increases, the robustness is bigger due to the fact that the tasks are sparser in time and they are able to absorb more incidences. However, it is not realistic to generate too sparse schedules, so generally a makespan bound is set and we try to find the more robust schedule for a given makespan threshold.

To carry out this study, the simulation carried out in the previous section gave us the amount of incidences that can be absorbed by modifying the energy consumption threshold. Figure 9.4 shows an approximate Pareto front for a set of 10 instances with 7 machines, 10 tasks per job and a maximum processing time of 100 time units. It can be observed that as the makespan increased the robustness also increased with a trigonometrical shape. Table 9.3 shows the makespan and robustness in different instances. It must be taken into account that the robustness is quite similar in all instances, so it is not directly dependent on the number of machines, number of tasks per job, neither the range of processing times. When the makespan threshold was set to the minimum possible (to achieve the optimal solution), these solutions were able to absorb an average of 29% of the incidences (first row of Table 9.3). This is due to the fact that natural buffers (black rectangles in Fig. 9.1) were able to absorb this percentage of incidences. Finally, when the makespan threshold set to an upper bound (obtained by minimizing energy consumption), the percentage of absorbed incidences was close to 100%. That means that the buffers are well distributed among all tasks and almost all disruptions were able to be absorbed.

The relationship between makespan and robustness can be obtained from the above formulas (9.8) and (9.9) to obtain formula (9.10):

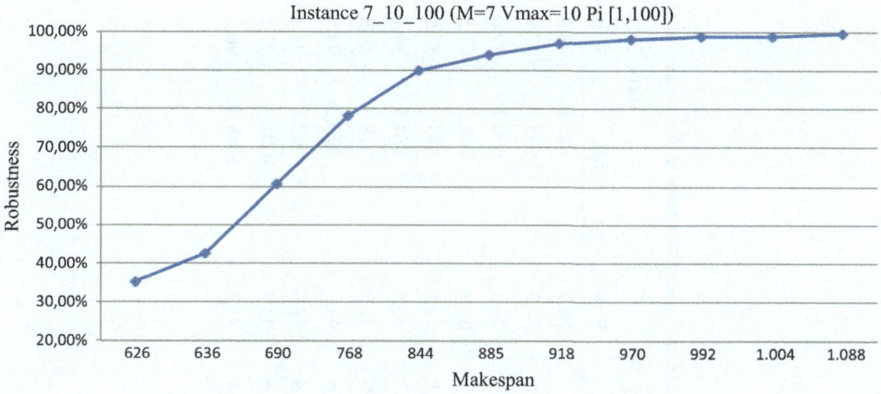

Fig. 9.4 Approximate Pareto front for the bicriteria makespan–robustness

Table 9.3 Makespan and robustness in instances $\langle m, v_{max}, p \rangle$

λ	5_10_50		7_10_100		3_20_50		3_25_100		3_30_200	
	Mk	Robust. (%)	Mk	Robust. (%)	Mk	Robust. (%)	Mk	Robust. (%)	Mk	Robust. (%)
1	317	26.7	625.9	29.2	707.8	26.8	1686.1	27.0	4048.9	25.6
0.9	322.8	37.2	635.7	37.9	728.3	36.3	1738.4	36.8	4182.9	39.7
0.8	343.1	48.2	690.1	49.0	777.8	50.6	1854.4	49.5	4530.8	48.9
0.7	384.7	68.1	768.2	67.5	854.6	66.0	2056.1	67.9	4953.3	62.5
0.6	410.1	79.8	843.7	82.6	946.8	79.6	2280.6	78.9	5452	80.9
0.5	454.3	86.6	884.5	89.1	1011.7	88.0	2421.8	86.3	5825	88.0
0.4	483.6	91.8	918.3	89.8	1079	93.4	2553	94.1	6158	92.7
0.3	502.6	94.2	970.2	94.8	1112.6	95.3	2667.1	94.9	6364.8	95.9
0.2	515.8	97.0	992.1	94.3	1145.7	96.6	2734.2	96.0	6513.2	96.8
0.1	524.4	93.7	1004.3	97.1	1168.7	97.3	2768.4	97.6	6600.8	96.3
0	565.4	97.2	1088.4	96.2	1296	99.0	31,601	98.9	7289.6	99.2

$$\frac{\text{Makespan}(\lambda)}{\text{Robustness}(\lambda, \%\text{incid})} \approx$$
$$\approx \frac{\text{ttasks} * p/2}{\left(6.3 - 6\frac{\%\text{incid}}{100}\right) * (8\cos(\lambda^2\pi) - \sin(\lambda\pi) + 11) * \left(e^{(0.25+1.2\lambda)} - \frac{\sin(\lambda\pi)}{2} + \frac{\cos(\lambda\pi)}{8}\right)} \tag{9.10}$$

Thus, given a makespan of a schedule with a given λ value and the duration of the incidence (%incid), we can estimate the robustness of this schedule. In the same way, given a robustness threshold, the duration of the incidence (%incid) and a λ value, we can estimate the makespan of this schedule. This formula can be refined by the operator according to the distribution of energy consumption of machines at different speeds.

9.10 General Analysis

In this section, a general analysis for all instance types was carried out. The main objective is to analyze the relationship among all relevant parameters around robustness and energy efficiency for all analyzed instances and different λ values (horizontal axe).

Figure 9.5 shows the results for disruptions of 40% of the maximum processing time (%incid = 40). The main vertical axe represents the robustness. Thus, the blue

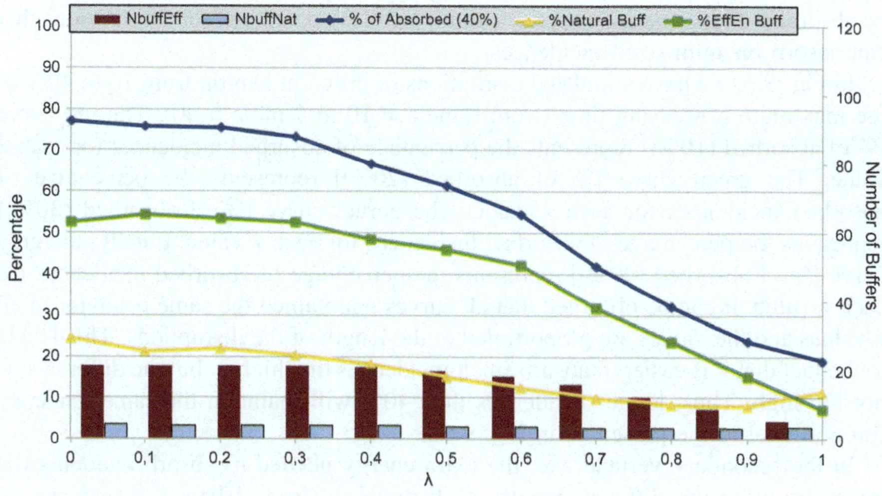

Fig. 9.5 General analysis for disruptions of 40% of maximum processing time (%incid = 40)

curve (% of Absorbed (40%)) represents the percentage of absorbed incidences for each λ value. The yellow curve (%Natural Buff) represents the percentage of incidences absorbed by a natural buffer. The green curve (%EffEn Buff) represents the percentage incidences absorbed by accelerating a machine. In this way, the robustness is % of Absorbed (40%) = %Natural Buff + %EffEn Buff.

In the secondary vertical axe, the garnet columns (NbuffEff) represent the average number of buffers generated by increasing the speed of machines and light blue columns (NbuffNat) represents the average number of natural buffers.

It can be observed that NbuffNat is mainly constant because they are independent of the objective (minimize makespan or energy consumption). However, the total amount of time involved in these natural buffers decreased as the value of λ increased. This is due to the fact that as λ increased the objective function gives more importance to minimize makespan, so the free slack is also minimized. Thus, the percentage of times that the incidence is absorbed by a natural buffer (%Natural Buff) also decreased. The same tendency is carried out by NbuffEff where the number of buffers generated decreased as λ increased. This is due to the fact that, as λ increases, the objective is to minimize makespan and more machines are assigned at maximum speed, so few buffer times can be generated by speeding up the machines.

The percentage absorbed incidences (%EffEn Buff) also decreased as the λ increased. However, it can be observed the difference between the percentage of absorbed incidences by the speeding up the machines (%EffEn Buff) against the percentage of absorbed incidences by natural buffers (%Natural Buff). Indeed, the main objective is represented by the blue curve [% of Absorbed (40%)] that represents the percentage of absorbed incidences for each λ value. It can be observed that for $\lambda = 0$ (minimizing energy consumption) almost all incidences can be absorbed. Thus, energy-aware schedules are also considered robust solutions that can absorb medium-sized incidences.

In Fig. 9.6, we have simulated disruptions of different length, from 10 to 40% of the maximum processing time (from %indic = 10 to %indic = 40). The red curve [% of absorbed (10%)] represents the percentage of absorbed incidences for each λ value. The green curve [% of absorbed (20%)] represents the percentage of absorbed incidences for each λ value. The garnet curve [% of absorbed (30%)] represents the percentage of absorbed incidences for each λ value. Finally, the gray curve [% of absorbed (40%)] represents the percentage of absorbed incidences for each λ value. It can be observed that all curves maintained the same behavior in all λ values and the values are proportional to the length of the disruptions. This is due to the fact that it is easier to absorb small incidences that higher, but the difference is not too high. Thus, longer incidences than 40% will maintain the same tendency (proportional to the presented in Fig. 9.6).

In the secondary vertical axe, the extra energy needed to absorb incidences is represented for the different length of disruptions (from 10 to 40%). It can be observed that although % of Absorbed (20%) was able to absorb less disruptions than % of Absorbed (10%), it needed more extra energy than the other in many cases. It must be taken into account that as the number of absorbed disruptions

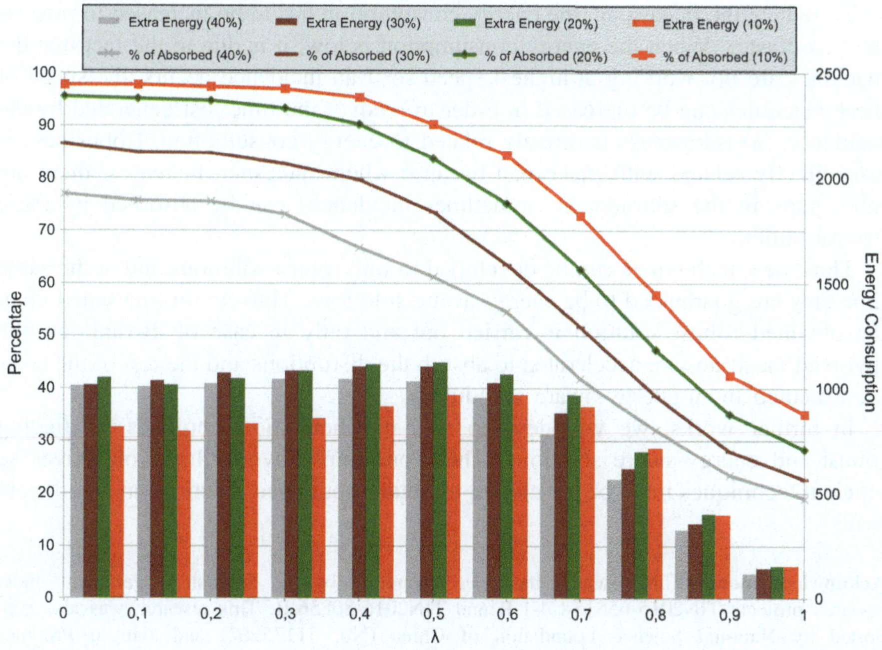

Fig. 9.6 General analysis for disruptions of 40% of maximum processing time (%incid = 40)

increased, the extra energy needed to absorb these disruptions also increased and the magnitude of needed energy is proportional to the size of the disruption. For instance, for $\lambda = 0$ (minimizing energy), the percentage of absorbed disruption of size 40% was around 77%, meanwhile the percentage of absorbed disruption of size 30% was around 87%. However, the extra energy needed to absorb these incidences was almost the same in both cases, due to the fact that larger disruptions generated larger needed of extra energy.

9.11 Conclusions

Many real-life problems can be modeled as a job-shop scheduling problem where machines can work at different speeds. It represents an extension of the classical job-shop scheduling problem, where each operation has to be executed by one machine and this machine has the possibility to work at different speeds. In this paper, we analyze the relationship among three important objectives that must be taken into consideration: Energy efficiency, robustness, and makespan. Analytical formulas are presented to estimate the relationship between these objectives in the analyzed instances. The results show the trade-off between makespan and robustness, and the direct relationship between robustness and energy efficiency.

To reduce the makespan, the energy consumption has to be increased to process the tasks faster. When the energy consumption is low, it is due to the fact that the machines are not working at highest speed so if an incidence occurs, the speed of these machines can be increased in order to recover the time lost generated by the incidence. So robustness is directly related to energy consumption. Robustness is also directly related with makespan because when makespan increases, there are more gaps in the solution, so sometimes incidences can be absorbed by these natural buffers.

Thus, new techniques can be developed to find robust solutions and at the same time they are guaranteed to be energy-aware solutions. Thus, in on-line scheduling, the obtained robust solution is carried out and only in case of disruptions, the involved machines are accelerated to absorb the disruptions and the rest of the tasks are executed in an energy-aware scheduling.

In further works, we will develop new metaheuristic techniques for finding robust and energy-aware solutions. These problems have multiple objectives so efficient techniques must be developed to obtain optimized solutions in an efficient way.

Acknowledgements This research has been supported by the Spanish Government under research projects TIN2015-65515-C4-1-R and TIN2016-80856-R. This research was also supported by National Science Foundation of China (No. 51175262) and Jiangsu Province Industry-Academy-Research Grant (No. BY201220116).

References

1. Agnetis, A., Flamini, M., Nicosia, G., & Pacifici, A. (2011). A job-shop problem with one additional resource type. *Journal of Scheduling, 14*(3), 225–237.
2. Billaut, J. C., Moukrim, A., & Sanlaville, E. (2008). *Flexibility and robustness in scheduling*. New York: Wiley.
3. Blazewicz, J., Cellary, W., Slowinski, R., & Weglarz, J. (1986). Scheduling under resource constraints-deterministic models. *Annals of Operations Research, 7,* 1–356.
4. BMWi. (2009). German federal ministry of economics and technology: Energy statistics.
5. Bruzzone, A. A. G., Anghinolfi, D., Paolucci, M., & Tonelli, F. (2012). Energy-aware scheduling for improving manufacturing process sustainability: A mathematical model for flexible flow shops. *CIRP Annals-Manufacturing Technology, 61*(1), 459–462.
6. Caplinskas, A., Dzemyda, G., Kiss, F., & Lupeikiene, A. (2012). Processing of undesirable business events in advanced production systems. *Informatica: International Journal., 23*(4), 563–579.
7. Dahmus, J., & Gutowski, T. (2004). An environmental analysis of machining. In *ASME International Mechanical Engineering Congress and RD\&D, Anaheim, California, USA*.
8. Dai, M., Tang, D. B., Giret, A., Salido, M. A., & Li, W. D. (2013). Energy-efficient scheduling for a flexible flow shop using an improved genetic-simulated annealing algorithm. *Robotics and Computer-Integrated Manufacturing., 29*(5), 418–429.
9. Duflou, J. R., Sutherland, J. W., Dornfeld, D., Herrmann, C., Jeswiet, J., Kara, S., et al. (2012). Towards energy and resource efficient manufacturing: A processes and systems approach. *CIRP Annals-Manufacturing Technology, 61*(2), 587–609.

10. Fang, K., Uhan, N., Zhao, F., & Sutherland, J. W. (2011). A new approach to scheduling in manufacturing for power consumption and carbon footprint reduction. *Journal of Manufacturing Systems., 30*(4), 234–240.
11. Garrido, A., Salido, M. A., Barbe, F., & López, M. A. (2000). Heuristic methods for solving job-shop scheduling problems. *In ECAI-2000 Workshop on New Results in Planning, Scheduling and Design, Berlin,* pp. 36–43.
12. Gutowski, T., Murphy, C., Allen, D., Bauer, D., Bras, B., Piwonka, T., et al. (2005). Environmentally benign manufacturing: observations from Japan, Europe and the United States. *Journal of Cleaner Production, 13*(1), 1–17.
13. Huang, K. L., & Liao, C. J. (2008). Ant colony optimization combined with taboo search for the job shop scheduling problem. *Computers & Operations Research, 35*(4), 1030–1046.
14. IBM. (2007). Modeling with IBM ILOG CP optimizer—Practical scheduling examples.
15. IBM ILOG CPLEX Optimizer. http://www-01.ibm.com/software/integration/optimization/cplex-optimizer/.
16. Kramer, L., Barbulescu, L., & Smith, S. (2007). Understanding performance tradeoffs in algorithms for solving oversubscribed scheduling. In *Proceedings 22nd Conference on Artificial Intelligence (AAAI-07).*
17. Laborie, P. (2009). IBM ILOG CP Optimizer for detailed scheduling illustrated on three problems. In *Proceedings of the 6th International Conference on Integration of AI and OR Techniques in Constraint Programming for Combinatorial Optimization Problems (CPAIOR09),* pp. 148–162.
18. Li, W., Zein, A., Kara, S., & Herrmann, C. (2011). An investigation into fixed energy consumption of machine tools. In *Glocalized solutions for sustainability in manufacturing,* pp. 268–273.
19. Mouzon, G., & Yildirim, M. B. (2008). A framework to minimise total energy consumption and total tardiness on a single machine. *International Journal of Sustainable Engineering., 1*(2), 105–116.
20. Mouzon, G., Yildirim, M. B., & Twomey, J. (2007). Operational methods for minimization of energy consumption of manufacturing equipment. *International Journal of Production Research, 45*(18–19), 4247–4271.
21. Neugebauer, R., Wabner, M., Rentzsch, H., & Ihlenfeldt, S. (2011). Structure principles of energy efficient machine tools. *CIRP Journal of Manufacturing Science and Technology, 4* (2), 136–147.
22. Nowicki, E., & Smutnicki, C. (2011). An advanced tabu search algorithm for the job shop problem. *Journal of Scheduling, 8*(145–159), 2005.
23. Seow, Y., & Rahimifard, S. (2011). A framework for modelling energy consumption within manufacturing systems. *CIRP Journal of Manufacturing Science and Technology, 4*(3), 258–264.
24. Szathmary, E. (2006). A robust approach. *Nature, 439,* 19–20.
25. Verfaillie, G., & Schiex T. (1994). Solution reuse in dynamic constraint satisfaction problems. In *Proceedings of the 12th National Conference on Artificial Intelligence (AAAI-94)*, pp. 307–312.
26. Weinert, N., Chiotellis, S., & Seliger, G. (2011). Methodology for planning and operating energy-efficient production systems. *CIRP Annals-Manufacturing Technology, 60*(1), 41–44.

Chapter 10
A Semantic Information Services Framework for Sustainable WEEE Management Toward Cloud-Based Remanufacturing

Kai Xia, Liang Gao, Lihui Wang, Weidong Li and Kuo-Ming Chao

Abstract Sustainable management of Waste Electrical and Electronic Equipment (WEEE) has attracted escalating concerns of researchers and industries. Closer information linking among the participants in the products' lifecycle should take place. How to interoperate among the distributed and heterogeneous information systems of various participants is a challenge faced. Targeting the cloud-based remanufacturing, this article aims to develop a semantic information services framework for sustainable WEEE management. In the proposed framework, an ontology-based approach is developed to integrate and represent the lifecycle information from multiple local data sources within an information services provider. Meanwhile, a semantic information services management platform is introduced for the advertisement, matchmaking, and retrieval of semantic information services. Some relevant techniques used to build the framework are introduced extensively. A demonstration case study on waste LCD TV is used to illustrate the effectiveness and significance of the proposed framework.

Keywords Cloud-based remanufacturing · Semantic information services
Waste electrical and electronic equipment · Sustainable management

---

K. Xia
Wuhan Second Ship Design and Research Institute, Wuhan, China

L. Gao (✉)
State Key Laboratory of Digital Manufacturing Equipment and Technology,
School of Mechanical Science and Engineering,
Huazhong University of Science and Technology, Wuhan, China
e-mail: gaoliang@mail.hust.edu.cn

L. Wang
Department of Production Engineering, KTH Royal Institute of Technology,
Stockholm, Sweden

W. Li · K.-M. Chao
Faculty of Engineering, Environment and Computing, Coventry University,
Coventry, UK

W. Li and S. Wang (eds.), *Sustainable Manufacturing and Remanufacturing Management*, https://doi.org/10.1007/978-3-319-73488-0_10

10.1 Introduction

Mass-customized production, rapid technology updating, and shortening lifespan of Electrical and Electronic Equipment (EEE) have resulted in enormous amount of Waste Electrical and Electronic Equipment (WEEE) in the world. It is estimated that 20–50 million tons of WEEE are discarded annually [1]. WEEE is harmful to the environment as it contains hazardous materials such as lead, mercury, cadmium, polybrominated biphenyls (PBBs), polybrominated diphenyl ethers (PBDEs), while it is rich in resource as it contains valuable materials including iron, copper, silver, gold, plastics.

The WEEE Directive (2002/96/EC) has been implemented in the EU since 2003 and has been updated in 2012 (2012/19/EU). Many other allied directives were implemented in other countries. They are intent to reduce the quantity of WEEE and encourage the industries to recover and remanufacture WEEE. As the legislative pressure is becoming stricter and the industries want to obtain better profit from end-of-life treatment, sustainable management of WEEE has attracted escalating concerns and developing technical solutions for sustainable WEEE management becomes a critical global concern [2]. Growing environmental consciousness has led manufacturing researchers and engineers to focus on sustainability increasingly [3]. Sustainable manufacturing requires considering simultaneously the economic, environmental and social implications associated with manufacturing products [4]. These considerations are also compliant to the considerations of sustainable WEEE management.

The closed-loop lifecycle of EEE/WEEE is illustrated in Fig. 10.1. In the beginning-of-life (BOL) stage, the raw materials are extracted firstly. Then, the components are fabricated using raw materials and assembled as new products. In the middle-of-life (MOL) stage, products are consumed by different users through distribution and retail. When the EEE come to the end-of-life (EOL) stage, different treatment options are selectable to recover the WEEE in the product level, component level, or materials level, including repair, reuse, remanufacturing, and recycling. The material flow is forward in the BOL and MOL stage while reverse in the EOL stage. Globally distributed participants, such as manufacturer, distributor, retailer, consumer, collector, recycler, remanufacturer, in the lifecycle of EEE/WEEE are involved and their information systems are usually heterogeneous.

WEEE remanufacturing and recycling count a great deal in WEEE management, and industries need to implement them in practices to tradeoff the environmental and economic objectives [6]. However, the information flows of EEE/WEEE between BOL/MOL stage and EOL stage have not yet been established effectively. Hence, it is hard to get efficient information support for applying sensible EOL treatment strategies to recover WEEE. The current information sharing mechanism in the lifecycle of EEE/WEEE is illustrated in Fig. 10.2. EEE manufacturer, distributor/retailer, and WEEE collector/importer report limited information to WEEE management authorities including product information, marketing information, and collecting information. All these types of information are very simple, for instance,

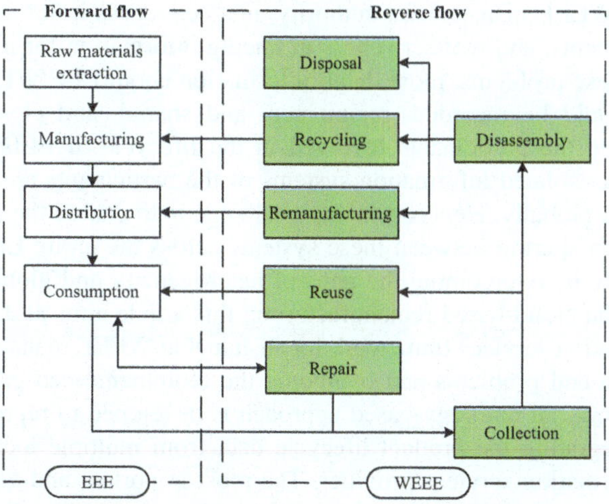

Fig. 10.1 Closed-loop lifecycle of EEE/WEEE developed from [5]

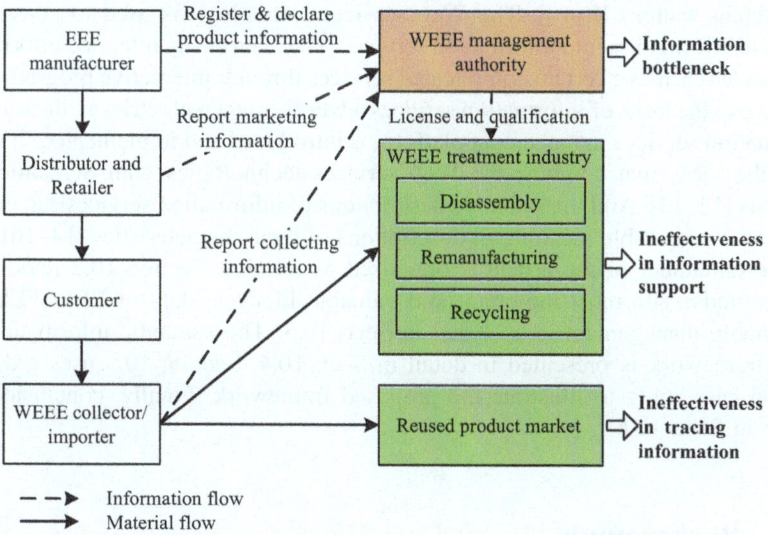

Fig. 10.2 Main problems in current EEE/WEEE information sharing mechanism

the product information just includes product type, name, amount and brief functional description and the marketing information and collection information just include product type, name, and amount. The WEEE management authority only provides licensing and qualification information to WEEE treatment industry without any decision supporting information. Hence, there are three main problems in current EEE/WEEE information sharing mechanism, including information bottleneck in WEEE management authority, ineffectiveness in information support for EOL treatments, and ineffectiveness in tracing information for reused products.

To solve these problems, more detail information about EEE/WEEE in different life stages should be recorded, maintained, and shared, and closer information linking between the participants involved in the lifecycle of EEE/WEEE should take place. The isolated information systems of the participants are heterogeneous and distributed globally. Hence, a facing challenge is to enable the interoperability and information sharing between these systems across the entire EEE/WEEE lifecycle efficiently by overcoming the gaps of heterogeneity and global distribution.

Targeting the cloud-based remanufacturing, this article aims at developing a semantic information services framework for sustainable WEEE management to solve the aforementioned problems and overcome the aforementioned gaps. In the proposed framework, an ontology-based approach is developed to represent the information by integrating the product lifecycle data from multiple local data sources within an information services provider. The most important and valuable product lifecycle data is modeled using the ontology-based approach. Ontology can provide an agile and dynamic map to the data contained in different sources within a domain [7], and ontology-based approaches can be used to extract and integrate data residing at different sources [8–10]. The Web services technology is used to convert the lifecycle information into information services. They can easily integrate information services and achieve certain complicated services through interactive programs [11]. To achieve the tasks of information services advertisement and retrieval, the semantic information services management platform is introduced and implemented. It comes from the ideas that combine the Web services technologies with Semantic Web methods [12, 13]. And the semantic descriptions of information services can provide the machine-readable meaningful descriptions of their characteristics [14–16].

The remainder of the article is organized as follows. Section 10.2 reviews the backgrounds. The most important and valuable lifecycle data of EEE/WEEE for sustainable management is analyzed in Sect. 10.3. The semantic information services framework is presented in detail in Sect. 10.4. Section 10.5 uses a demonstration case study to illustrate the proposed framework. Finally, conclusions are drawn in Sect. 10.6.

10.2 Backgrounds

10.2.1 Lifecycle Information Management

Product lifecycle data can be classified as static data and dynamic data [17]. Static data is related to the characteristics of products and gives details about the contained materials, configuration options, and operation instructions. It occurs at the BOL stage and rarely changes. Dynamic data is obtained at the MOL and EOL stages. Usage data, including consumers, conditions, and serving actions, forms the major part of dynamic data [18].

To obtain and store the dynamic data, product embedded information devices have been introduced including: the intelligent data unit (IDU) [18], the lifecycle units (LCUs) [19]. Radio-frequency identification (RFID) technologies and sensor networks are implemented in the generation and sharing of product information among lifecycle participants [20]. To manage and use the product lifecycle data, some lifecycle information management systems have been studied and introduced. Reference [21] developed a component-based software framework for lifecycle information management of consumer products. Reference [22] proposed a reverse logistics decisions conceptual framework with a reverse logistics information system. Reference [23] introduced a service-oriented information frameworks for global EEE/WEEE information management. Some other works studied the impacts or benefits of sharing product lifecycle information [24–26].

10.2.2 Cloud-Based Remanufacturing

Cloud manufacturing is based on the ideas of networked manufacturing and cloud computing and developed for collaborative manufacturing in a distributed environment [27]. It is web-based, service-oriented, highly efficient, and knowledge-based. Remanufacturing enables the potential for recovering the economic value and reducing the environmental impact of waste products simultaneously [28]. Cloud-based remanufacturing is a domain implementation of cloud manufacturing on EOL treatment of waste products [29]. The cloud concept is introduced into the remanufacturing context, and it will reshape the remanufacturing industry and transform their business model, with the characteristics of customization and distribution [30]. Several kinds of architectures have been proposed in previous work, such as the main framework of the cloud-based remanufacturing system proposed in [29], the three-layer WEEE Remanufacturing Cloud (WRCloud) proposed in [30], and the service-oriented remanufacturing platform proposed in [31]. In these proposed architectures, the lifecycle information exchange and management frameworks are discussed. However, the methods of integrating and representing the lifecycle information in each information provider and the approaches to retrieving the lifecycle information services have not been studied or introduced.

Addressing the aforementioned main problems in current EEE/WEEE information sharing mechanism and the gaps in lifecycle information management, this paper proposes a semantic information services framework for sustainable WEEE management toward cloud-based remanufacturing.

10.3 Lifecycle Data for Sustainable WEEE

The important and valuable product lifecycle data for sustainable WEEE management is studied and analyzed in the view of EOL treatment. There exist different EOL treatment opinions for WEEE, including repair, reuse, disassembly, remanufacturing, recycling. In order to choose an EOL treatment option that is technically feasible, cost-effective, and environment-friendly, the detail data about the products, the tracing data, the technical data for EOL treatment and the legal, economic, and ecological data should be provided to support decision making. The detail categories of the important lifecycle data are summarized and listed in Table 10.1.

The product data generated at the BOL stage of products belongs to static data. The product data mainly includes identification data, bill of materials (BOM), material properties, and assembly data. Figure 10.3 gives a general concept data model for the product data. In the concept data model, the BOM is the key part and represented as a tree structure, in which, the root node represents a product, the intermediate node represents a subassembly and the leaf node represents a kind of materials. The BOM provides abundant information of a product, including the structure tree of the product, the identification, quantity, and mass of the contained materials. Hence, it is very valuable for supporting reuse of parts, recycling of

Table 10.1 Categories of important lifecycle data for sustainable management of WEEE

Categories	Subcategories	Detail data
Product data	Identification data	Product's id, name, type, category, mass, manufacturer, etc.
	BOM	Product structure tree, quantity and mass of the contained materials, etc.
	Material properties	Material's id, name, type, category, composition, hazardousness, supplier, etc.
	Assembly data	Assembly operation instruction, assembly tools, assembly precedence constraints, etc.
Tracing data	Utilization history	Served consumers, served time, repairing and updating records, etc.
	Performance data	Performance parameters, condition assessment, etc.
Technical data for EOL treatment	Technical data for disassembly	Process capabilities and resource availability, disassembly precedence constraints and disassembly sequences, etc.
	Technical data for remanufacturing	Process capabilities and resource availability, remanufacturing strategy and process, etc.
	Technical data for recycling	Process capabilities and resource availability, recycling strategy and process, etc.
The other data	Legal data	WEEE directive, RoHS directive, etc.
	Economic data	Price of product/component/materials, cost, etc.
	Ecological data	Carbon footprint, environmental impact, etc.

materials, and remanufacturing of products. Meanwhile, the assembly data, including the assembly operation instructions, assembly tools, assembly precedence constraints, is very useful to help and support disassembly planning, which is one of the most important processes in EOL treatment of WEEE.

The tracing data occurred at the MOL stage of products belongs to dynamic data. The tracing data mainly includes the utilization history and performance data of product. The utilization history data includes the product's served consumers, served time, and repairing and updating records. The performance data records the product's performance parameters and condition assessments. As discussed in the introduction section, it is infeasible to record the tracing data in the current management mechanism of EEE/WEEE. To implement a cloud-based information management mechanism, the distributor, retailer, and repairer should take more responsibilities to record the tracing data when a product is sold, repaired, or

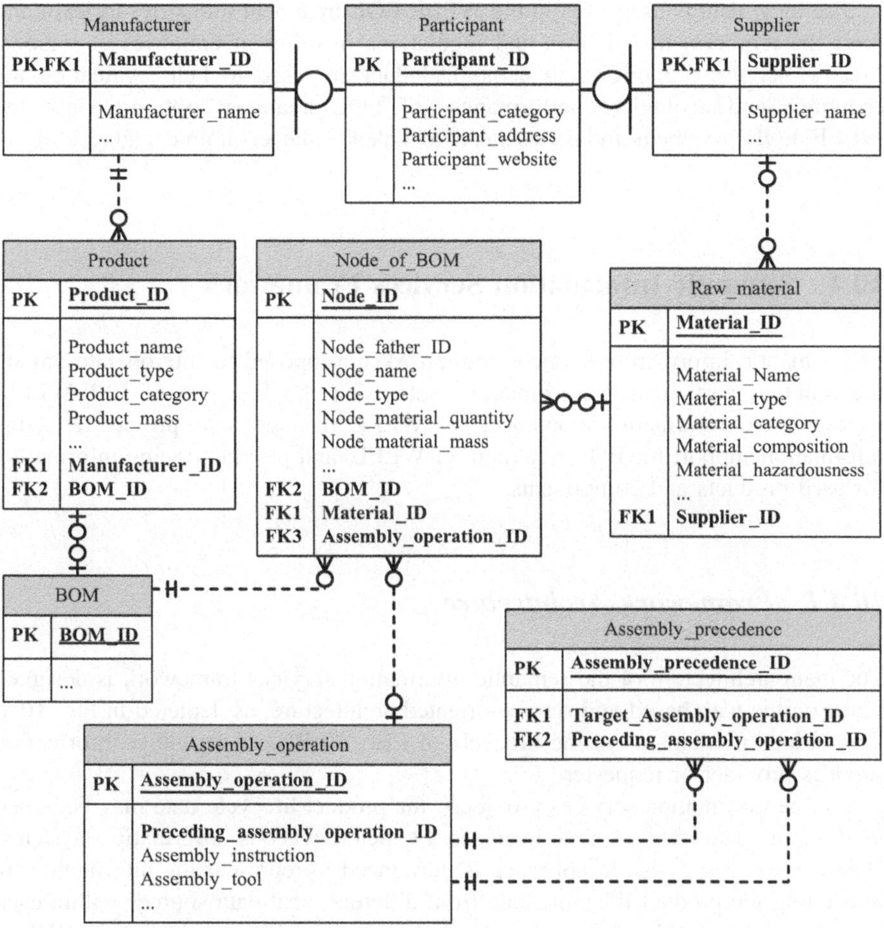

Fig. 10.3 Concept data model for the product data

updated by new components. Each product should have a unique identification and the tracing data is stored in the cloud storage. When new tracing data is generated, the users can find the tracing data updating service according to the product identification and update the data in the cloud storage.

Technical data for EOL treatment is implemented at the EOL stage of products. In this paper, three kinds of EOL treatment options are considered: disassembly, remanufacturing, and recycling. For disassembly, the disassembly precedence constraints, which determine the disassembly precedence relations among disassembly operations, are especially important to support disassembly sequence planning. The suggested disassembly sequences and disassembly operation instructions from the manufacturer or EOL treatment industries are also useful. For remanufacturing and recycling, suggested strategies and processes, and operation instructions from the manufacturer or EOL treatment industries are significant and valuable supports.

The legal data is used to help the WEEE EOL treatment industries to learn and meet the requirements of laws and regulations in different countries or regions. Meanwhile, the economic and ecological data are used to help evaluating the economic and ecological performances of EOL treatment options, since the WEEE EOL treatment industries need to balance the economic and ecological objectives.

10.4 Semantic Information Services Framework

The semantic information services framework is proposed to interoperate among the different participants' information systems in the lifecycle of EEE/WEEE, eliminate the information bottleneck of WEEE management, provide effective information support for EOL treatment of WEEE, and provide tracing information for used products and components.

10.4.1 Framework Architecture

The main architecture of the semantic information services framework is designed based on the web-based and service-oriented architecture, as depicted in Fig. 10.4. The various participants in the lifecycle of EEE/WEEE can be either information services provider or requester.

For the information services providers, the product lifecycle data may be stored in different data sources and managed by heterogeneous information systems. Hence, the ontology-based approach is introduced to represent the information by integrating the product lifecycle data from different local data sources within each participant. The local ontology is extracted from the relational databases (RDBs). The path mapping function maintains the links between the ontology and the data

Fig. 10.4 Concept data model for the product data

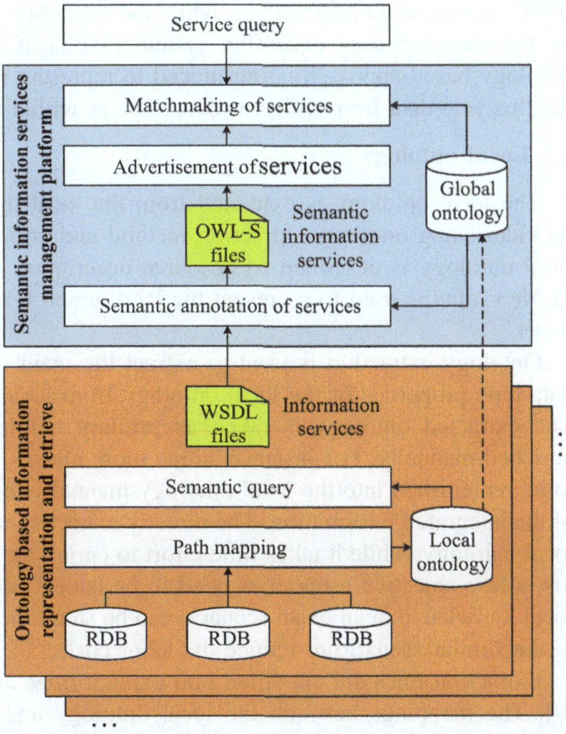

sources. And the semantic query function is used for retrieving the lifecycle information.

The retrieve of the lifecycle information is published as information services using Web services technology and described using the Web Services Definition Language (WSDL) [32]. Then the published information services are described using the OWL-S [33] via semantic annotation. OWL-S is a Web service ontology built on the Ontology Web Language (OWL) [34]. The semantic information services management platform provides a registry where the advertisements of information services (OWL-S files) are stored. And the retrieval of the semantic information services is performed by submitting quires described by OWL-S and by means of matchmaking processes.

10.4.2 Ontology-Based Information Representation and Retrieve

The lifecycle data of EEE/WEEE may be stored in different data sources and managed by heterogeneous information systems within an information provider. For instance, the BOM of a product is managed by the enterprise resource planning

(ERP) system of a manufacturer, while the assembly data of the product is managed by the manufacturing execution system (MES) of the manufacturer. Hence, the ontology-based approach is introduced to represent the information by integrating the lifecycle data from different data sources within an information provider.

(1) **Local ontology**

The local ontology is extracted from the local data sources of an information provider using ontology extraction method and enriched manually afterward. The local ontology is described by resource description framework (RDF) graph, and Turtle syntax is used to represent the RDF graph textually in the examples of this paper.

Ontology extraction is used to extract the main classes, object properties, and data type properties of the local ontology from local data sources, such as RDBs. The extracted ontology is taken as primary ontology and needs to be further enriched manually. For instance, some more classes and objective properties need to be replenished into the local ontology manually to represent the whole structure of the integrated information. The directly extracted ontology composes most of the local ontology, while it takes little effort to enrich the ontology manually. Actually, just a few objective proprieties need to be added. Moreover, the ontology enrichment knowledge in an exact scenario can be stored in a knowledge base and reused in the similar scenario to reduce the labor cost.

Extraction rules are classified into explicit rules and implicit rules according to [9]. The mappings between the local ontology and the local RDBs are directly generated using explicit rules. For example, a table is mapped to a class, and a column is mapped to a property. In addition, the implicit rules are used to explore the hidden information that is not explicitly appeared, e.g., using similarity to extract owl:sameAs property (sameAs belongs to the OWL vocabularies), and using hierarchy relationship to discover rdfs:subClassOf property (subClassOf belongs to the RDF Schema (RDFS) vocabularies). Some extraction rules are summarized in Table 10.2 according to [9].

According to these rules, the local ontology for the product data shown in Fig. 10.3 is extracted and the ontology graph is illustrated in Fig. 10.5, where the extracted classes come from the tables in different data sources, and the extracted object properties come from the foreign keys of the tables. In addition, the extracted

Table 10.2 Ontology extraction rules according to [9]

Type	RDB	Ontology
Explicit	Table	owl: Class
	Column excluding keys	owl: DatatypeProperty
	Primary key	owl: InversFunctionalProperty
	Foreign key	owl: ObjectProperty
	Row	owl: NamedIndividual
Implicit	Similarity check	owl: sameAs
	Hierarchy mining	rdfs: subClassOf

object properties are elaborated by manual adjustment and the ontology extracted from different local data sources are integrated together by some of the objective properties. For instance, the object property manufacturedBy, which's domain and range are Class Product and Class Manufacturer, respectively, is introduced to represent that Class Product is connected with Class Manufacturer via the Foreign Key Manufacturer_ID. Moreover, the ontology about the manufacturer and the ontology about the product are integrated in such way.

After the classes, objective properties and data type properties are extracted; the path mapping function is used to generate the individuals to enrich the local ontology.

(2) **Path mapping**

The links between the local ontology and local data sources need to be maintained to make upper information services able to retrieve information from the data sources through the local ontology. There are some technologies used to represent the mappings between local ontology and local RDBs, such as RDB to RDF Mapping Language (R2RML) [35], and D2RQ Mapping Language [36]. Both R2RML mapping and D2RQ mapping are an RDF document written in Turtle syntax.

R2RML is used to express the mappings from RDBs to RDF datasets. Such mappings make it possible to represent the data structure in RDBs using the RDF data model. R2RML enables different types of mapping implementations, such as mapping a simple table, linking two tables, mapping many-to-many tables. A mapped table is a logical table that can be a base table or a view. A triples map is used to map a logical table to RDF. The rows of the logical table are mapped to RDF triples. A RDF triple is formed by combining a subject map with multiple

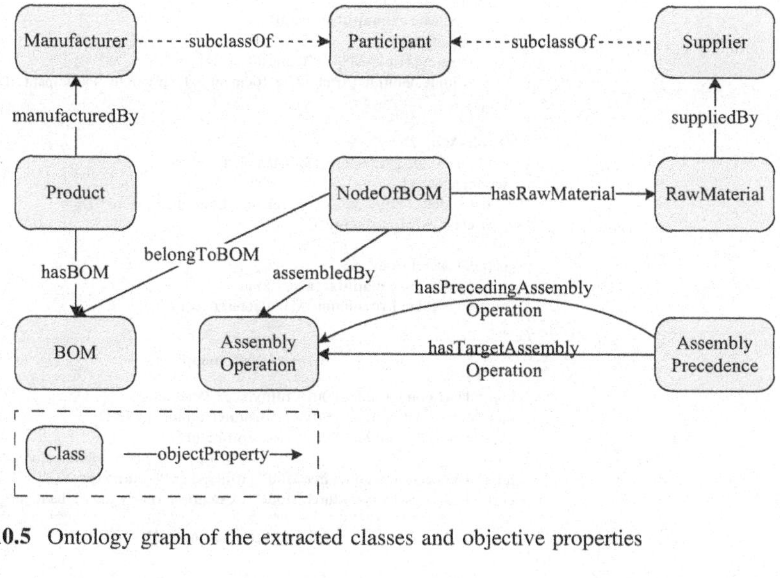

Fig. 10.5 Ontology graph of the extracted classes and objective properties

predicate-object maps. A subject map maps the individuals of a class with the rows of the logical table, and a predicate-object map maps the property values of an individual with the column values in the relevant row [35].

An example for using the R2RML to express the mappings between RDB and RDF/OWL is shown in Fig. 10.6: firstly, the input local RDB including two tables, product and manufacturer, is given in (a); secondly, the local ontology is extracted

Fig. 10.6 An example of R2RML mapping

Product				Manufacturer	
Product_ID	Product_name	Participant_ID		Participant_ID	Manufacturer_name
P001	LCD TV	M001		M001	Changhong Ltd

(a): Input local RDB

```
@prefix owl: <http://www.w3.org/2002/07/owl#>.
@prefix rdf: <http://www.w3.org/1999/02/22-rdf-syntax-ns#>.
@prefix ex: <http://example.com/>.
<http://example.com> rdf:type owl:Ontology.
ex:manufacturedBy rdf:type owl:ObjectProperty.
ex:manufacturerName rdf:type owl:DatatypeProperty.
ex:participantID rdf:type owl:DatatypeProperty.
ex:productName rdf:type owl:DatatypeProperty.
ex:Manufacturer rdf:type owl:Class.
ex:Product rdf:type owl:Class.
```

(b): Extracted local ontology (in Turtle syntax)

```
@prefix rr: <http://www.w3.org/ns/r2rml#>.
@prefix ex: <http://example.com/>.
<#TriplesMap_1>
  rr:logicalTable [ rr:tableName "Product" ];
  rr:subjectMap [
    rr:template "http://data1.com/product/{Product_ID}";
    rr:class ex:Product;
  ];
  rr:predicateObjectMap [
    rr:predicate ex:productName;
    rr:objectMap [ rr:column "Product_name" ];
  ];
  rr:predicateObjectMap [
    rr:predicate ex:manufacturedBy;
    rr:objectMap [
      rr:parentTriplesMap <#TriplesMap_2>;
      rr:joinCondition [ rr:child "Participant_ID"; rr:parent "Participant_ID"; ];
    ];
  ].
<#TriplesMap_2>
  rr:logicalTable [ rr:tableName "Manufacturer" ];
  rr:subjectMap [
    rr:template "http://data2.com/manufacturer/{Participant_ID}";
    rr:class ex:Manufacturer;
  ];
  rr:predicateObjectMap [
    rr:predicate ex:manufacturerName
    rr:objectMap [ rr:column "Manufacturer_name" ];
  ].
```

(c) R2RML mapping

```
<http://data1.com/product/P001> rdf:type ex:Product.
<http://data1.com/product/P001> ex:productName  "LCD TV".
<http://data1.com/product/P001> ex:manufacturedBy
                                  <http://data2.com/Manufacturer/M001>.
<http://data2.com/manufacturer/M001> rdf:type ex:Manufacturer.
<http://data2.com/manufacturer/M001> ex:manufactuerName "Changhong Ltd".
```

(d) Output mapping RDF

as shown in (b); and then, the customized R2RML mapping is built in (c); lastly, the output mapping RDF is shown in (d), according to which, the individuals can be generated.

(3) **Semantic query**

Semantic query languages, e.g., Simple Protocol and RDF Query Language (SPARQL) [37], can be used to execute queries toward the mapping RDF generated according to the customized R2RML mapping. A SPARQL query toward the output mapping RDF in Fig. 10.6d is illustrated in Fig. 10.7.

The semantic query function is service-oriented and used for retrieving lifecycle information from the local databases of the information services provider through the Internet. Figure 10.8 gives a general framework of the service-oriented SPARQL query, where the SPARQL endpoint refers to the endpoint of the information services and the RDF datasets refer to the storage of the local ontology. The SPARQL endpoint is used to provide the information services using the SPARQL Protocol over the Internet. SPARQL Protocol is a means of conveying SPARQL queries from clients to SPARQL processors [38]. When the clients invoke the information services, they send HTTP requests to SPARQL endpoint that provide the information services, and after that, the endpoint executes the

```
PREFIX ex: <http://example.com/>
SELECT ?Product ?Product_name ?Manufacturer ?Manufacturer_name
WHRE { ?Product a ex:Product.
        ?Product ex:productName ?Product_name.
        ?Product ex:manufacturedBy ?Manufacturer.
        ?Manufacturer ex:manufacturerName ?Manufacturer_name.
}
```

(a): SPARQL query

Product	Product_name	Manufacturer	Manufacturer_name
P001	LCD TV	M001	Changhong Ltd

(b): Query result

Fig. 10.7 An example of SPARQL query

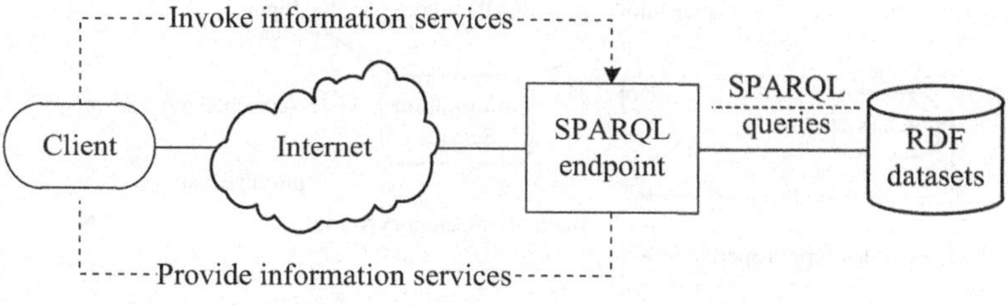

Fig. 10.8 General framework of the service-oriented semantic query

relevant SPARQL queries against the RDF datasets and sends HTTP responses back to the originating client.

10.4.3 Semantic Information Services Management Platform

The semantic information services management platform is developed to manage the information services published by the globally distributed and various information providers. Existing service description languages, e.g., WSDL, are used to describe services to help a client making a decision on whether and how to invoke the published services. Global distribution and huge amount of the EEE/WEEE lifecycle participants will lead to that thousands of services will have to be searched and retrieved, and automation will be the key requirement. In order to fulfill these challenges, existing service specifications need to be augmented with semantic descriptions. The developed platform provides three key functions: semantic annotation, advertisement, and matchmaking of the published information services.

(1) **Global ontology**

The semantic annotation of the information services for sustainable WEEE management concerns the name, categories, inputs, and outputs of the information services. The global ontology represents the knowledge about the different information services provided by various information providers in a globally distributed environment. Figure 10.9 illustrates the global ontology with selected classes and properties.

The classes include information service, parameter, input, output. The parameter refers to a class in the local ontology of the relevant information service provider, which represents a kind of lifecycle data. The input and output are subclasses of parameter. Inputs specify the data that the information service requires for its

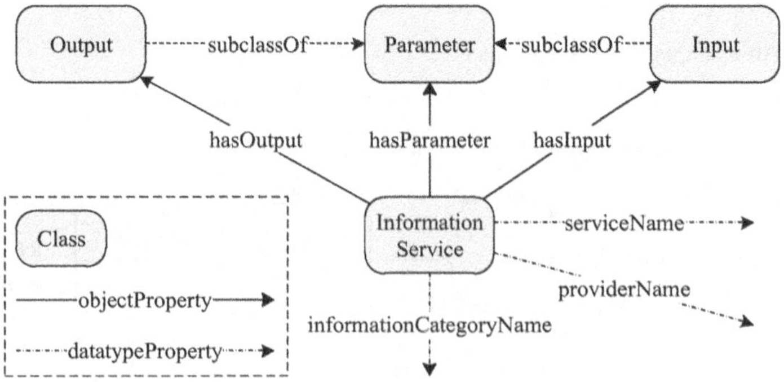

Fig. 10.9 Global ontology

execution. And outputs specify the data that the information service transforms to the client.

The object properties include hasParameter, hasInput, hasOutput. The hasParameter ranges over the instances of parameters. The hasInput and hasOutput both are the subproperties of the hasParameter. The hasInput ranges over the instances of Input, and the hasOutput ranges over the instances of Output.

The data type properties include serviceName, serviceProviderName, informationCategoryName. The serviceName is an identifier of a service. The serviceProviderName refers to an identifier of a service provider. The informationCategoryName represents the category of the information transformed by the service, which has been described in Table 10.1 in Sect. 10.3.

(2) **Semantic annotation**

The purpose of semantic annotation is to generate the semantic description for the information services based on the global ontology. We deal with the information services' name, information category, and the input/output parameters. The information services are described by WSDL and defined using elements of type, message, porttype, binding, port, and service. The characteristics of a service specify a number of input and output messages. Each message consists of some parameters and gives the transmitted data's definition.

The semantic annotation function provides mappings of parameter of WSDL input/output messages to ontology classes. For each parameter in WSDL, there is only one ontology class mapped to. As described in Sect. 10.4.2, the lifecycle data of an information provider is integrated and represented using an ontology-based approach and the local ontology is extracted from the local data sources. Each kind of data is mapped to a local ontology class. Hence, it is possible to generate the mappings of WSDL parameters to global ontology classes automatically, when the relevant local ontology is intergraded to the global ontology.

Given the semantic annotations of the parameters of the WSDL input/output messages, the signature of information services is constructed in OWL-S files. In addition, the properties of service name and information category are also represented in OWL-S files. The service name property refers to the name of the service that is being offered and can be used as an identifier of the service. And the information category property is inserted by human annotators. An example of the semantic annotation is given in Fig. 10.10.

(3) **Advertisement and matchmaking**

When an information service provider submits a service description in WSDL to the semantic information services management platform, the semantic annotation function generates an OWL-S file to represent the signature of the service. Then, the OWL-S file is inserted into the platform registry for advertisement. Information service requesters submit queries to search for required information services advertised in the platform registry.

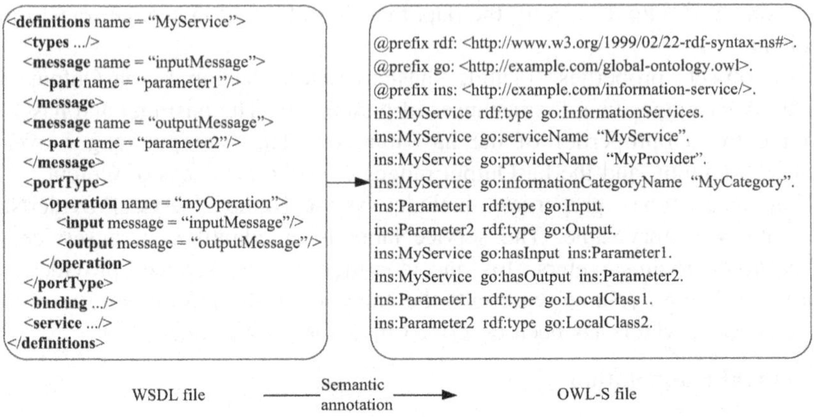

Fig. 10.10 Example of semantic annotation

Similar to the semantic annotation of information services, a query can be also specified and described by input/output parameters, service provider, and information category of the required service, which are defined by the global ontology. The matchmaking function is used to explore and find the semantic relations between the semantic description of the request and the semantic description of registered information services. In the matchmaking process, the matching of a submitted query to an advertised semantic information service can be spread to the matching of service provider, information category, input parameters, and output parameters. The query result is a list of identifications of the matched information services.

SPARQL query is used to retrieve the semantic information services matching with the given requirements. An example of using a SPARQL query to retrieve the semantic information services is illustrated in Fig. 10.11. After executing a SPARQL query, the exactly mapped and near matched services are retrieved and their names of identifications are outputted.

10.4.4 Process of Information Services Retrieval and Invoking

In the proposed semantic information services framework, a typical process of information services retrieval is illustrated in Fig. 10.12. The information service providers publish services binding to their SPARQL endpoints to the semantic information services management. Then, the semantic descriptions of the services are generated by semantic annotation service and inserted into the registry using advertisement service. After the semantic information services are registered, the providers get some responses. Similarly, the information service requesters submit

```
PREFIX rdf: <http://www.w3.org/1999/02/22-rdf-syntax-ns#>.
PREFIX go: <http://example.com/global-ontology.owl>.
PREFIX ins: <http://example.com/information-service/>.
SELECT ?InformationService ?ServiceName
WHRE { ?InformationService rdf:type go:InformationService.
       ?InformationService go:providerName “MyProvider”.
       ?InformationService go:informationCategoryName “MyCategory”.
       ?InformationService go:hasInput ?InputParameter.
       ?InputParameter rdf:type go:Input.
       ?InputParameter rdf:type go:LocalClass1.
       ?InformationService go:hasOutput ?OutputParameter.
       ?OutputParameter rdf:type go:Output.
       ?OutputParameter rdf:type go:LocalClass2.
}
```

Fig. 10.11 Example of SPARQL query for matchmaking of information services

queries to the matchmaking service. After the matchmaking service executes SPARQL requests against the registered OWL-S files based on the global, the service requesters get the responses with the names of information services that match with the requirements. Finally, the requesters can invoke the required information services automatically according to the query results.

The process of invoking an information service is illustrated in Fig. 10.13. When a service requester sends a request of invoking an information service to a SPARQL endpoint of a service provider, the SPARQL endpoint executes the relevant SPARQL queries against the RDF datasets of local ontology. The query results are

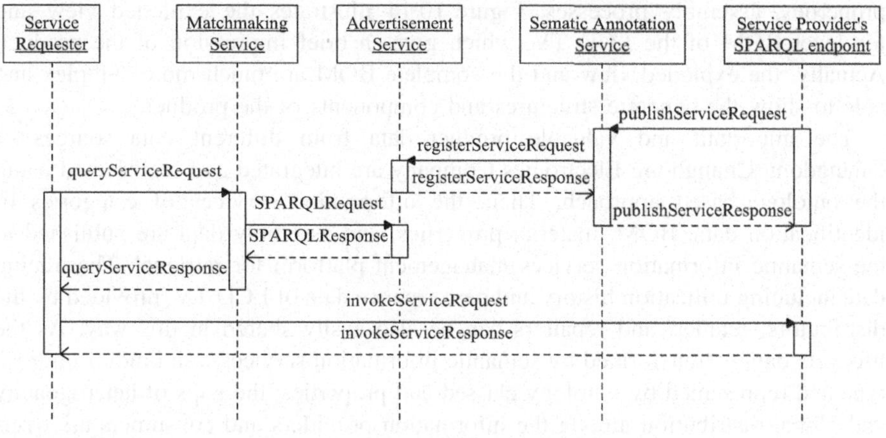

Fig. 10.12 Sequence diagram of information services retrieval

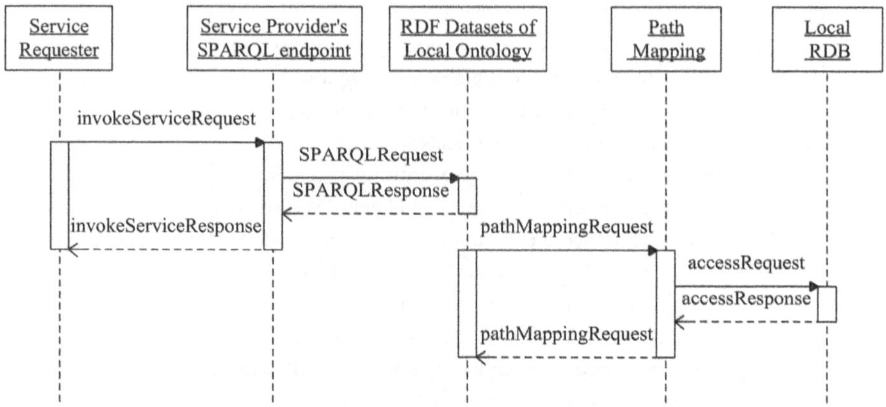

Fig. 10.13 Sequence diagram of invoking information services

sent back to the information service resister. The path mapping function is used to update the local ontology and maintained the links between the local ontology and local RDBs.

10.5 Demonstration Case Study

To illustrate the effectiveness and significance of the proposed framework, a demonstration case study on waste LCD TV is used. The LCD TV studied in this paper is produced by a Chinese television producer named Guangdong Changhong Electronics Company, Ltd. The case study is implemented in the authors' previous work published in [6, 30]. The product data of the LCD TV is acquired from the company, including the exploded view, bill of materials (BOM), components' properties, assembly processes. Figure 10.14 illustrates the exploded view and top-level BOM of the LCD TV, which gives a brief impression of the product. Actually, the exploded view and the complete BOM are much more complex and able to show the concrete structures and components of the product.

The important and valuable product data from different data sources of Guangdong Changhong Electronics Company are integrated and represented using the ontology-based approach. Then, the information services of categories of identification data, BOM, material properties, and assembly data are published to the semantic information services management platform for retrieval. The tracing data including utilization history and performance data of LCD TV, provided by the distributors, retailer, and repairers can be also easily shared in this way. As the lifecycle data is transformed by semantic information services in a kind of uniform type and represented by ontology classed and properties, the gaps of heterogeneity and global distribution among the information providers and consumers are overcame. In this way, the previously mentioned information bottleneck in WEEE

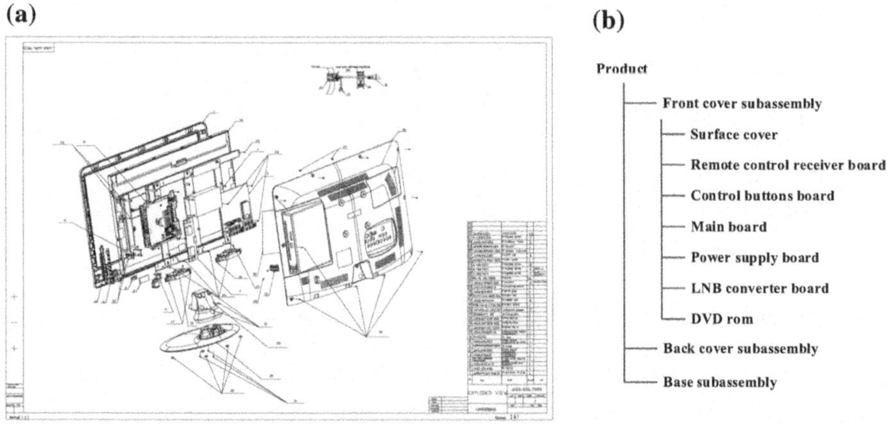

Fig. 10.14 Production information about the LCD TV. **a** Exploded view and **b** top-level BOM

Table 10.3 Properties of the disassembly operations [6]

#	Operation	Time (s)	Potential value (CNY)	Potential Environmental Impact(10^{-3} Pt)
1	Unscrew and remove base part	86.4	0.02244	2.54
2	Unscrew and remove cover plate	86.4	0.26188	10.66
3	Remove back cover part	43.2	0.1248	6.92
4	Disassemble back cover part	21.6	6.0184	328.9
5	Remove wire with pin	86.4	0	0
6	Remove power switch part	43.2	0	0
7	Remove control button part	43.2	0	0
8	Unscrew and remove main board	129.6	0.024	2.24
9	Unscrew and remove loudspeaker part	86.4	0.0132	1.49
10	Unscrew and remove power supply board	86.4	0.15926	5.84
11	Unscrew and remove metal support	86.4	1.22364	138.31
12	Unscrew	86.4	0.20592	23.27
13	Remove loudspeaker	43.2	0	0
14	Remove Remote control receiver board	21.6	0	0
15	Separate surface frame and LCD screen	21.6	7.92736	591.52
16	Disassemble power switch part	64.8	0	0

(continued)

Table 10.3 (continued)

#	Operation	Time (s)	Potential value (CNY)	Potential Environmental Impact(10^{-3} Pt)
17	Disassemble loudspeaker part	64.8	0.64284	72.66
18	Disassemble base part	86.4	0.24684	27.9
19	Disassemble brace part	86.4	0.58084	42.64
20	Disassemble seat part	64.8	0.33	37.3

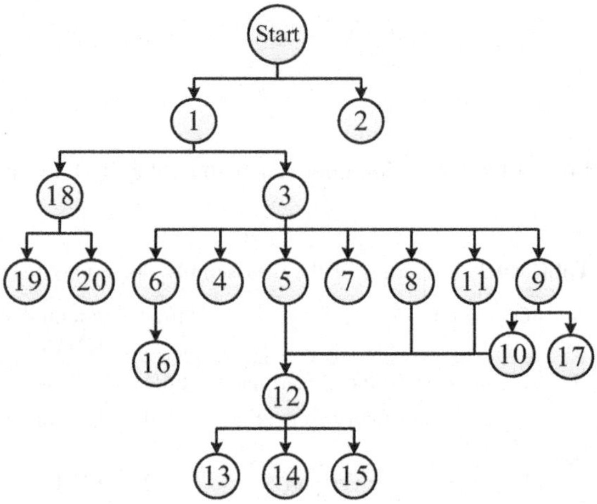

Fig. 10.15 Disassembly precedence graph

management authority and ineffectiveness in tracing information for reused products are eliminated.

Furthermore, when the information services of the category of product data, economic data, and ecological data are easily published and retrieved for sharing, technical data for supporting EOL treatment, including disassembly, recycling, and remanufacturing, can be analyzed and generated by EOL treatment industries. For instance, the disassembly operations and their properties for disassemble the studied LCD TV are given in Table 10.3 and the disassembly precedence graph (DPG) for these disassembly operations is illustrated in Fig. 10.15. In the DPG, the circled numbers refer to the operation sequence numbers listed in Table 10.3, and arcs represent precedence relationships between disassembly operations. According to this information, optimized disassembly sequence planning can be made. The detailed description of disassembly sequence planning optimization can be found in [6, 39]. The technical data for EOL treatment can also be easily shared in the proposed semantic information services framework. In this way, the previously mentioned ineffectiveness in information support for EOL treatments is eliminated.

10.6 Conclusions

This paper pointed out three main problems in current EEE/WEEE information sharing mechanism, including information bottleneck in WEEE management authority, ineffectiveness in information support for EOL treatments, and ineffectiveness in tracing information for reused products. The important and valuable product lifecycle data for sustainable WEEE management is studied and analyzed in the view of EOL treatment. Facing the challenge of enabling the interoperability and information sharing between the global distributed and heterogeneous information systems across the entire EEE/WEEE lifecycle efficiently, and targeting the cloud-based remanufacturing, this article proposed a semantic information services framework for sustainable WEEE management. In the framework, the ontology-based approached is used to integrate and represent the lifecycle data in information service providers. And the semantic information service management platform is proposed for the semantic annotation, advertisement, and retrieval of information services. A demonstration case study on waste LCD TV is used to illustrate the effectiveness and significance of the proposed framework.

In the future, more advanced technologies of cloud manufacturing will be introduced to the file of remanufacturing, e.g., cloud-based remanufacturing planning, cloud-based disassembly sequence planning, and cloud-based reverse logistics planning. A full-featured cloud-based remanufacturing for sustainable management of WEEE will be achieved.

References

1. Robinson, B. H. (2009). E-waste: An assessment of global production and environmental impacts. *Science of the Total Environment, 408*(2), 183–191.
2. Georgiadis, P., & Besiou, M. (2008). Sustainability in electrical and electronic equipment closed-loop supply chains: A system dynamics approach. *Journal of Cleaner Production, 16* (15), 1665–1678.
3. Linke, B. S. (2013). Sustainability indicators for grinding applied to dressing strategies. *Journal of Manufacturing Science and Engineering, 135*(5), 054502.
4. Haapala, K. R., Zhao, F., Camelio, J., Sutherland, J. W., Skerlos, S. J., Dornfeld, D. A., et al. (2013). A review of engineering research in sustainable manufacturing. *Journal of Manufacturing Science and Engineering, 135*(4), 041013.
5. Zeng, X., Li, J., Stevels, A. L. N., & Liu, L. (2013). Perspective of electronic waste management in china based on a legislation comparison between China and the EU. *Journal of Cleaner Production, 51,* 80–87.
6. Li, W. D., Xia, K., Gao, L., & Chao, K.-M. (2013). Selective disassembly planning for waste electrical and electronic equipment with case studies on liquid crystal displays. *Robotics and Computer-Integrated Manufacturing, 29*(4), 248–260.
7. Gardner, S. P. (2005). Ontologies and semantic data integration. *Drug Discovery Today, 10* (14), 1001–1007.
8. Panetto, H., Dassisti, M., & Tursi, A. (2012). Onto-pdm: Product-driven ontology for product data management interoperability within manufacturing process environment. *Advanced Engineering Informatics, 26*(2), 334–348.

9. Song, F., Zacharewicz, G., & Chen, D. (2013). An ontology-driven framework towards building enterprise semantic information layer. *Advanced Engineering Informatics, 27*(1), 38–50.
10. Chang, X., & Terpenny, J. (2009). Ontology-based data integration and decision support for product e-Design. *Robotics and Computer-Integrated Manufacturing, 25*(6), 863–870.
11. Yang, S.-Y. (2013). A novel cloud information agent system with web service techniques: Example of an energy-saving multi-agent system. *Expert Systems with Applications, 40*(5), 1758–1785.
12. McIlraith, S. A., Tran, Cao S., & Honglei, Z. (2001). Semantic web services. *IEEE Intelligent Systems, 16*(2), 46–53.
13. Breslin, J. G., O'Sullivan, D., Passant, A., & Vasiliu, L. (2010). Semantic web computing in industry. *Computers in Industry, 61*(8), 729–741.
14. Sycara, K., Paolucci, M., Ankolekar, A., & Srinivasan, N. (2003). Automated discovery, interaction and composition of semantic web services. *Web Semantics: Science, Services and Agents on the World Wide Web, 1*(1), 27–46.
15. Puttonen, J., Lobov, A., Soto, M. A. C., & Lastra, J. L. M. (2010). A semantic web services-based approach for production systems control. *Advanced Engineering Informatics, 24*(3), 285–299.
16. Vouros, G. A., Papasalouros, A., Tzonas, K., Valarakos, A., Kotis, K., Quian-Ruiz, J.-A., et al. (2010). A semantic information system for services and traded resources in grid e-markets. *Future Generation Computer Systems, 26*(7), 916–933.
17. Seliger, G., Buchholz, A., & Kross, U. (2003). Enhanced product functionality with life cycle units. *Proceedings of the Institution of Mechanical Engineers Part B-Journal of Engineering Manufacture, 217*(9), 1197–1202.
18. Yang, X., Moore, P., & Chong, S. K. (2009). Intelligent products: From lifecycle data acquisition to enabling product-related services. *Computers in Industry, 60*(3), 184–194.
19. Kim, H. J., Ciupek, M., Buchholz, A., & Seliger, G. (2006). Adaptive disassembly sequence control by using product and system information. *Robotics and Computer-Integrated Manufacturing, 22*(3), 267–278.
20. Parlikad, A. K., & McFarlane, D. (2010). Quantifying the impact of AIDC technologies for vehicle component recovery. *Computers & Industrial Engineering, 59*(2), 296–307.
21. Yang, X., Moore, P. R., Wong, C.-B., & Pu, J.-S. (2007). A component-based software framework for product lifecycle information management for consumer products. *IEEE Transactions on Consumer Electronics, 53*(3), 1195–1203.
22. Lambert, S., Riopel, D., & Abdul-Kader, W. (2011). A reverse logistics decisions conceptual framework. *Computers & Industrial Engineering, 61*(3), 561–581.
23. Li, W. D., Chao, K.-M., Jin, G., Xia, K., Gao, L. (2012). Sustainable information management for waste electrical and electronic equipment. In *Proceedings of 2012 IEEE 16th International Conference on Computer Supported Cooperative Work in Design (CSCWD)* (pp. 875–881).
24. Huang, G. Q., Lau, J. S. K., & Mak, K. L. (2003). The impacts of sharing production information on supply chain dynamics: A review of the literature. *International Journal of Production Research, 41*(7), 1483–1517.
25. Huang, G. Q., Lau, J. S. K., Mak, K. L., & Liang, L. (2005). Distributed supply-chain project rescheduling: Part I-Impacts of information-sharing strategies. *International Journal of Production Research, 43*(24), 5107–5129.
26. Parlikad, A. K., & McFarlane, D. (2007). RFID-based product information in end-of-life decision making. *Control Engineering Practice, 15*(11), 1348–1363.
27. Lin, Q., Xia, K., Wang, L., & Gao, L. (2014). Cloud manufacturing in China: A literature survey. *International Journal of Manufacturing Research, 9*(4), 369–388.
28. Krill, M., & Thurston, D. L. (2005). Remanufacturing: Impacts of sacrificial cylinder liners. *Journal of Manufacturing Science and Engineering, 127*(3), 687–697.
29. Xia, K., Gao, L., Li, W. D., Wang, L. H., Chao, K.-M. (2014). A q-learning based selective disassembly planning service in the cloud based remanufacturing system for WEEE. In

Proceedings of ASME 2014 International Manufacturing Science and Engineering Conference (pp. V001T04A012–V001T04A012).

30. Wang, L., Wang, X. V., Gao, L., & Vncza, J. (2014). A cloud-based approach for weee remanufacturing. *CIRP Annals-Manufacturing Technology, 63*(1), 409–412.
31. Wang, X. V., Wang, L. H., Gao, L. (2013). From cloud manufacturing to cloud remanufacturing: A cloud-based approach for WEEE. In *Proceedings of 2013 IEEE 10th International Conference on e-Business Engineering (ICEBE)* (pp. 399–406).
32. Christensen E., Curbera F., Meredith G., Weerawarana S., 2001. *Web services description language (WSDL) 1.1*. Technical report, W3C Note.
33. Martin, D., Burstein, M., Hobbs, J., Lassila, O., McDermott, D., McIlraith, S., et al. (2004). *OWL-S: Semantic markup for web services*. Technical report, W3C Member Submission.
34. McGuinness, D. L., Harmelen, F. V. (2004). *Owl web ontology language*. Technical report, W3C Recommendation.
35. Das, S., Sundara, S., Cyganiak, R. (2012). *R2RML: RDB to RDF mapping language*. Technical report, W3C Recommendation.
36. Cyganiak, R., Bizer, C., Garbers, J., Maresch, O., Becker, C. (2012). *The D2RQ mapping language*, Technical report.
37. Prud'hommeaux, E., Seaborne, A. (2008). *SPARQL query language for RDF*. Technical report, W3C Recommendation.
38. Feigenbaum, L., Williams, G. T., Clark, K. G., Torres, E. (2013). *SPARQL 1.1 protocol*. Technical report, W3C Recommendation.
39. Xia, K., Gao, L., Li, W. D., & Chao, K.-M. (2014). Disassembly sequence planning using a simplified teaching-learning-based optimization algorithm. *Advanced Engineering Informatics, 28*(4), 518–527.

36. [illegible] (2012). *OWL 2 Web Ontology Language* [illegible] Technical report, W3C [illegible]
38. [illegible] (2005). [illegible] Technical report, W3C [illegible]
39. [illegible] (2005). [illegible] Technical report, W3C Member Submission.
41. Farrell, J., Lausen, H. [illegible] (2007). *SAWSDL*. Technical report, W3C Recommendation.
42. [illegible] (2016). [illegible] service-based [illegible] *Procedia CIRP* [illegible]

Chapter 11
Selective Disassembly Planning for Waste Electrical and Electronic Equipment with Case Studies on Liquid Crystal Displays

W. D. Li, K. Xia, L. Gao and K. M. Chao

Abstract Waste Electrical and Electronic Equipment (WEEE) is one of the most significant waste streams in modern societies. In the past decade, disassembly of WEEE to support remanufacturing and recycling has been growingly adopted by industries. With the increasing customization and diversity of Electrical and Electronic Equipment (EEE) and more complex assembly processes, full disassembly of WEEE is rarely an ideal solution due to high disassembly cost. Selective disassembly, which prioritizes operations for partial disassembly according to the legislative and economic considerations of specific stakeholders, is becoming an important but still challenging research topic in recent years. In order to address the issue effectively, in this chapter, a particle swarm optimization (PSO)-based selective disassembly planning method embedded with customizable decision-making models and a novel generic constraint handling algorithm has been developed. With multi-criteria and adaptive decision-making models, the developed method is flexible to handle WEEE to meet the various requirements of stakeholders. Based on the generic constraint handling and intelligent optimization algorithms, the developed research is capable to process complex constraints and achieve optimized selective disassembly plans. Industrial cases on liquid crystal display (LCD) televisions have been used to verify and demonstrate the effectiveness and robustness of the research in different application scenarios.

Keywords Disassembly planning · Waste Electrical and Electronic Equipment
Electrical and Electronic Equipment · Particle swarm optimization
Liquid crystal display

W. D. Li (✉) · K. M. Chao
Faculty of Engineering, Environment and Computing, Coventry University, Coventry, UK
e-mail: weidong.li@coventry.ac.uk

K. Xia · L. Gao
School of Mechanical Engineering and Sciences, Huazhong University of Science and Technology, Wuhan, China

W. Li and S. Wang (eds.), *Sustainable Manufacturing and Remanufacturing Management*, https://doi.org/10.1007/978-3-319-73488-0_11

11.1 Introduction

The mounting demand for new products has brought more production activities worldwide in recent years. The rapid development, however, has been hindered by the increasing concerns of the scarcity of natural resources and environmental issues. Statistics show that from 1985 the resource consumption on the global level has been higher than the ecological capability of the Earth. It has been estimated that the required bio-capacity of two Earths is necessary to satisfy the need of the development in 2050 according to current production and consumption trends [1]. On the other hand, more and more products after services are filled up in landfills. Among them, Electrical and Electronic Equipment (EEE) after services, that is, Waste Electrical and Electronic Equipment (WEEE), is becoming one of the major and challenging waste streams in terms of quantity and toxicity. For instance, there are approximately seven million tons of WEEE generated in Europe per year [2]. In China, 1.1 million tons of WEEE are generated per year [3]. Due to the rapid technical innovations and shorter usage lifecycle of EEE, WEEE is growing much faster than any other municipal waste streams. In order for the Earth to be cleaner, end-of-life (EoL) recovery strategies are critical to shape the future of WEEE lifecycle management patterns. Among the strategies, remanufacturing is viewed as a "hidden green giant" and attracting escalating attentions of researchers and practitioners [4–7]. Remanufacturers seek to bring some components of products after their services back into "as new" conditions by carrying out necessary disassembly, overhaul, and/or repairing operations for reuse to extend lifecycles. There are two driving forces for industries in adopting the relevant technologies and practices, i.e., stricter legislative pressure for environmental protection and better profit margins from remanufacturing. The explanations are expanded below.

- The WEEE Directive has been enacted and implemented from 2003 in Europe, and the equivalent directives have been developed in different countries of the world. Further proposals for the tighter WEEE Directives have been suggested to regulation bodics with an aim to make products and components after services more recyclable, reusable, and remanufactureable. According to the WEEE Directives, a producer (manufacturer, brand owner, or importer)'s responsibility is extended to the post-consumer stage of WEEE, instead of stopping at selling and maintenance (i.e., Extended Producer Responsibility—EPR [8, 9]). The EPR is aimed at encouraging producers especially manufacturers to provide cradle-to-grave support to reduce environmental impacts, such that they work closely with remanufacturing industries to recover maximum values and reduce environmental toxicity/hazardousness. For instance, the remanufacturing legislative initiatives are underway in the EU and USA to ensure Original Equipment Manufacturers (OEMs) and suppliers to provide free access to remanufacturing information facilities in global chains [10].
- Good remanufacturing planning and management can effectively balance economic and environmental targets and close gaps between the shorter innovation cycles of EEE and the extended lives of components of WEEE.

Remanufacturing industries in the EU and worldwide have been recently growing quickly because of better economic return values. There are a number of successful cases in industries, including single-use cameras (Eastman Kodak and Fuji Film), toner cartridges (Xerox), personal computers (IBM, HP, Toshiba, Reuse network—Germany), photocopiers (Fuji Xerox—Australia, Netherlands, and UK), commercial cleaning equipment (Electrolux), washing machines (ENVIE—France), mobile phones (Nokia, ReCelluar—USA, Greener solution—UK).

Disassembly planning, which is used to determine sensible disassembly operations and sequencing, is critical in remanufacturing. Effective disassembly planning can significantly improve the recycling and reuse rates of components and materials from WEEE to ensure maximum value recovery. For a set of WEEE, there could be a number of different sequences of disassembly operations constrained technically and geometrically between the components of the WEEE, leading to the different decision-making models according to the perspectives and criteria of stakeholders [11]. As thus, it becomes difficult for remanufacturers to solely depend upon their experiences to plan disassembly operations so as to recover a larger proportion of components and fulfill environmental targets at a reasonable cost. In the past years, research has been carried out to address the issues of disassembly. The previous research can be generally summarized as the following two categories:

- Disassembly for design. Disassembly approaches for EEE such as consumer electronic products have been developed to use smart materials like shape-memory polymers (SMPs) in the design of embedded releasable fasteners to facilitate the disassembly processes of the products [12–17]. Design for remanufacturing/disassembly principles have been spread among Japanese manufacturers since products with the principles are more profitable in this context than those that were not designed with this purpose [5, 18, 19].
- Disassembly planning and operation sequencing. Typical disassembly operations based on manual, semi-automatic, and automatic processes and the associated tool kits were summarized [5]. Based on disassembly operations and the precedence constraint relationships among the disassembly operations, sequencing rules and intelligent and/or meta-heuristic reasoning algorithms were applied to deduce an optimal plan from a large pool of candidate solutions [11, 20–22]. In recent years, remanufacturers are facing many challenges to disassemble WEEE due to their high customization and diversity, high integration level, and more complex assembly processes. Current economic analyses have demonstrated that full disassembly is rarely an optimal solution and necessary owing to high disassembly cost. Selective disassembly, which prioritizes operations to implement partial dismantling of WEEE so as to take account of the legislative and economic considerations and meet the specific requirements of stakeholders, is a promising alternative and has therefore become a new research trend [5, 23, 24].

Attributing to booming personalized and mass-customized EEE, there is still challenging to apply the developed methods to the increasingly diversified and personalized WEEE to make sensible decisions and meet different stakeholders' perspectives. In this chapter, a particle swarm optimization (PSO)-based selective disassembly planning method with customizable decision-making models and a novel constraint handling algorithm has been developed. The method is adaptive to various types of WEEE, flexible for customized decision modeling and making for different stakeholders, and capable for handling complex constraints and achieving optimized solutions during disassembly planning. Industrial cases on liquid crystal display (LCD) televisions have been used to verify and demonstrate the effectiveness of the developed method in different application scenarios.

11.2 Selective Disassembly Planning Approach

11.2.1 Customizable Decision-Making Modeling for Selective Disassembly

Disassembly of WEEE involves different stakeholders, such as environmental regulators and remanufacturers. The different levels of targets will lead them to adopt or develop different decision-making models. For instance, according to the WEEE Directive, WEEE regulators will check whether remanufacturing companies are able to recycle at least 75% of WEEE by weight and remove/recover all the hazardous materials. In other words, at least 75% of WEEE are required to be dismantled to a component level, and all the components containing hazardous materials need to be taken apart from WEEE for further recycling and processing. Apart from fulfilling these fundamental environmental targets, remanufacturers would also improve the economic efficiency by prioritizing valued components during disassembly. In Fig. 11.1, an example of LCD WEEE is used to illustrate the above scenario.

To develop a selective disassembly planning method that is suitable for stakeholders to process various types of WEEE and meet their specific requirements, it is imperative to define customizable decision-making models. The models (disassembly indices and objective) developed in this research are described below.

Disassembly Indices

In the following formulas, several symbols will be used frequently and they are explained here first.

n	The number of the total disassembly operations in a plan of a set of WEEE
m	The number of the disassembly operations in a selective disassembly plan

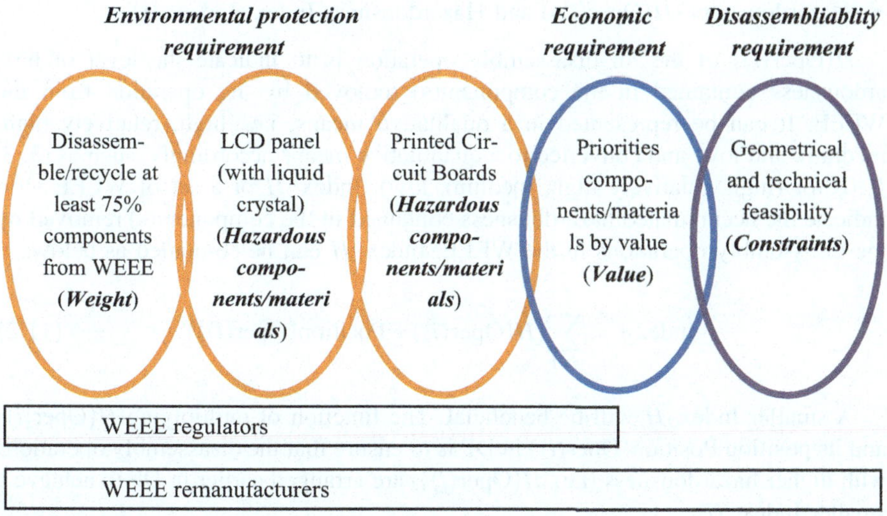

Fig. 11.1 Criteria used to develop different decision-making models to address various users' needs

Position(Oper(i)) The position (sequence) of the ith disassembly operation in a disassembly plan

- Selective Disassembly Plan (DP) and Disassembly Operation (Oper(i))

A set of WEEE can be fully disassembled using a disassembly plan. The number of all the operations in the plan is n. A selective disassembly plan (DP) consists of a set of disassembly operations, which is a part of the above complete operations. The number of the selected operations is m, and the ith operation is denoted as Oper(i). DP can be represented as:

$$\mathrm{DP} = \mathop{Y}_{i=1}^{m} (\mathrm{Oper}(i), \mathrm{Position}(\mathrm{Oper}(i))) \tag{11.1}$$

where Y represents the set of disassembly operations, and $m \leq n$.

For instance, there are a set of disassembly operations Oper(1), Oper(2), Oper(3), Oper(4), and their positions in DP are (e.g., Position(Oper(1) = 4) , so that the sequence of the operations in DP is Oper(3), Oper(2), Oper(4), Oper(1).

Meanwhile, Oper(i) has some properties related to the environmental and economic targets defined as follows.

- Hazardousness ($H(\text{Oper}(i))$) and Hazardousness Index (Index_*H*)

$H(\text{Oper}(i))$ of the *i*th disassembly operation is to indicate the level of hazardousness contained in the component(s) removed by the operation from the WEEE. It can be represented in a qualitative means, i.e., high, relatively high, medium, and low, and converted to a quantitative means accordingly, such as (5, 3, 1, 0) for (high, relatively high, medium, low). Index_*H* of a set of WEEE is to indicate the accumulated hazardousness contained in the component(s) removed by the disassembly operations in the WEEE. Index_*H* can be computed as below:

$$\text{Index}_H = \sum_{i=1}^{m} (H(\text{Oper}(i)) * \text{Position}(\text{Oper}(i))) \tag{11.2}$$

A smaller Index_*H* will be beneficial. The function of multiplying $H(\text{Oper}(i))$ and its position $\text{Position}(\text{Oper}(i))$ in DP is to ensure that the disassembly operations with higher hazardousness (i.e., $H(\text{Oper}(i))$) are arranged earlier in DP to achieve a smaller Index_*H*.

For instance, the hazardousness of $\text{Oper}(1), \text{Oper}(2), \text{Oper}(3), \text{Oper}(4)$ are high, low, medium, relatively high, respectively, which can be converted to (5, 0, 1, 3). The positions of the operations in DP are (4, 2, 1, 3). Therefore, the hazardousness index of DP is (5 * 4 + 0 * 2 + 1 * 1 + 3 * 3) = 30. If the positions of the operations are rearranged as (1, 4, 3, 2), then the hazardousness index is (5 * 1 + 0 * 4 + 1 * 3 + 3 * 2) = 14. The latter is lower than the earlier since the operations with higher hazardousness are arranged earlier in the latter. In objective defined later on, a weighted minimum hazardousness index will be pursued to ensure the operations to remove the most hazardous components will be arranged as early as possible to improve the efficiency of hazardousness removal in a selective disassembly plan.

- Potential Recovery Value ($V(\text{Oper}(i))$), Disassembly Time ($T(\text{Oper}(i))$) and Potential Value Index (Index_*V*)

$V(\text{Oper}(i))$ of the *i*th disassembly operation is to indicate the potential recovery value of the component(s) disassembled from the WEEE by the operation. The disassembled component(s) could be reusable so that $V(\text{Oper}(i))$ can be represented as the depreciation value of the equivalent new component(s). $T(\text{Oper}(i))$ represents the time spent for the disassembly operation $\text{Oper}(i)$. Index_*V* of a set of WEEE is to indicate the accumulated potential value index by the disassembly operations in the WEEE. Index_*V* can be computed as below:

$$\text{Index\_}V = \sum_{i=1}^{m} (V(\text{Oper}(i))/T(\text{Oper}(i)) * \text{Position}(\text{Oper}(i))) \tag{11.3}$$

A smaller Index_*V* will be beneficial. $V(\text{Oper}(i))/T(\text{Oper}(i))$ represents the potential value recovery efficiency of $\text{Oper}(i)$. The function of multiplying $V(\text{Oper}(i))/T(\text{Oper}(i))$ and its position $\text{Position}(\text{Oper}(i))$ in DP is to ensure that the disassembly operations with higher $V(\mathrm{Oper}(i))/T(\mathrm{Oper}(i))$ are arranged earlier to achieve a smaller Index_V so as to achieve a higher efficiency of potential value recovery for a selective disassembly plan.

- Weight Removal ($W(\mathrm{Oper}(i))$) and Weight Removal Index (Index_W)

$W(\mathrm{Oper}(i))$ is to indicate the level of the removed weight by the ith disassembly operation from the WEEE. It can be represented by the weight of the component(s) disassembled by the operation. Index_W of a set of WEEE is to indicate the accumulated weight removal index by the disassembly operations in the WEEE. Index_W can be computed as below:

$$\mathrm{Index\_}W = \sum_{i=1}^{m} (W(\mathrm{Oper}(i)) * \mathrm{Position}(\mathrm{Oper}(i))) \quad (11.4)$$

Similarly, a smaller Index_W will be beneficial. The function of multiplying $W(\mathrm{Oper}(i))$ and its position $\mathrm{Position}(\mathrm{Oper}(i))$ in DP is to ensure that the disassembly operations with higher $W(\mathrm{Oper}(i))$ are arranged earlier to achieve a smaller Index_W in order to improve the efficiency of weight removal in a selective disassembly plan.

Disassembly Constraints

During the process of disassembly, there are some geometrical or technical constraints to specify precedent relationships between disassembly operations. Three examples in Fig. 11.2 are used to illustrate the concept.

In (a) and (b), there are two disassembly directions for Components A and B, respectively. Due to the geometrical blocking relationship, the first disassembly operation (denoted as Oper(1)) is to disassemble the joining mechanism associated with Component B and housing, and the second disassembly operation (Oper(2)) is to dismantle the joining mechanism between Components A and B. Therefore, Oper(1) is constrained to be prior to Oper(2) geometrically.

In (c), there is a single disassembly direction for Components A and B. Geometrically, it can dismantle either the joining mechanism between

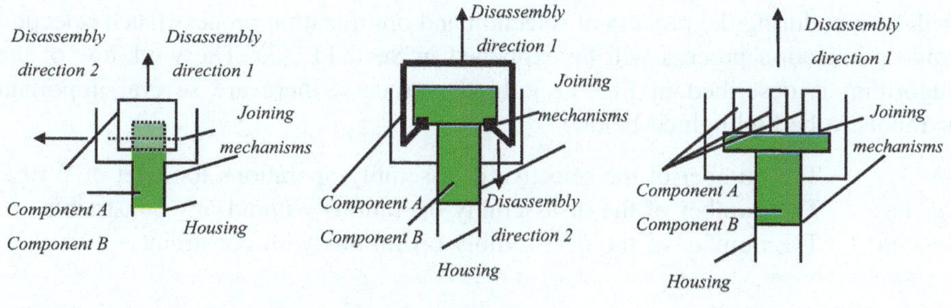

Fig. 11.2 Examples of constraints during disassembly

Component B and housing first (Oper(1)) or the joining mechanism between Components A and B first (Oper(2)) first. However, from the technical point of view, it is recommended to remove the joining mechanism between Component B and housing first, considering that the disassembly of the second joining mechanism needs more operation space. Therefore, Oper(1) is constrained to be prior to Oper(2) technically.

Decision-Making Objective

Disassembly decision-making will be modeled as a constraint-based optimization problem. The objective can be customized to address the different requirements of stakeholders through providing weight setting by users. The objective is represented below:

$$\begin{aligned}\text{Minimis}(\text{Index\_}H, \text{Index\_}V, \text{Index\_}W) = \text{Minimise}(&\omega_1 * \text{Index\_}H \\ &+ \omega_2 * \text{Index\_}V + \omega_3 * \text{Index\_}W)\end{aligned} \tag{11.5}$$

where $\omega_1 - \omega_3$ are the weights. The setting of weights can be used to reflect importance. A higher weight means more attentions will be paid to that index, and a zero value means such the index will not be considered. In order to rationalize the model, the three indices are required to be normalized to be in the same measurement scale. The late case studies can illustrate the normalization process.

11.2.2 A Generic Constraint Handling Algorithm

There could be a number of precedence constraints between the disassembly operations for a set of WEEE. Under the situation, it is usually difficult to generate a valid disassembly plan. In order to address complex constraints in WEEE disassembly dynamically and adaptively, a new constraint handling algorithm, which employs a generic process to handle various constraints, has been developed. The manipulation operations of the algorithm, which are based on data structure and double-linked list design, can ensure that all the constraints in a disassembly plan will be met during the process of selection and optimization process (such selection and optimization process will be explained in Sect. 11.2.3). The workflow of the algorithm is described in Fig. 11.3. In the process, there are several important symbols to be highlighted below.

m	The number of the selective disassembly operations for a set of WEEE
m_1	The number of the disassembly operations without any constraints
$m - m\_1$	The number of the disassembly operations with constraints

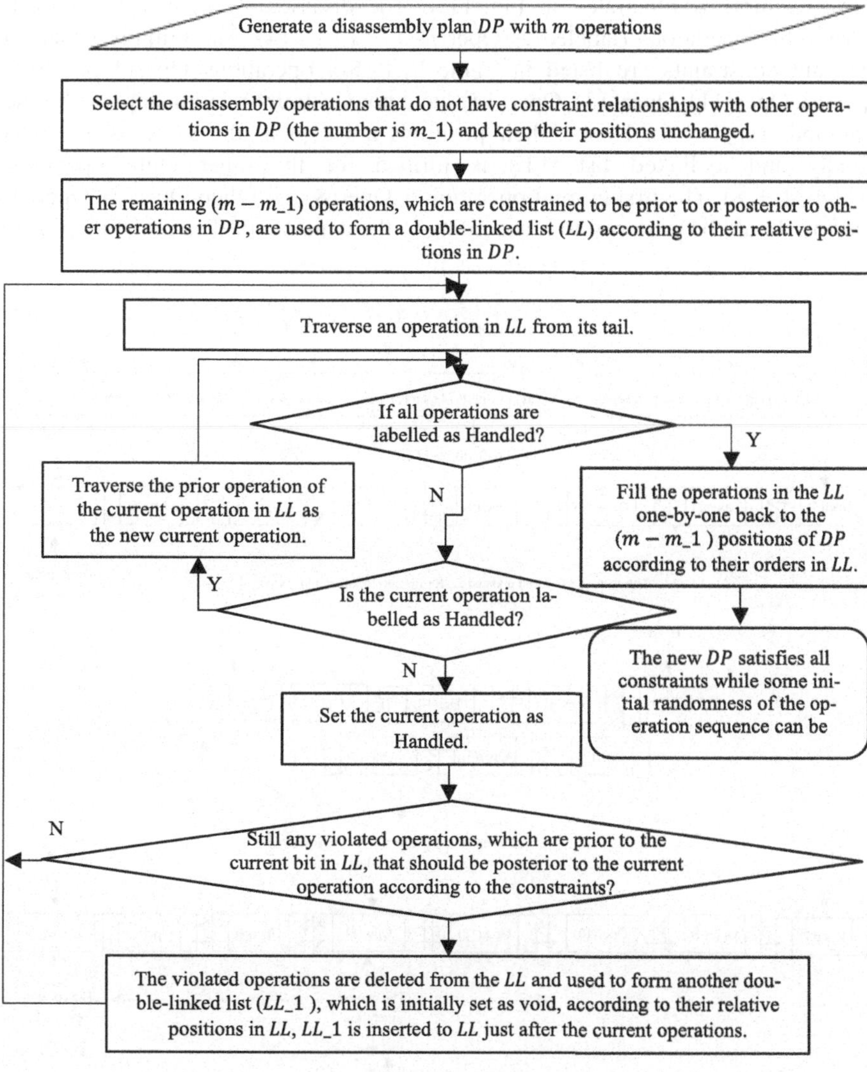

Fig. 11.3 Workflow of a generic constraint handling algorithm

LL	A double-linked list for the disassembly operations with constraints
LL_1	A double-linked list to store immediate results during the algorithm manipulation
Current operation	The working operation during the manipulation of the algorithm

An example is illustrated in Fig. 11.4. The disassembly plan consists of 14 disassembly operations (labeled as Oper(i), $i = 1, \ldots, 14$). The initially generated plan and constraints are listed in Table 11.1. Six operations Oper(1), Oper(4), Oper(6), Oper(11), Oper(13), Oper(14) have no constraint relationships with other operations ($m\_1 = 6$). Hence, their positions are kept as the same as generated initially and a linked list (LL) is formed for the other eight operations ($m - m\_1 = 8$). The first current operation is Oper(8), and then Oper(3), Oper(9), Oper(5) should be posterior to it according to the constraints. The updating process

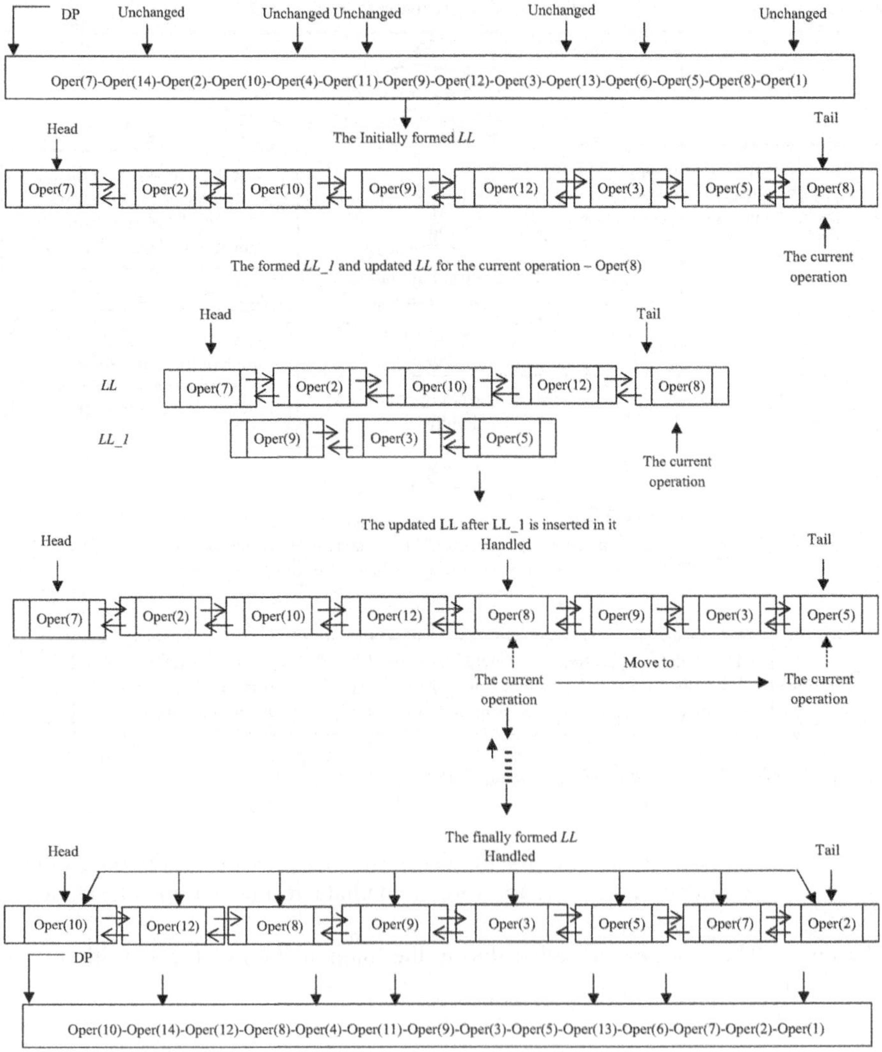

Fig. 11.4 An example process of the generic constraint handling algorithm

Table 11.1 Constraints in a disassembly plan

Initial process plan	Oper(7)-Oper(14)-Oper(2)-Oper(10)-Oper(4)-Oper(11)-Oper(9)-Oper(12)-Oper(3)-Oper(13)-Oper(6)-Oper(5)-Oper(8)-Oper(1)
Constraint 1	Oper(5) and Oper(9) should be prior to Oper(2) and Oper(7)
Constraint 2	Oper(12) and Oper(8) should be prior to Oper(3), Oper(5) and Oper(9)
Constraint 3	Oper(3) should be prior to Oper(5)
Constraint 4	Oper(10) should be prior to Oper(7)

of LL is illustrated. After Oper(8) has been handled, the reference to the current operation is moved to the tail and the same procedure is continued until all operations are assigned as handled. The final updated disassembly plan satisfies all the constraints.

11.2.3 Improved Particle Swarm Optimization Algorithm

The different selection and optimization sequencing of disassembly operations for a set of WEEE usually brings forth a large search space. Conventional algorithms are often incapable of optimizing the problem. To address it effectively, some modern optimization algorithms, such as genetic algorithm (GA) and simulated annealing (SA), have been developed to quickly identify an optimized solution in a large search space through some evolutional or heuristic strategies. In this research, an improved algorithm based on a modern intelligent algorithm, i.e., PSO, has been applied to facilitate the search process. Moreover, the improved PSO has been also compared with GA and SA for this disassembly planning problem to show the characteristics of the algorithms. More details of GA and SA implementation can refer to [25, 26].

A classic PSO algorithm was inspired by the social behavior of bird flocking and fish schooling [27]. Three aspects will be considered simultaneously when an individual fish or bird (particle) makes a decision about where to move: (1) its current moving direction (velocity) according to the inertia of the movement; (2) the best position that it has achieved so far; and (3) the best position that all the particles have achieved so far. In the algorithm, the particles form a swarm, and each particle can be used to represent a potential disassembly plan of a problem. The velocity and position of a particle (disassembly plan) will be computed below.

$$V_i^{t+1} = w * V_i^t + c_1 * \mathrm{Rand}(1) * (P_i^t - X_i^t) + c_2 * \mathrm{Rand}(1) * \left(P_g^t - X_i^t\right) \quad (11.6)$$

$$X_i^{t+1} = X_i^t + V_i^{t+1} \quad (11.7)$$

$$X_i = (X_{i1}, X_{i2}, \ldots, X_{iN}) \quad (11.8)$$

$$V_i = (V_{i1}, V_{i2}, \ldots, V_{iN}) \quad (11.9)$$

Here, i is the index number of particles in the swarm; t is the iteration number; V and X are the velocity vector and the position vector of a particle, respectively. For an N-dimensional problem, V and X can be represented by N particle dimensions as Formulas (11.3) and (11.4) show. p_i is the local best position that the ith particle has achieved so far; p_g is the global best position that all the particles have achieved so far; w is the inertia weight to adjust the tendency to facilitate global exploration (smaller w) and the tendency to facilitate local exploration to fine-tune the current search area (larger w); Rand(1) returns a random number in [0, 1]; c_1 and c_2 are two constant numbers to balance the effect of p_i and p_g.

In each iteration, the position and velocity of a particle can be adjusted by the algorithm that takes the above three considerations into account. After a number of iterations, the whole swarm will converge at an optimized position in the search space. A classic PSO algorithm can be applied to optimize the disassembly planning models in the following steps:

(1) Initialization

- Set the size of a swarm, e.g., the number of particles "Swarm_Size" and the max number of iterations "Iter_Num".
- Initialize all the particles (a particle is a disassembly plan DP) in a swarm. Calculate the corresponding indices and objective of the particles according to formulas (11.1)–(11.5) (the result of the objective is called fitness here).
- Set the local best particle and the global best particle with the best fitness.

(2) Iterate the following steps until "Iter_Num" is reached.

- For each particle in the swarm, update its velocity and position values.
- Decode the particle into a disassembly plan in terms of new position values, and calculate the fitness of the particle. Update the local best particle and the global best particle if a lower fitness is achieved.

(3) Decode global best particle to get the optimized solution.

However, the classic PSO algorithm introduced above is still not effective in resolving the problem. There are two major reasons for it:

- Due to the inherent mathematical operators, it is difficult for the classic PSO algorithm to consider the different arrangements of operations, and therefore, the particle is unable to fully explore the entire search space.
- The classic algorithm usually works well in finding solutions at the early stage of the search process (the optimization result improves fast), but is less efficient during the final stage. Due to the loss of diversity in the population, the particles move quite slowly with low or even zero velocities and this makes it hard to

reach the global best solution. Therefore, the entire swarm is prone to be trapped in a local optimum from which it is difficult to escape.

To solve these two problems and enhance the capability of the classic PSO algorithm to find the global optimum, new operations, including crossover and shift, have been developed and incorporated in an improved PSO algorithm. Some modification details are depicted below.

(1) New operators in the algorithm

- Crossover. Two particles in the swarm are chosen as parent particles for a crossover operation. In the crossover, a cutting point is randomly determined, and each parent particle is separated as left and right parts of the cutting point. The positions and velocities of the left part of Parent 1 and the right part of Parent 2 are reorganized to form Child 1. The positions and velocities of the left part of Parent 2 and the right part of Parent 1 are reorganized to form Child 2.
- Shift. This operator is used to exchange the positions and velocities of two operations in a particle in a random position so as to change their relative positions in the particle.

(2) Escape method

During the optimization process, if the iteration number of obtaining the same best fitness is more than 10, then the crossover and shift operations are applied to the best particle to escape from the local optima.

A general diagram to show the above flow is shown in Fig. 11.5.

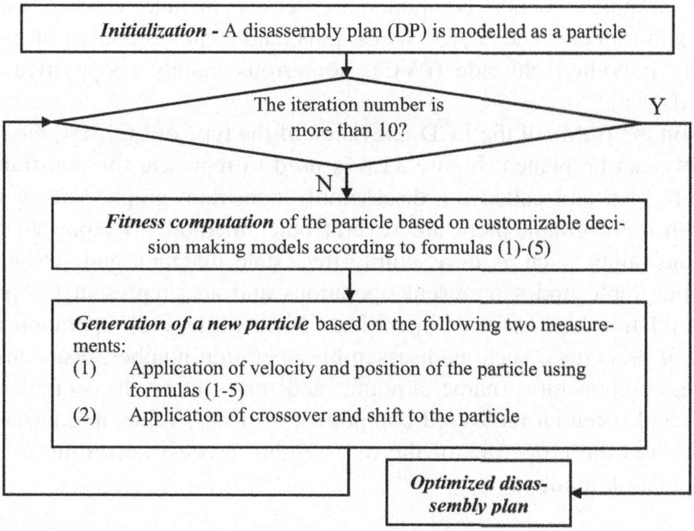

Fig. 11.5 General workflow of the PSO-based disassembly plan optimization

11.3 Case Studies in Selective Disassembly Planning

11.3.1 Background

Televisions can be generally classified into five groups: CRT, LCD, PDP, OLED, and RP. The LCD televisions have been developed quickly over the past decades, and they are now sharing the biggest market (e.g., the global market figures for the LCD televisions are forecasted to surpass $80 billion in 2012 [24]). A LCD television produces a black and colored image by selectively filtering a white light. The light is typically provided by a series of cold cathode fluorescent lamps (CCFLs) at the back of the screen, although some displays use white or colored LED. The LCD televisions studied here are produced by the Changhong Electronics Company, Ltd. from China, which is the biggest television producer in China. The company provides information about LCD televisions of the type of LC24F4, such as the Bill of Materials (BoMs), exploded view, mass of each part, and the detailed assembly processes. The structure of the LCD television is shown in Fig. 11.6a, b. The typical exploded view of a LCD television is shown in (c). As shown in (d), a LCD television is typically assembled by three main parts: front cover assembly part, back cover assembly part, and base assembly part. Among them, the front cover assembly part is composed of a surface frame, a remote control receiver board, a control button board, a main board, a power supply board, a low-noise block (LNB) converter board (optional), and a DVD ROM (optional). The mass of the LC24F4 LCD television is 5963.8 g, and the main component/material composition is shown in Fig. 11.7, in which the percentage is represented in terms of the ratio of mass. Among the component/material composition, the printed circuit boards (PCBs, which are mainly main boards and power supply boards) and LCD screens are quite complex. Other components/materials include cables, wires, pins, switches, and rubbers. The cables, wires, pins, and switches consist of plastics that are usually polyvinyl chloride (PVC), nonferrous mainly copper (Cu) and aluminum (Al).

Based on the BoMs of the LCD television of the type of LC24F4, the process of disassembly can be planed. Figure 11.8 is used to represent the constraints of the disassembly plan and called the disassembly constraint graph. Except the disassembly constraint graph, there are several other methods to represent the disassembly constraints, such as disassembly tree, state diagram, and AND/OR graph [28]. In the graph, nodes represent operations and arcs represent the precedence constraint relationships between operations. Meanwhile, each operation is defined with several properties, such as disassembly operation number, disassembly operation time, component(s) (name, amount, and mass) to be disassembled by each operation, and potential recovered component(s)' mass, value, and hazardousness. Table 11.2 lists the properties of the disassembly process according to the disassembly operation number.

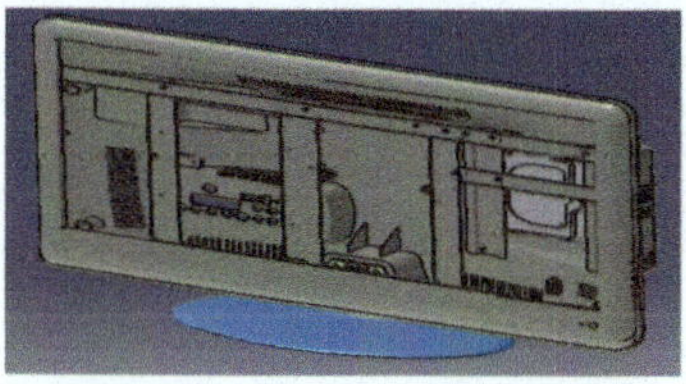

(a) Front view of the LCD television framework

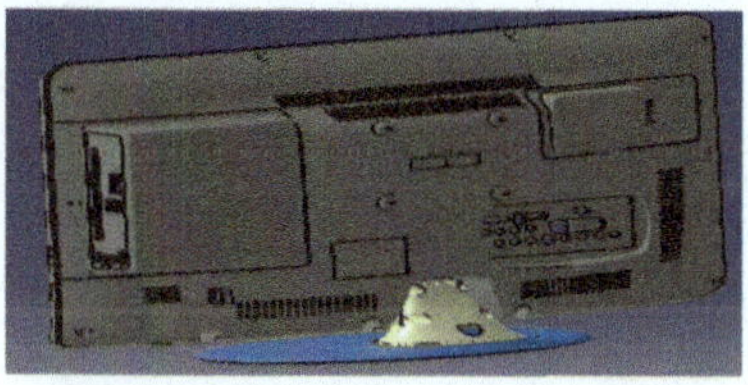

(b) Back view of the LCD television framework

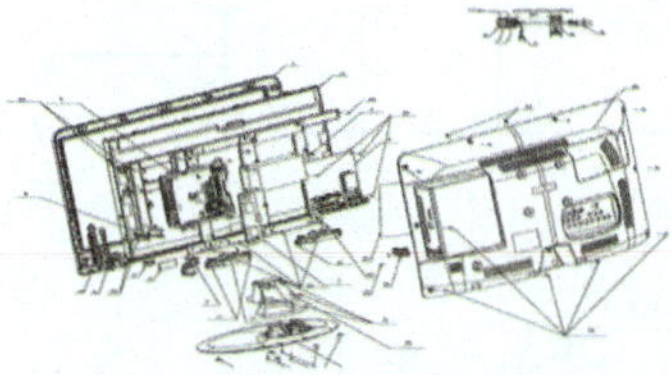

(c) Typical exploded view of the LCD television structure

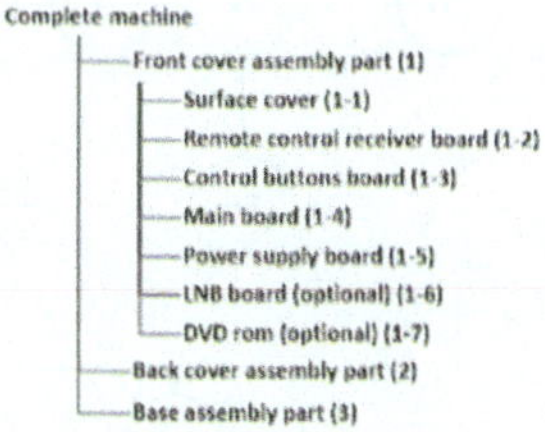

(d) Part of the BoMs of the LCD television

Fig. 11.6 LCD television and its structures

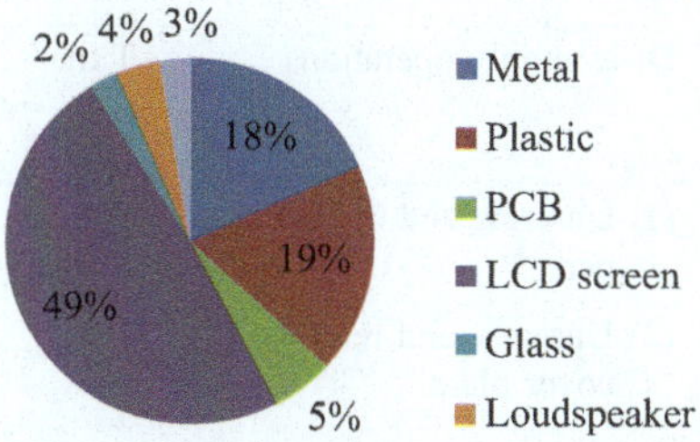

Fig. 11.7 Component/material composition of the LCD television

11.3.2 Selective Optimizations and Comparisons

An Initial Plan

According to the constraints, different disassembly plans can be created. One of these chosen is (1, 2, 3, 4, 5, 6, 7, 8, 9, 10, 11, 12, 13, 14, 15, 16, 17, 18, 19, 20). It physical disassembly process is shown in Fig. 11.9. This plan is called "an initial plan" to be used in the following scenarios for the comparisons with an optimized plan for a better understanding of the optimization process.

Scenario 1 for Selective Optimization

It is aimed to determine a selective optimization disassembly plan (part of the full disassembly plan) to meet the environmental protection targets (100%

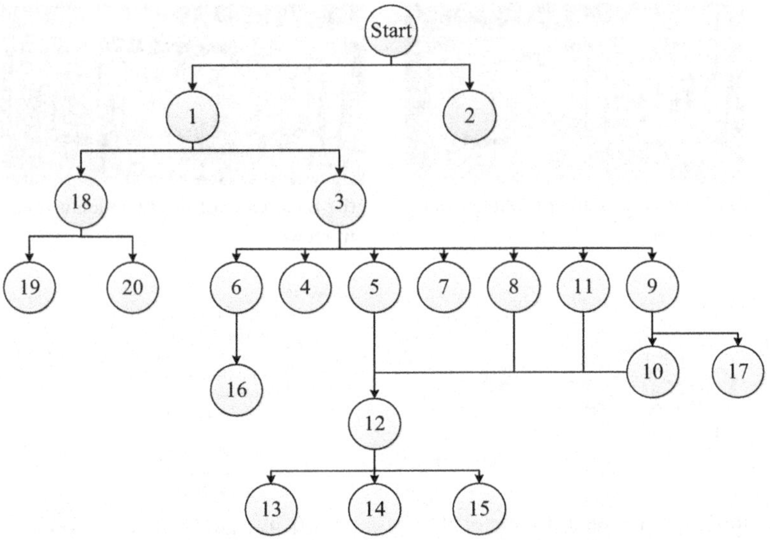

Fig. 11.8 Disassembly constraint graph

Table 11.2 Disassembly operations and some properties of the LCD television

Disassembly operations	Time (s)	Components	Mass (g)	Potential value (Yuan)	Hazardousness removal
(1) Unscrew and remove base part	86.4	Base part	1.8	0.0119	Low
		M4x12	1.6	0.0106	
(2) Unscrew and remove cover plate	86.4	4x10BTECh	11.2	0.0739	Low
		Cover plate	23.0	0.1840	
		3x10KTHCh	0.6	0.0004	
(3) Remove back cover part	43.2	Support structure	15.6	0.1248	Low
(4) Disassembly back cover part	21.6	Back cover	723.8	1.7904	Low
		Insulation board	25.0	0.2280	
(5) Remove wire with pin	86.4	Wire with pin	50.0	0.1000	Low
(6) Remove power switch part	43.2	Power switch part	5.0	0.0100	Low
(7) Remove control button part	43.2	Control button	3.7	0.0050	Low
		Control button part	5.5	0.0050	

(continued)

Table 11.2 (continued)

Disassembly operations	Time (s)	Components	Mass (g)	Potential value (Yuan)	Hazardousness removal
(8) Unscrew and remove main board	129.6	Main board	196.0	0.7908	Relatively high
		M3x8GB/T9074.4	3.0	0.0021	
		Insulating washer	3.0	0.0100	
(9) Unscrew and remove loudspeaker part	86.4	Loudspeaker part	60.0	1.3000	Low
		M3x8GB/T9074.4	2.0	0.0040	
(10) Unscrew and remove power supply board and insulating board	86.4	Power supply board	118.0	0.6466	Medium
		Insulating board	25.0	0.1520	
		M3x8GB/T9074.4	0.5	0.0033	
		M4x8GB/T9074.4	0.6	0.0004	
(11) Unscrew and remove metal support	86.4	Metal support	183.0	1.2078	Low
		M4x8GB/T818	2.4	0.0158	
(12) Unscrew	86.4	4x8BTHCh	7.2	0.0475	Low
		Clamping bush	24.0	0.1584	
(13) Remove loudspeaker	43.1	Loudspeaker	77.8	0.0600	Low
(14) Remove remote control receiver board	21.6	Remote control receive board	3.0	0.4000	Medium
(15) Separate surface frame and LCD screen	21.6	Surface frame	270.8	1.1000	High
		LCD screen	2900.0	9.6684	
		Metal mounting plate	639.0	1.2170	
(16) Disassemble power switch part	64.8	Power switch	5.0	0.0100	Low
		Power wire	75.5	0.1000	
		Wire with pin	5.0	0.0100	
(17) Disassemble loudspeaker part	64.8	Loudspeaker	152.0	0.6000	Low
		Support	95.0	0.0200	
		Washer	2.0	0.0070	
		4x8BTHCh	2.4	0.0158	

(continued)

Table 11.2 (continued)

Disassembly operations	Time (s)	Components	Mass (g)	Potential value (Yuan)	Hazardousness removal
(18) Disassemble base part	86.4	Metal washer 1	10.0	0.0660	Low
		Metal washer 2	10.0	0.0660	
		Metal fixing plate	15.0	0.0990	
		M4x12GB/T818	2.4	0.0158	
(19) Disassemble brace part	86.4	Metal support	25.0	0.1650	Low
		Plastic support 1	30.0	0.2400	
		Plastic support 2	20.0	0.1600	
		M4x12GB/T818	2.4	0.0158	
(20) Disassemble seat part	64.8	Toughened grass	150.0	0.3300	Low
		seat	50.0	0.0640	
		Steel plate	20.0	0.0200	
		Rubber gasket			

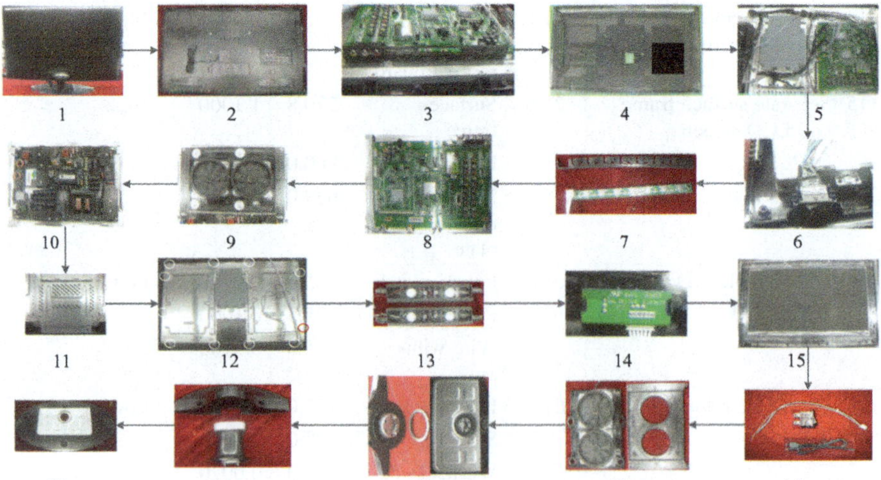

Fig. 11.9 A disassembly plan of the LCD television (an initial plan)

hazardousness removal and 75% component disassembled for the whole WEEE) and achieve the optimized potential recovery value (all the three weights in Formulas (11.5) were set 1). The input data is shown in Table 11.2.

In Fig. 11.10a, the disassembly planning selection and optimization process are shown. During the computation process, results were normalized; i.e., the index result of each operation was converted as the percentage of the overall results of all the operations. The results in the *Y*-axis were also accumulated for the operations.

The hazardousness removal, weight removal, and potential recovery value for the initial plan and an optimized plan are shown in (b)–(d), respectively. In (b), a 100% hazardousness removal target will be achieved after 13 disassembly operations for the optimized plan, In (c), a target to achieve 75% component disassembled by weight (of the total weight of the WEEE) took six operations for the optimized plan. In (d), the result of potential recovery value divided by spent time for each operation is shown, which is a target to achieve the most potential recovery

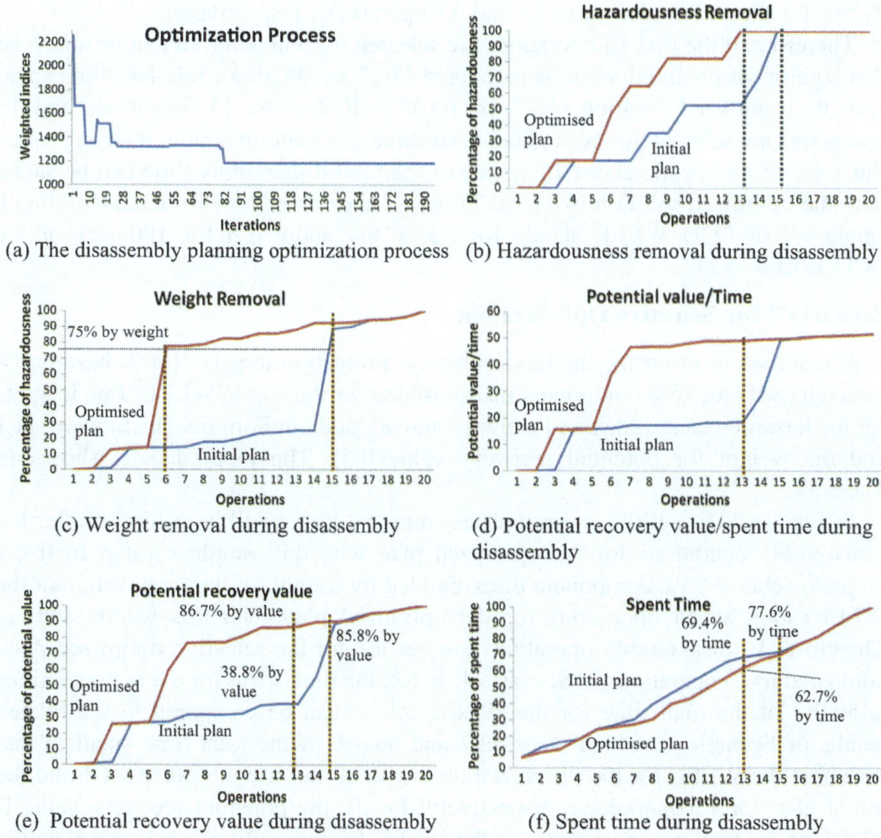

(a) The disassembly planning optimization process (b) Hazardousness removal during disassembly

(c) Weight removal during disassembly (d) Potential recovery value/spent time during disassembly

(e) Potential recovery value during disassembly (f) Spent time during disassembly

Fig. 11.10 Disassembly planning optimization with customizable decision-making models (all weights are 1)

value within the shortest time. To meet the environmental protection targets of removing 100% components with hazardous materials and 75% components by weight to be disassembled, the first 13 disassembly operations were selected from the optimized plan as the selective optimized plan. Meanwhile, the potential recovery value and spent time for this plan were optimized in this selective plan.

In (b) and (c), it can show that the initial plan will take 15 disassembly operations to achieve 100% hazardousness removal and also 15 operations for 75% components by weight to be disassembled. Therefore, 15 operations are necessary to achieve the environmental protection targets. Therefore, the optimized plan will have two less operations. The potential value/time in (d) can be separated and interpreted in (e) and (f). It shows that with the selective optimized plan, the potential recovery values during the disassembly process are 86.7% (of the total potential value of all the disassembled components in the WEEE) for 13 operations and 38.8 and 85.8% for the initial plan after 13 and 15 operations, respectively. With the selective optimized plan, the time spent during the process was 62.7% (of the total time spent to disassemble the WEEE) for 13 operations and 69.4 and 77.6% for the initial plan after 13 and 15 operations, respectively.

Therefore, if the first 13 operations are selected for both plans, it can be observed that significant potential value is recovered (86.7 vs. 38.3%) while less time spent with the optimized solution (62.7 vs. 69.4%). If the first 13 operations and 15 operations are selected for both plans, respectively, a better potential recovery value (86.7 vs. 85.8%) while about 15% time of the total disassembly time can be saved with the optimized solution (62.7 vs. 77.6%). Labor time (15%) of disassembling a single set of LCD WEEE stands for 200 s and about 6 h for 100 sets of the LCD WEEE.

Scenario 2 for Selective Optimization

It is aimed to prioritize the environmental protection targets (100% hazardousness removal and 75% component disassembled for the whole WEEE) (the weights for the hazardousness index and weight removal index in Formula (11.5) were set 1 and the weight for potential recovery value 0.5). The input data is shown in Table 11.2.

In Fig. 11.11a, a 100% hazardousness removal target will be achieved after ten disassembly operations for the optimized plan with this weight setting. In (b), a target to achieve 75% component disassembled by weight (of the total weight of the WEEE) took seven operations for the optimized plan with this weight setting. Therefore, 10 disassembly operations are needed for the selective optimized plan, compared to 13 operations in Scenario 1. In (c), the time spent for the ten operations is 50.0% of the total time for the WEEE, which can be compared to the related results of Scenario 1, which were 62.7 and 69.4% of the total time spent to disassemble the WEEE for the optimized plan with all the weights were set 1 and the initial plan for 13 operations, respectively. In (d), the potential recovery value is 77.4% of the total potential value of the WEEE for this setting, while the potential recovery values are 86.7 and 38.8% of the total potential value of all the

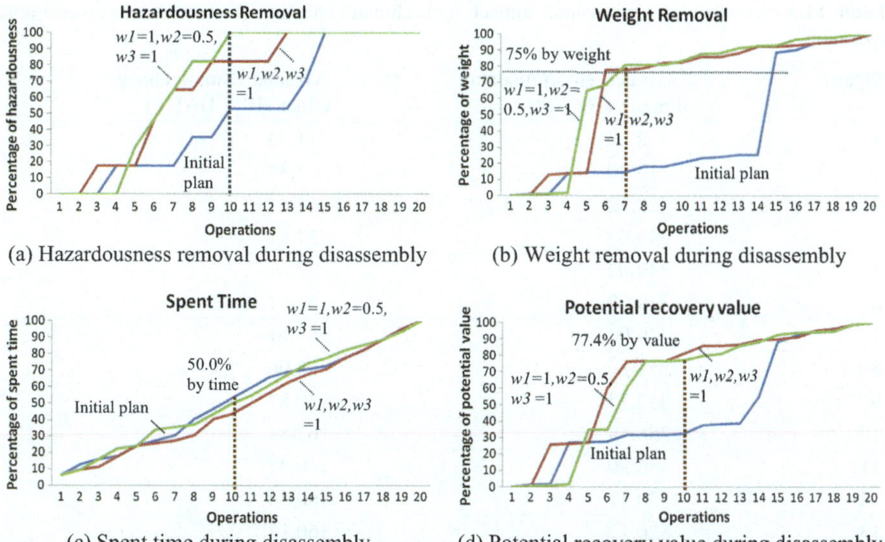

(a) Hazardousness removal during disassembly (b) Weight removal during disassembly

(c) Spent time during disassembly (d) Potential recovery value during disassembly

Fig. 11.11 Disassembly planning optimization with customizable decision-making models (weights are 1, 0.5, and 1)

disassembled components in the WEEE for the optimized plan and the initial plan in Scenario 1, respectively. It can be clearly observed that with the prioritized considerations of hazardousness and weight removal, less operations and time are needed accordingly while the potential recovery value has to be traded off (from 86.7 to 77.4%).

Environmental Impact Evaluation

Disassembled components can potentially generate values through component reuse/remanufacturing and material recycling, and they can therefore reduce the environmental impact and climate change effect without needing to make the components from raw materials. The environmental impact and climate change effect of each operation are shown in Table 11.3. The results of the optimized plan under Scenario 1 (all the weights were set 1) and the initial plan are shown in Fig. 11.12. It can be observed that significant improvements can be made with the optimization process compared to the initial plan (69.4 and 155.3% improvement in the two aspects).

Algorithm Comparisons

The generic constraint handling method in this research was compared with a classic penalty method [29], which is a popular method applicable to complex constraints. The results are shown in Fig. 11.13a. It can be concluded that the

Table 11.3 Avoided environmental impact and climate change effect of each disassembly operation

Oper(i)	Avoided environmental impact (10^{-3} Pt)	Avoided climate change effect (10^{-8} DALY)
1	2.54	1.53
2	13.20	3.37
3	20.12	4.44
4	349.02	57.61
5	349.02	57.61
6	349.02	57.61
7	349.02	57.61
8	351.26	58.96
9	352.75	59.87
10	358.59	62.85
11	496.90	146.47
12	520.17	160.54
13	520.17	160.54
14	520.17	160.54
15	1111.69	465.03
16	1111.69	465.03
17	1184.35	508.96
18	1212.25	525.82
19	1254.89	541.60
20	1292.19	564.15

(a) Accumulated avoided environmental impact (b) Accumulated avoided climate change effect

Fig. 11.12 Optimized accumulated avoided environmental impact/climate change effect

developed generic constraint handling method ensures that the computational process can be conducted in a smoother and more efficient way, and all the generated plans are valid.

The GA, SA, and improved PSO algorithms were also used for optimization shown in (b). All of them can yield good results but the SA and the improved PSO both outperform the GA in the case studies, while the improved PSO algorithm was better than the SA. Each iteration of the improved PSO algorithm uses mainly

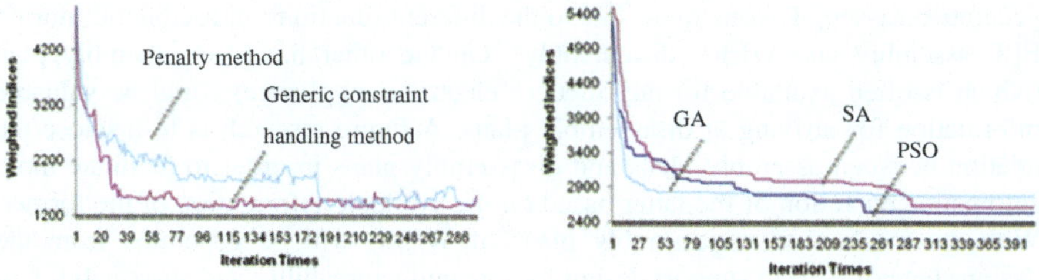

(a) Comparison on constraint handling methods

(b) Comparison on intelligent methods

Fig. 11.13 Algorithm comparisons

simple mathematical operators that can be finished in a shorter time than those for the GA and the SA algorithms with mainly complex position changing operators so that the improved PSO algorithm is also more efficient to achieve the best value generally.

11.4 Conclusions

WEEE has been increasingly customized and diversified, and the selective disassembly planning of WEEE to support remanufacturing decision making is an important but challenging research issue. In this chapter, an effective selective disassembly planning method has been developed to address the issue systematically. The characteristics and contributions of the research include:

- An improved PSO algorithm-based selective disassembly planning method with customizable decision-making models and a novel constraint handling algorithm has been developed in a systematic means. In the method, the customizable decision-making models embedded with adaptive multi-criteria to meet different stakeholders' requirements have been designed to enable the method flexible and customizable in processing WEEE effectively.
- Based on the constraint handling and intelligent optimization algorithms, the developed method is capable to process complex constraints for different types of WEEE based on a generic and robust process and achieve selective optimized disassembly plans efficiently.
- Industrial cases on LCD WEEE have been successfully carried out to verify the effectiveness and generalization of the developed research. Different application scenarios and targets have been set to validate and demonstrate that this research is promising for practical problem-solving.

In the future, a more intelligent mechanism needs to be developed to generate disassembly constraints from the functions and semantics of the BoMs of EEE automatically and accurately (e.g., not all the assembly constraints will be used to generate disassembly constraints due to the different functions and semantics during EEE assembly and WEEE disassembly). On the other hand, an assembly plan (which is often available for any modern electronic appliance) could be valuable information for arriving at disassembly plans. A future research is to disclose the relation between assembly plans and disassembly plans in order to facilitate more automatic generation of the latter based on the available information of the former. With the mechanism, disassembly plans of WEEE will be generated from the design stage of EEE to support design for remanufactureability and sustainability in a more efficient means.

Acknowledgements This research was carried out as a part of the GREENet and CASES projects which are supported by a Marie Curie International Research Staff Exchange Scheme Fellowship within the 7th European Community Framework Programme under the grant agreement No 269122 and No 294931. The authors would also appreciate Mr Qiang Peng, the Technical Director of the Guangdong Changhong Electronics Company, Ltd., and his team for their strong support during the project in terms of technical consultancy/discussions and raw data providing/ explanations.

References

1. Jovane, F., Yoshikawa, H., Alting, L., Boer, C. R., Westkamper, E., Williams, D., et al. (2008). The incoming global technological and industrial revolution towards competitive sustainable manufacturing. *CIRP Annals Manufacturing Technology, 75,* 641–659.
2. Walther, G., Steinborn, J., Spengler, T. S., Luger, T., & Herrmann, C. (2010). Implementation of the WEEE-directive—Economic effects and improvement potentials for reuse and recycling in Germany. *International Journal of Advanced Manufacturing Technology, 47,* 461–474.
3. Hicks, C., Dietmar, R., & Eugster, M. (2005). The recycling and disposal of electrical and electronic waste in China—Legislative and market responses. *Environmental Impact Assessment Review, 25,* 459–471.
4. Kopacek, B., & Kopacek, P. (1999). Intelligent disassembly of electronic equipment. *Annual Reviews in Control, 23,* 165–170.
5. Duflou, J. R., Seliger, G., Kara, S., Umeda, Y., Ometto, A., & Willems, B. (2008). Efficiency and feasibility of product disassembly: A case-based study. *CIRP Annals Manufacturing Technology, 57,* 583–600.
6. Kernbaum, S., Heyer, S., Chiotellis, S., & Seliger, G. (2009). Process planning for IT-equipment remanufacturing. *CIRP Journal of Manufacturing Science and Technology, 2,* 13–20.
7. Hatcher, G. D., Ijomah, W. L., & Windmill, J. F. C. (2011). Design for remanufacturing: A literature survey and future research needs. *Journal of Cleaner Production, 19,* 2004–2014.
8. Mayers, C. K. (2007). Strategic, financial, and design implications of extended producer responsibility in Europe: A producer case study. *Journal of Industrial Ecology, 11,* 113–131.
9. Sander, K., Schilling, S., Tojo, N., van Rossem, C., Vernon, J., & George, C. (2007). *The producer responsibility principle of the WEEE Directive, DG ENV.* Study Contract N° 07010401/2006/449269/MAR/G4, https://ec.europa.eu/environment/waste/weee/pdf/final_rep_okopol.pdf (accessed on 01.07.2017).
10. Giuntini, R., & Gaudette, K. (2003). Remanufacturing: The next great opportunity for boosting US productivity. *Business Horizons,* 41–48.

11. Kara, S., Pornprasitpol, P., & Kaebernick, H. (2006). Selective disassembly sequencing: A methodology for the disassembly of end-of-life products. *CIRP Annals Manufacturing Technology, 55*(1), 37–40.
12. Masui, K., Mizuhara, K., Ishii, K., & Rose, C. (1999). Development of products embedded disassembly process based on end-of-life strategies. In *Proceedings of the EcoDesign'99: 1st International Symposium on Environmentally Conscious Design and Inverse Manufacturing*, Tokyo, pp. 570–575.
13. Chiodo, J. D., Harrison, D. J., & Billett, E. H. (2001). An initial investigation into active disassembly using shape memory polymers. *Proceedings of the Institution of Mechanical Engineers, Part B: Journal of Engineering Manufacture, 215*(5), 733–741.
14. Jones, N., Harrison, D., Billett, E., & Chiodo, J. (2004). Electrically self-powered active disassembly. *Proceedings of the Institution of Mechanical Engineers, Part B: Journal of Engineering Manufacture, 218*(7), 689–697.
15. Braunschweig, A. (2004). Automatic disassembly of snap-in joints in electromechanical devices. In *Proceedings of the 4th International Congress Mechanical Engineering Technologies'04,* Varna, pp. 48–56.
16. Hussein, H., & Harrison, D. (2008). New technologies for active disassembly: Using the shape memory effect in engineering polymers. *International Journal of Product Development, 6*(3/4), 431–449.
17. Ijomah, W. L., & Chiodo, J. D. (2010). Application of active disassembly to extend profitable remanufacturing in small electrical and electronic products. *International Journal of Sustainable Engineering, 3*(4), 246–257.
18. Sundin, E., Lindahl, M., & Ijomah, W. (2009). Product design for product/service systems—Design experiences from Swedish industry. *Journal of Manufacturing Technology Management, 20*(5), 723–753.
19. Dindarian, A., Gibson, A. A. P., & Quariguasi-Frota-Neto, J. (2012). Electronic product returns and potential reuse opportunities: A microwave case study in the United Kingdom. *Journal of Cleaner Production, 32,* 22–31.
20. Santochi, M., Dini, G., & Failli, F. (2002). Computer aided disassembly planning: State of the arts and perspectives. *CIRP Annals Manufacturing Technology, 51*(2), 507–529.
21. Lambert, A. J. D. (2002). Determining optimum disassembly sequences in electronic equipment. *Computers & Industrial Engineering, 43*(3), 553–575.
22. Kuo, T. C. (2012). Waste electronics and electrical equipment disassembly and recycling using Petri net analysis: Considering the economic value and environmental impacts. *Computers & Industrial Engineering, 65*, 54.
23. Renteria, A., Alvarez, E., Perez, J., & Pozo, D. (2011). A methodology to optimize the recycling process of WEEE: Case of television sets and monitors. *International Journal of Advanced Manufacturing Technology, 54,* 789–800.
24. Ryan, A., O'Donoghue, L., & Lewis, H. (2011). Characterising components of liquid crystal displays to facilitate disassembly. *Journal of Cleaner Production, 19,* 1066–1071.
25. Li, W. D., Ong, S. K., & Nee, A. Y. C. (2002). Hybrid genetic algorithm and simulated annealing approach for the optimization of process plans for prismatic parts. *International Journal of Production Research, 40*(8), 1899–1922.
26. Li, W. D., & McMahon, C. A. (2007). A simulated annealing-based optimization approach for integrated process planning and scheduling. *International Journal of Computer Integrated Manufacturing, 20*(1), 80–95.
27. Kennedy, J., & Eberhart, R. (1995). Particle swarm optimization. In *Proceedings of IEEE International Conference on Neural Networks* (Vol. IV, pp. 1942–1948).
28. Lambert, A. J. D., & Gupta, S. M. (2005). *Disassembly modelling for assembly, maintenance, reuse, and recycling*. Boca Raton, FL: CRC Press.
29. Reddy, S. V. B., Shunmugam, M. S., & Narendran, T. T. (1999). Operation sequencing in CAPP using genetic algorithm. *International Journal of Production Research, 37,* 1063–1074.

[illegible]

Chapter 12
A Systematic Selective Disassembly Approach for Waste Electrical and Electronic Equipment (WEEE)

G. Q. Jin, W. D. Li, S. Wang and S. M. Gao

Abstract Waste Electrical and Electronic Equipment (WEEE) is one of the major waste streams in terms of quantity and toxicity, and a critical step in WEEE end-of-life (EOL) processing is through disassembly. Compared with full disassembly, which is a sub-optimal solution due to its high operational cost, selective disassembly is more economic and practical as only selected parts with recycling potential are considered. In this paper, a systematic selective disassembly approach for handling WEEE with a maximum disassembly profit in accordance with the WEEE and Restriction of Hazardous Substances (ROHS) Directives has been developed. Firstly, a space interference matrix is generated based on the interference relationship between individual components in the 3D CAD model of WEEE. A matrix analysis algorithm is then applied to obtain all the feasible disassembly sequences. Secondly, an evaluation and decision-making method is developed to find out an optimal selective disassembly sequence from the obtained feasible disassembly sequences. The evaluation takes into account the disassembly profit and requirements of the WEEE and ROHS Directives, which regulate on recycling rates of different types of products and removal requirements of (i) hazardous, (ii) heavy, and (iii) high-value components. Thus, an optimal solution is a selective disassembly sequence that can achieve the maximum disassembly profit, while complying with the WEEE and ROHS restrictions based on a brute-force search method. Finally, an industrial case on Changhong Liquid Crystal Display Televisions (LCD-TVs) of the type LC24F4 is used to demonstrate the effectiveness of the developed approach.

G. Q. Jin
Robotics and Microsystems Center, School of Mechanical and Electric Engineering, Soochow University, Suzhou, People's Republic of China

W. D. Li (✉) · S. Wang
Faculty of Engineering, Environment and Computing, Coventry University, Coventry, UK
e-mail: weidong.li@coventry.ac.uk

S. M. Gao
State Key Lab of CAD&CG, Zhejiang University, Hangzhou, People's Republic of China

W. Li and S. Wang (eds.), *Sustainable Manufacturing and Remanufacturing Management*, https://doi.org/10.1007/978-3-319-73488-0_12

Keywords Selective disassembly · The WEEE directive · The ROHS directive LCD-TVs

12.1 Introduction

Due to the huge market demand and shorter usage lifecycle of Electrical and Electronic Equipment (EEE), the mounting WEEE is posing a severe threat to the environment and sustainable economy. To tackle this issue, the WEEE Directive together with the ROHS Directive was enacted and became a European Environment Law in February 2003 [1]. They are aimed at encouraging manufacturers to provide cradle-to-grave support to recycle maximum values and reduce environmental hazardousness by reducing the amount of WEEE destined for landfill and eradicating certain hazardous substances from WEEE.

Disassembly is a key process of recycling strategies in the treatment of WEEE. Generally, the approach can be classified as full disassembly and selective disassembly. Full disassembly is to completely disassemble every single part in a waste product. On the other hand, selective disassembly aims to minimize the effort required to separate parts selected for separation or replacement. Unlike full disassembly, selective disassembly allows a partial and non-sequential disassembly procedure [2–4]. In the past years, there have been many research articles published on full disassembly of WEEE. In the literatures [5–7], some detailed reviews on the full disassembly research were given. However, the high cost of disassembly has impeded the cost-effectiveness of the full disassembly approach and has resulted in a research trend toward selective disassembly. Although there are a number of papers reporting research on selective disassembly, a systematic selective disassembly approach for handling WEEE with environmental and economic considerations is still unavailable and highly desirable. In this paper, the focus is on the development of a systematic selective disassembly approach that can maximize the disassembly profit and meet the environmental restrictions simultaneously. The approach incorporates the environment requirements of WEEE/ROHS Directives into decision making to ensure the restricted recycling rates of different products are achieved and all hazardous components from WEEE are removed for further EOL processes. Meanwhile, the cost-effectiveness of the disassembly operation is maximized by selecting the optimal disassembly planning based on the proposed evaluation and decision-making method. Figure 12.1 shows the main flow of the developed approach. A summary of the developed approach is given below:

- A feasible solution space method is developed to generate all the candidate solutions for further evaluation and decision making. A space interference matrix is used to represent the space relationships of components of WEEE in six directions in a 3D Cartesian coordinate system. In this manner, all the space interference relationships between components of WEEE can be digitally

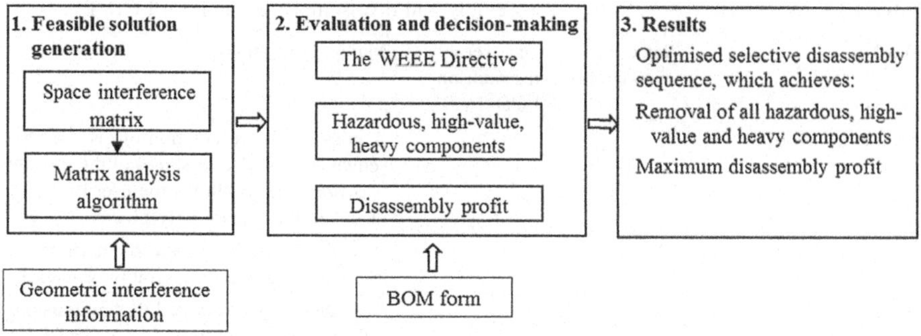

Fig. 12.1 Main flow of the developed approach

presented and analyzed by a matrix analysis algorithm to find out all the feasible disassembly sequences of WEEE.
- An evaluation and decision-making methods are developed to identify an optimized selective disassembly sequence for maximizing the disassembly profit by considering the recycling rate requirement of the WEEE Directive and removal of all hazardous, high-value and heavy components from WEEE.

A selective disassembly optimization system based on the above method was implemented and an industrial case study on Changhong Liquid Crystal Display Televisions (LCD-TVs) was performed to validate the developed method. The LCD-TV is selected as the case study due to the fact that it is a typical product of EEE and occupies a significant portion of WEEE (e.g., more than 200 million units are shipped in the global market in 2014 [8]). The performance results on a LCD-TV (type LC24F4) proved the effectiveness of the developed approach.

12.2 Related Research

Selective disassembly is targeting on singling out hazardous and valuable components of end-of-life (EOL) products. Compared with full disassembly, it is a more economical approach in the practice of recycling WEEE. Recently, attention is being paid toward selective disassembly research, such as LCD-TVs EOL processing. A summary of the previous work is given in Table 12.1, and detailed discussions are presented below.

Kara et al. [3, 9] developed a selective disassembly method by modifying the typical rule-based question–answer method proposed by Nevins and Whitney [10] for assembly sequence generation. It provided a graphical representation of disassembly sequences at the different stages of the process, which allows the user to visualize e the disassembly process. Two cases studies, i.e., a single-hole punch and a washing machine, were used to explain the concept and efficiency of the methodology. Garcia et al. [11] presented a method to determine low-cost selective

Table 12.1 A summary of selective disassembly and disassembly research on LCD-TVs

Works	Selective disassembly	Disassembly research on LCD-TVs	Major characteristics
Kara et al. [3, 9]	O		A graphical representation of disassembly sequences at different stages of the process was provided. It allows the user to visualize the disassembly process
Garcia et al. [11]	O		A method was developed to determine low-cost selective disassembly sequences
Behdad et al. [12]	O		A method was developed to simultaneous selective disassembly and EOL decision making for products
ElSayed et al. [13]	O		AGenetic Algorithm was utilized to obtain economically and environmentally sustainable selective disassembly sequences for EOL products
Srinivasan and Gadh. [14, 15]	O		A geometric algorithm was used to find out the optimal disassembly sequence for the selected component with minimal component removals
Smith and Chen [16, 17]	O		The developed method can eliminate unrealistic and uncommon disassembly sequences and find out optimal selective disassembly sequences for complex assemblies effectively
Ryan et al. [18]		O	An overview was presented for the LCD assembly and detailed material composition of the LCD structure
Umeda et al. [19, 20]		O	A recyclability evaluation method was developed to evaluate LCD-TVs recyclability in product life cycles at the design stage
Chiodo et al. [21]		O	An LCD bracket made from shape memory polymer was used to separate LCD screens from PCBs
Li et al. [2, 22]	O	O	The developed method is capable of handling complex constraints to achieve better economic value and environmental protection requirements

disassembly sequences. The algorithm computes the minimum distances from the exterior components of the assembly to the rest of components. A set of partial disassembly sequences is obtained by finding minimum spanning trees in the precedence graph. Behdad et al. [12] presented a method on simultaneous selective disassembly and EOL decision making for products. It integrates a transition matrix with mixed integer linear programming to determine the extent to which products should be disassembled and the optimal EOL strategy for each resultant component. Two cell phone products were used to illustrate the research. ElSayed et al. [13] presented an evolutionary algorithm for generating optimal sequences for selective disassembly of EoL products. A Genetic Algorithm was utilized to obtain economically and environmentally sustainable disassembly sequences. Srinivasan and Gadh [14, 15] used a geometric algorithm to determine an optimal disassembly sequence for selected components with minimal component removals. The search space of the algorithm is reduced by analyzing a subset of components in the assembly. Smith and Chen [16, 17] presented a rule-based recursive method for finding an optimal selective disassembly sequence to increase products' recyclability and maintainability. Based on four matrices and five disassembly rules, the method can eliminate unrealistic and uncommon disassembly sequences and find optimal selective disassembly sequences for complex assemblies effectively.

LCD-TVs are one of the most important WEEE. The requirement for recycling LCD-TVs is increasing rapidly as the huge amount of LCD-TVs is to be replaced or in the end of their useful life sooner or later. There are several papers reporting disassembly research on LCD-TVs. Ryan et al. [18] presented an overview of the LCD assembly and detailed material composition of the LCD structure. It investigated the best approach to recycle or disassemble LCD with a hybrid system of manual and automated processes. Umeda et al. [19, 20] presented a recyclability evaluation method to evaluate LCD-TVs recyclability in product lifecycles at the design stage. Firstly, it describes an EoL scenario of the product, and then to calculate the recyclability rate of the product based on the described EOL scenario. A case study of LCD-TVs was used, and the result shows that recyclability of LCD-TVs depends on its EOL processes and material constitution. Chiodo et al. [21] investigated the technical feasibility of removing LCD screens using the smart materials technology. An LCD bracket made from shape memory polymer was used to separate LCD screens from printed circuit boards (PCBs). Li et al. [2, 22] developed a particle swarm optimization (PSO)-based selective disassembly planning method embedded with customizable decision-making models and a novel generic constraint handling algorithm. The method is flexible for customized decision modeling and is capable of handling complex constraints to achieve better economic value and environmental protection requirements. LCD-TVs have been used to demonstrate the effectiveness and robustness of the developed method. Further research is expected to develop a set of more systematic criteria to evaluate the different stages of remanufacturing in terms of environment and economy.

As mentioned in the introduction section, there still lacks a systematic selective disassembly approach to handle WEEE to meet the environmental and economic requirements. The work in this paper is to develop a systematic selective disassembly approach to achieve better economic value and meet the environmental protection requirements of the WEEE/ROHS Directives.

12.3 Methodology and Characteristics

The overview of the developed methods is presented as follows:

- A feasible solution generation method is developed to find out all the feasible disassembly sequences of WEEE by analyzing the space interference matrices in a 3D environment.
- An evaluation and decision-making method is devised to identify the optimized disassembly sequence in achieving better economic value and environmental protection requirements.
- An industrial case study on LCD-TVs is carried out to verify and demonstrate the performance of the developed methods.

The developed methods are shown in Fig. 12.2. The details are described below.

12.3.1 A Feasible Solution Generation Method

The development of the feasible solution generation method is carried out in two phases.

- Phase 1 is to generate a space interference matrix based on a 3D CAD model of WEEE. It can be used to represent the space interference relationship between components of the WEEE.
- Phase 2 is to obtain all the feasible disassembly sequences with the developed matrix analysis algorithm.

The details of each phase are explained below.

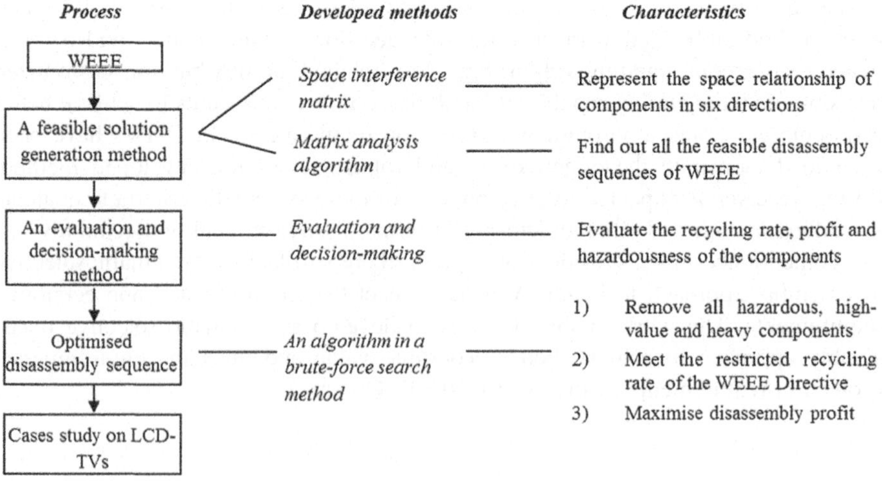

Fig. 12.2 Process of the developed methods

12.3.1.1 Phase 1—Space Interference Matrix

Firstly, based on a CAD model of WEEE, row-major six space interference matrices are generated in six directions separately in a 3D environment. It can be used to represent the space interference relationship of components of the waste product.

$$\begin{array}{c} \\ C_1 \\ C_2 \\ \vdots \\ C_n \end{array} \begin{array}{c} \begin{array}{cccc} C_1 & C_2 & \cdots & C_n \end{array} \\ \begin{bmatrix} r_{11} & r_{12} & \cdots & r_{1n} \\ r_{21} & r_{22} & & r_{2n} \\ \vdots & \vdots & \ddots & \vdots \\ r_{n1} & r_{n2} & \cdots & r_{nn} \end{bmatrix} \end{array} \tag{12.1}$$

In the matrix, the element in each row and column denotes one of the components in the product. The element represents the space interference relationship between components *i* (to be removed component) and *j* (interfering component) in six directions (*X*+ , *X*−, *Y*+ , *Y*−, *Z*+ , *Z*−) in the 3D environment. If space interference exists between components *i* and *j* in one direction, the element in the matrix corresponding to this specific direction is set ‘1’. Otherwise, it is ‘0’.

An example is used here to explain the space interference relationship between ‘*A*’ and ‘*B*’ components (shown in Fig. 12.3). As the component ‘*B*’ is in the *X* + direction of the component ‘*A*’, and ‘*A*’ is in the *X*− direction of ‘*B*’, the element in the *X*+ direction matrix is therefore ‘1’, and the element in the *X*− direction matrix is ‘1’. All the other elements are ‘0’.

A four-component product (shown in Fig. 12.4) will be used as an example to explain the matrix analysis method (Phase 2). The space interference matrices are first obtained as given in Eqs. (12.2)–(12.7).

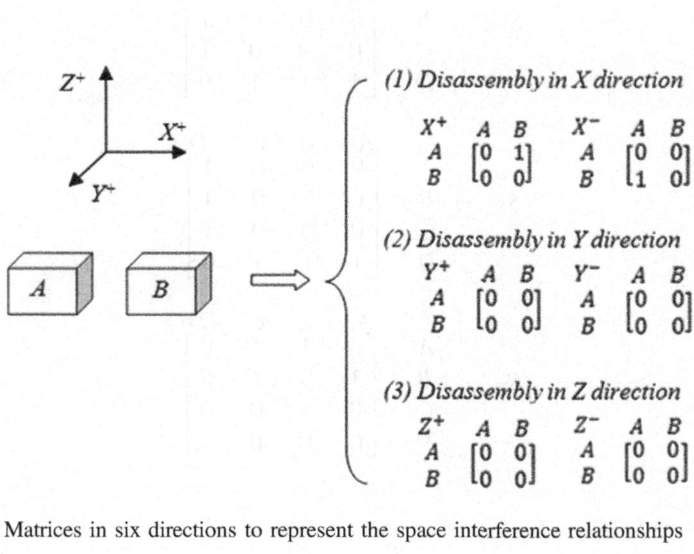

Fig. 12.3 Matrices in six directions to represent the space interference relationships

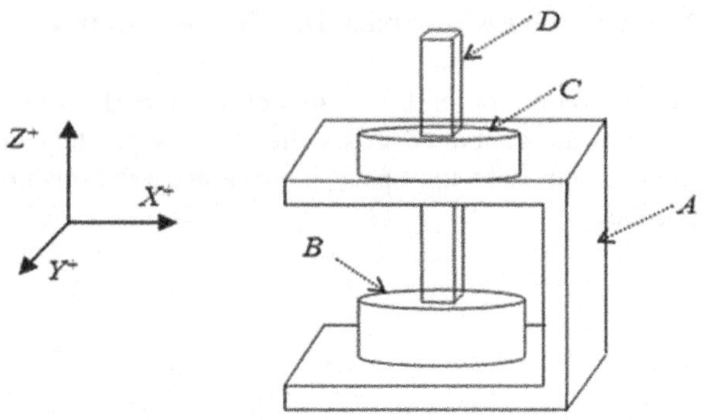

Fig. 12.4 Product with four components

$$S_{X^+} = \begin{array}{c} \\ A \\ B \\ C \\ D \end{array} \begin{array}{c} \begin{array}{cccc} A & B & C & D \end{array} \\ \begin{bmatrix} 0 & 0 & 0 & 1 \\ 1 & 0 & 0 & 1 \\ 0 & 0 & 0 & 0 \\ 1 & 1 & 1 & 0 \end{bmatrix} \end{array} \tag{12.2}$$

$$S_{X^-} = \begin{array}{c} \\ A \\ B \\ C \\ D \end{array} \begin{array}{c} \begin{array}{cccc} A & B & C & D \end{array} \\ \begin{bmatrix} 0 & 1 & 0 & 1 \\ 0 & 0 & 0 & 1 \\ 0 & 0 & 0 & 1 \\ 1 & 1 & 1 & 0 \end{bmatrix} \end{array} \tag{12.3}$$

$$S_{Y^+} = \begin{array}{c} \\ A \\ B \\ C \\ D \end{array} \begin{array}{c} \begin{array}{cccc} A & B & C & D \end{array} \\ \begin{bmatrix} 0 & 0 & 0 & 1 \\ 0 & 0 & 0 & 1 \\ 0 & 0 & 0 & 1 \\ 1 & 1 & 1 & 0 \end{bmatrix} \end{array} \tag{12.4}$$

$$S_{Y^-} = \begin{array}{c} \\ A \\ B \\ C \\ D \end{array} \begin{array}{c} \begin{array}{cccc} A & B & C & D \end{array} \\ \begin{bmatrix} 0 & 0 & 0 & 1 \\ 0 & 0 & 0 & 1 \\ 0 & 0 & 0 & 1 \\ 1 & 1 & 1 & 0 \end{bmatrix} \end{array} \tag{12.5}$$

$$S_{Z^+} = \begin{array}{c} \\ A \\ B \\ C \\ D \end{array} \begin{array}{c} \begin{array}{cccc} A & B & C & D \end{array} \\ \begin{bmatrix} 0 & 1 & 1 & 0 \\ 1 & 0 & 1 & 0 \\ 0 & 0 & 0 & 0 \\ 0 & 0 & 0 & 0 \end{bmatrix} \end{array} \tag{12.6}$$

$$S_{Z^-} = \begin{array}{c} \\ A \\ B \\ C \\ D \end{array} \begin{array}{c} \begin{array}{cccc} A & B & C & D \end{array} \\ \begin{bmatrix} 0 & 1 & 0 & 0 \\ 1 & 0 & 0 & 0 \\ 1 & 1 & 0 & 0 \\ 0 & 0 & 0 & 0 \end{bmatrix} \end{array} \tag{12.7}$$

12.3.1.2 Phase 2—Matrix Analysis Algorithm

Based on the obtained space interference matrices in six directions, a matrix analysis algorithm is then developed to obtain all the feasible disassembly sequences of the product. The aforementioned example is used here to explain the details of the developed matrix analysis algorithm. Firstly, Eq. (12.8) is generated by combining Eqs. (12.2)–(12.7) in six directions:

$$S = \begin{array}{c} \\ A \\ B \\ C \\ D \end{array} \begin{array}{c} \begin{array}{cccc} A & B & C & D \end{array} \\ \begin{bmatrix} 000000 & 010011 & 000010 & 111100 \\ 100011 & 000000 & 000010 & 111100 \\ 000001 & 000001 & 000000 & 111100 \\ 111100 & 111100 & 111100 & 000000 \end{bmatrix} \end{array} \tag{12.8}$$

The Boolean operator 'OR' is used here for the above equation for each component in the row direction. For instance, in a row, if there is any '1' in a direction, the final result in that direction after the 'OR' operation will be '1'. Otherwise, the final result in the direction will be '0'. Equation (12.9) is obtained below:

$$S = \begin{array}{c} \\ A \\ B \\ C \\ D \end{array} \begin{array}{c} \begin{array}{cccc} A & B & C & D \end{array} \\ \begin{bmatrix} 000000 & 010011 & 000010 & 111100 \\ 100011 & 000000 & 000010 & 111100 \\ 000001 & 000001 & 000000 & 111100 \\ 111100 & 111100 & 111100 & 000000 \end{bmatrix} \end{array} \begin{array}{c} \text{Result} \\ 111111 \\ 111111 \\ 111101 \\ 111100 \end{array} \tag{12.9}$$

The result '111111' represents the relationship between one component and all the other remaining components of the product in six directions. If the result is always '1', it means the component could not be disassembled in any direction; if the result includes '0', it means the component can be disassembled from that direction. The example in Fig. 12.5 can be used to explain the concept. In Eq. (12.9), components '*A*' and '*B*' could not be disassembled in any direction as the results are all '1'; component '*C*' can be disassembled in *Z*+ direction as the result is '0' in this direction; component '*D*' can be disassembled in *Z*+ and *Z*− directions.

If component '*D*' is disassembled in the *Z*+ direction firstly, the remaining combined space interference matrix is updated as shown below:

Boolean operator 'OR' in rows

$$S=\begin{array}{c} \\ A \\ B \\ C \\ D \end{array}\begin{array}{c} \begin{array}{cccc} A & B & C & D \end{array} \\ \begin{bmatrix} 000000 & 010011 & 000010 & 111100 \\ 100011 & 000000 & 000010 & 111100 \\ 000001 & 000001 & 000000 & 111100 \\ 111100 & 111100 & 111100 & 000000 \end{bmatrix} \end{array}\begin{array}{l} Result \\ 111111 \Rightarrow \text{'A' could not be disassembled} \\ 111111 \Rightarrow \text{'B' could not be disassembled} \\ 111101 \Rightarrow \text{'C' could be disassembled in } Z^+ \\ 111100 \Rightarrow \text{'D' could be disassembled in } Z^+Z^- \end{array}$$

$$Direction: X^+X^-Y^+Y^-Z^+Z^-$$

Fig. 12.5 Feasible disassembly sequence analysis for the product

$$S=\begin{array}{c} \\ A \\ B \\ C \end{array}\begin{array}{c} \begin{array}{ccc} A & B & C \end{array} \\ \begin{bmatrix} 000000 & 010011 & 000010 \\ 100011 & 000000 & 000010 \\ 000001 & 000001 & 000000 \end{bmatrix} \end{array}\begin{array}{c} \text{Result} \\ 010011 \\ 100011 \\ 000001 \end{array} \tag{12.10}$$

From Eq. (12.10), components '*A*' and '*B*' can be disassembled in three directions, and the component '*C*' can be disassembled in five directions. If component '*C*' is disassembled in the *Z*+ direction, then the remaining combined space interference matrix is shown below:

$$S=\begin{array}{c} \\ A \\ B \end{array}\begin{array}{c} \begin{array}{cc} A & B \end{array} \\ \begin{bmatrix} 000000 & 010011 \\ 100011 & 000000 \end{bmatrix} \end{array}\begin{array}{c} \text{Result} \\ 010011 \\ 100011 \end{array} \tag{12.11}$$

From Eq. (12.11), components '*A*' and '*B*' can be disassembled in three directions. After '*B*' is disassembled in the *Y*+ direction, the product has been disassembled completely. Loop the above analysis process until all the feasible disassembly sequences of the product are obtained. Based on the above analysis, the total feasible disassembly sequences for the product is 192 (30 + 30 + 30 + 30 + 30 + 30 + 6 + 6) (shown in Fig. 12.6).

The obtained all feasible disassembly sequences with geometrical constraints are then evaluated based on the evaluation and decision-making method to get the optimized selective disassembly sequence of a product within an acceptable time. Details on the evaluation and decision-making method are explained in the next section.

12.3.2 An Evaluation and Decision-Making Method

According to the WEEE/ROHS Directives, the restricted recycling rate of WEEE is required to be met, and all the components containing hazardous materials need to be taken apart from WEEE for further recycling and processing. Apart from fulfilling these fundamental environmental targets, disassembly factories would also

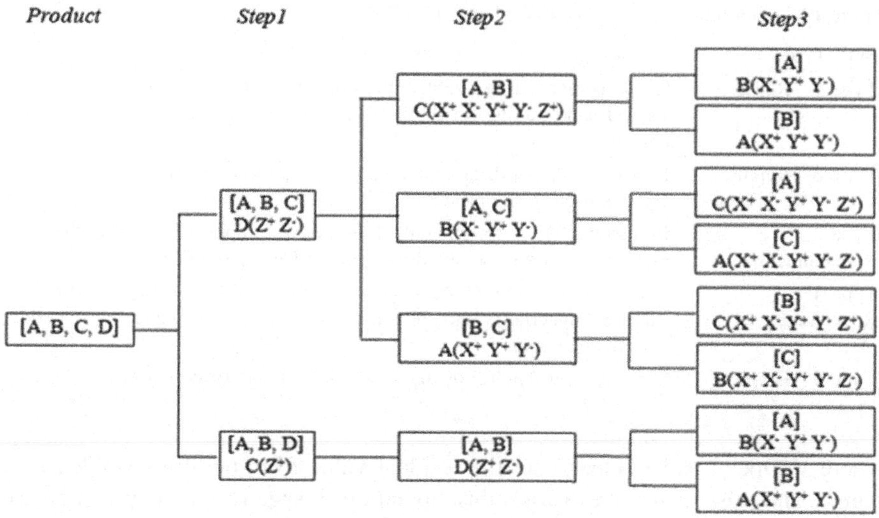

Fig. 12.6 All feasible disassembly sequences for the product

improve the disassembly profit by prioritizing heavy and valued components during disassembly. Based on the above scenario, an evaluation and decision-making method is developed to selectively disassemble WEEE to meet the above requirements. It is composed by three steps: (1) to calculate the profit of the disassembly operation, (2) to identify hazardous, heavy and high-value components, and (3) to calculate the recycling rate of WEEE. Meanwhile, the developed methods are incorporated with a brute-force search method [23] to find out the optimized sequence to meet the economic and environmental requirements, including: (1) to remove all hazardousness, high-value and heavy components from WEEE before shredding and disposal, (2) to meet the restricted recycling rate of the WEEE Directive, and (3) to maximize the disassembly profit.

12.3.2.1 Profit calculation of disassembly operation

The profit of disassembly operation changes with the different EOL treatments of WEEE. Generally, there are five EOL options available for WEEE, as described in Table 12.2 [24, 25].

The profit of the *i*th disassembly operation (Profit (Oper_i)) can be computed as below:

$$\mathrm{Profit}(\mathrm{Oper}_i) = V(\mathrm{Oper}_i) - C(\mathrm{Oper}_i) \qquad (12.12)$$

Table 12.2 Definitions of the five EOL Types [24, 25]

EOL Types	Characters
Type 1—Reuse	Reuse is that a disassembled component can be reused in the second-hand trading for the product without any physical or chemical change
Type 2—Repair	Repair is when a disassembled component needs to be repaired before being reused in the product
Type 3—Remanufacture	Remanufacture is to remanufacture a new component by using the materials of a disassembled component from a product
Type 4—Recycling	Recycling is to reduce the material size to facilitate sorting and the shredded material is separated and recycled with magnetic, air and eddy current separation
Type 5—Disposal	Disposal is to landfill or incinerate the product or component

where $V(\text{Oper}_i)$ and $C(\text{Oper}_i)$ are the residual value from the disassembled component and cost of the *i*th disassembly operation, respectively. They can be calculated as follows:

(1) The residual value of the disassembled component of the *i*th disassembly operation $(V(\text{Oper}_i))$

- If the disassembled component is recycled in Type 1 or Type 2, $V(\text{Oper}_i)$ is calculated with the residual value of component in the second-hand trading market.
- If the disassembled component is recycled in Type 3 or Type 4, $V(\text{Oper}_i)$ is calculated with the residual value of materials of the component.
- If the disassembled component belongs to Type 5, $V(\text{Oper}_i)$ is zero as no material is recycled from the component.

Equation (12.13) is used to explain the above concept:

$$\begin{cases} V(\text{Oper}_i) = V(\text{Component}_i) & \text{if Comp}_i = \text{Type 1 or Type 2} \\ V(\text{Oper}_i) = V\left(\sum_{j=1}^{n}\left(w_{ji} \times r_{ji}^{\text{cyc}}\right)\right) & \text{if Comp}_i = \text{Type 3 or Type 4} \\ V(\text{Oper}_i) = 0 & \text{if Comp}_i = \text{Type 5} \end{cases} \quad (12.13)$$

where Comp_i represent the *i*th disassembled component,w_{ji} and r_{ji}^{cyc} represent the weight and recycling rate of the *j*th material in the ith component, respectively.

(2) The cost of the *i*th disassembly operation $(C(\text{Oper}_i))$

The result of $C(\text{Oper}_i)$ varies with the different EOL Types. It can be computed as below:

$$\begin{cases} C(\text{Oper}_i) = C^{\text{disassembly}}(\text{Oper}_i) + C^{\text{clean}}(\text{Oper}_i) & \text{if Comp}_i = \text{Type 1} \\ C(\text{Oper}_i) = C^{\text{disassembly}}(\text{Oper}_i) + C^{\text{repair}}(\text{Oper}_i) & \text{if Comp}_i = \text{Type 2} \\ C(\text{Oper}_i) = C^{\text{disassembly}}(\text{Oper}_i) & \text{if Comp}_i = \text{Type 3} \\ C(\text{Oper}_i) = C^{\text{shredding}}(\text{Oper}_i) + C^{\text{dseparation}}(\text{Oper}_i) & \text{if Comp}_i = \text{Type 4} \\ C(\text{Oper}_i) = W_i \times C^{\frac{\text{landfill}}{\text{incinerate}}}(\text{Oper}_i) & \text{if Comp}_i = \text{Type 5} \end{cases} \tag{12.14}$$

where W_i is the weight of the disassembled component; $C^{\text{disassembly}}(\text{Oper}_i)$, $C^{\text{clean}}(\text{Oper}_i)$, $C^{\text{repair}}(\text{Oper}_i)$, $C^{\text{shredding}}(\text{Oper}_i)$, $C^{\text{separation}}(\text{Oper}_i)$, and $C^{\text{landfill/incinerate}}(\text{Oper}_i)$ represent the cost of disassembly, cleaning, repair, shredding, separation, landfill, and incineration of the ith component, respectively. They can be calculated based on the time spent on the disassembly operation $\left(T^{\text{disassembly}}(\text{Oper}_i)\right)$, the wage of labor ($W^{\text{labor}}$), the expense of factory (E^{factory}) per day (indirect cost), and the number of workers in each factory (N^{labor}). For instance, $C^{\text{disassembly}}(\text{Oper}_i)$ can be computed in Eqs. (12.15)–(12.17) as follows:

$$C^{\text{disassembly}}(\text{Oper}_i) = C^{\text{labor}}(\text{Oper}_i) + C^{\text{factory}}(\text{Oper}_i) \tag{12.15}$$

$$C^{\text{labor}}(\text{Oper}_i) = T^{\text{disassembly}}(\text{Oper}_i) \times W^{\text{labor}} \tag{12.16}$$

$$C^{\text{factory}}(\text{Oper}_i) = T^{\text{disassembly}}(\text{Oper}_i) \times E^{\text{factory}}/N^{\text{labor}} \tag{12.17}$$

Based on the above analysis, total profit (*TP*) of a product after the disassembly operation can be computed as below:

$$TP = \sum_{i=1}^{n} \text{Profit}(\text{Oper}_i) \tag{12.18}$$

12.3.2.2 Identification of hazardous, heavy and high-value components of WEEE

(1) Hazardous components

According to the environment law, all the components containing hazardous materials need to be taken apart from WEEE for further recycling and processing. The hazardous components of WEEE could be identified with the ROHS Directive in Europe [26] and the Code of Federal Regulations in USA with Title 40: Protection of Environment [27].

- *The ROHS Directive restricts the following six substances*: (1) lead, (2) mercury, (3) cadmium, (4) hexavalent chromium, (5) polybrominated biphenyls, and (6) polybrominated diphenyl ether. The maximum permitted concentrations in non-exempt products are 0.1% or 1000 ppm (except for cadmium, which is limited to 0.01% or 100 ppm) by weight.

Table 12.3 Maximum concentration of contaminants for the toxicity characteristic [27]

Contaminant	Regulatory Level (mg/L)	Contaminant	Regulatory Level (mg/L)
Arsenic	5.0	Hexachlorobenzene	30.13
Barium	100.0	Hexachlorobutadiene	0.5
Benzene	0.5	Hexachloroethane	3.0
Cadmium	1.0	Lead	5.0
Carbon tetrachloride	0.5	Lindane	0.4
Chlordane	0.03	Mercury	0.2
Chlorobenzene	100.0	Methoxychlor	10.0
Chloroform	6.0	Methyl ethyl ketone	200.0
Chromium	5.0	Nitrobenzene	2.0
o-Cresol	4200.0	Pentrachlorophenol	100.0
m-Cresol	4200.0	Pyridine	35.0
p-Cresol	4200.0	Selenium	1.0
Cresol	4200.0	Silver	5.0
2,4-D	10.0	Tetrachloroethyl-ene	0.7
1,4-Dichlorobenzene	7.5	Toxaphene	0.5
1,2-Dichloroethane	0.5	Trichloroethyl-ene	0.5
1,1-Dichloroethylene	0.7	2,4,5-Trichlorophenol	400.0
2,4-Dinitrotoluene	30.13	2,4,6-Trichlorophenol	2.0
Endrin	0.02	2,4,5-TP (Silvex)	1.0
Heptachlor (and its epoxide)	0.008	Vinyl chloride	0.2

- *The Code of Federal Regulations with Title 40*: Protection of Environment identifies the hazardousness by calculating the component that contains any of the contaminants listed in Table 12.3 at the concentration equal to or greater than the respective value given in this table.

Here, Eq. (12.19) is used to identify the hazardous components ($C^{\text{hazardous}}$) of WEEE as below:

$$\text{Comp}_i = C^{\text{hazardous}} \text{ if } \text{Comp}_i\left(\text{Material}_j\right) \geq \text{Restricted}\left(\text{Material}_j\right) \qquad (12.19)$$

Some major hazardous components of WEEE are also listed in Table 12.4 [28], among them, printed circuit boards and liquid crystal display screen are hazardous components in LCD-TVs.

(2) Heavy and high-value components

The heavy components (C^{heavy}) and high-value components ($C^{\text{high-value}}$) can be identified as below:

Table 12.4 Major hazardous components of WEEE [28]

Components	Characters
Cathode ray tubes (CRTs)	Fluorescent coating covers the inside of panel glass and leads in the cone glass
Printed circuit boards	Cadmium in certain components, such as SMD chip resistors, infrared detectors, and semiconductors
Liquid crystal displays	LCD screens greater than 100 cm^2 have to be removed from WEEE
Gas discharge lamps	Mercury has to be removed
Component containing mercury such as switches and thermostats	Mercury is used in some thermostats, sensors, relays, and switches
Component containing chlorofluorocarbon (CFC), Hydrochlorofluorocarbons (HCFCs) and Hydrofluorocarbons (HFCs)	CFCs, HCFCs, and HFCs present in the foam and the refrigerating circuit must be properly extracted and destroyed

$$\mathrm{Comp}_i = C^{\mathrm{hazardous}} \quad \text{if } W(\mathrm{Comp}_i) \geq \mathrm{Set}(\text{weight of } \mathrm{Comp}_i) \tag{12.20}$$

$$\mathrm{Comp}_i = C^{\mathrm{high-value}} \quad \text{if } V(\mathrm{Comp}_i) \geq \mathrm{Set}(\text{value of } \mathrm{Comp}_i) \tag{12.21}$$

If the weight/value of the disassembled component is greater than the setting weight/value by the customer, the component is identified as heavy/high-value component.

12.3.2.3 Recycling rate of WEEE

The recycling rate ($R_{\mathrm{recycling}}$) is defined in below:

$$R_{\mathrm{recycling}} = \frac{\sum_{j=1}^{N} \sum_{i=1}^{j} \left(w_{ji} \times r_{ji}^{\mathrm{cyc}} \right)}{W_{\mathrm{total}}} \times 100\% \tag{12.22}$$

where W_{total} is the total weight of a waste product, w_{ji} and r_{ji}^{cyc} are the weight and recycling rate of its *j*th material of the *i*th component, respectively. Table 12.5 shows the recycling rate of different materials [19, 20]. The restricted recycling rate changes with different categories of the WEEE Directive. Table 12.6 shows ten different categories of the WEEE Directive [29]. Table 12.7 shows the minimum targets applicable for different categories in the WEEE Directive [29].

For instance, LCD-TVs belong to category 4 'consumer equipment', and the restricted recycling rate of LCD-TVs is 75% which can be obtained in Table 12.7. Meanwhile, the recycling rate of LCD-TVs can be calculated based on the mass of each material in components and the related recycling rate as shown in Table 12.5.

Table 12.5 A part of recycling rate of product materials [19, 20]

Material name	Recycling rate	Recovering rate	Material name	Recycling rate	Recovering rate
ABS	74	90	PS	62	90
PC	0	90	PVC	0	90
PC/ABS	0	90	Steel	91	91
PMMA	0	90	Aluminum	91	91
PET	90	91	Copper	85	85
PP	90	91	Iron	94	94

Notes if component is comprised of a single material, all values are set as 100%

Table 12.6 Categories in the WEEE Directive [29]

1. Large household appliances	6. Electrical and electronic tools
2. Small household appliances	7. Toys, leisure, and sports equipment
3. IT and telecommunications equipment	8. Medical devices
4. Consumer equipment	9. Monitoring and control instruments
5. Lighting equipment	10. Automatic dispensers

Table 12.7 Recycling rates for different categories [29]

Categories	1, 10	3, 4	2, 5, 6, 7, 8, 9
Recycling (%)	80	75	70

12.3.2.4 Objective function of the selective disassembly optimization

Based on the evaluation and decision-making method, the choice of a selective disassembly plan is then converted into a constrained optimization problem. The constraints are represented in Eqs. (12.23)–(12.26), and the objective function is represented in Eq. (12.27).

(1) Remove all the hazardous, high-value and heavy components

$$\text{Remove}\left(\sum_{r=1}^{m} C_r^{\text{hazardous}}\right) \tag{12.23}$$

$$\text{Remove}\left(\sum_{s=1}^{p} C_s^{\text{high-value}}\right) \tag{12.24}$$

$$\text{Remove}\left(\sum_{t=1}^{q} C_t^{\text{heavy}}\right) \tag{12.25}$$

(2) Meet the restricted recycling rate of the WEEE Directive

$$\sum_{i=1}^{n} R_{\text{recycling}}(\text{Oper}_i) \geq \text{WEEEDirective}\left(R_{\text{recycling}}\right) \tag{12.26}$$

(3) Maximize the disassembly profit

$$\text{Maximize}\left(TP = \sum_{i=1}^{n} \text{Profit}(\text{Oper}_i)\right) \tag{12.27}$$

A software package was developed in Java language to obtain the feasible solution space as described in the Sect. 12.3.1 and to incorporate the decision-making method for WEEE. A brute-force search method [23] is employed initially to find out the optimized selective disassembly sequence from the feasible solution space. Although the search time of the brute-force method for LCD-TVs is affordable, when the number of components in the space interference matrices increases, the search time increases exponentially, and therefore a more efficient optimization method that can handle products with much more components is desirable. Further research on the aspect is ongoing.

12.4 Industrial Cases Study on Changhong LCD-TVs

The LCD-TVs studied here are produced by the Changhong Electronics Company, Ltd., China, which is the biggest television producer in China. The company provides information about LCD-TVs of the type of LC24F4, such as the bill of materials, exploded view, mass of each component, and the detailed assembly processes. The structure of the LCD-TV is shown in Fig. 12.7a, b. The exploded view of a LCD-TV is shown in Fig. 12.7c. As shown in Fig. 12.7d, a LCD-TV is typically assembled by three main parts: (1) base assembly part, (2) front cover assembly part, and (3) back cover assembly part.

12.4.1 Feasible Solution Generation on LCD-TV

Base assembly part

The base assembly part of the LCD-TV is shown in Fig. 12.8. It is composed of nine components: (*A*) metal fixing plate, (*B*) metal washer 1, (*C*) metal washer 2, (*D*) top metal support, (*E*) cylindrical metal support 1, (*F*) cylindrical metal support 2, (*G*) toughened glass seat, (*H*) steel plate, and (*I*) rubber gasket. The space interference matrices to represent the base assembly part in six directions are below.

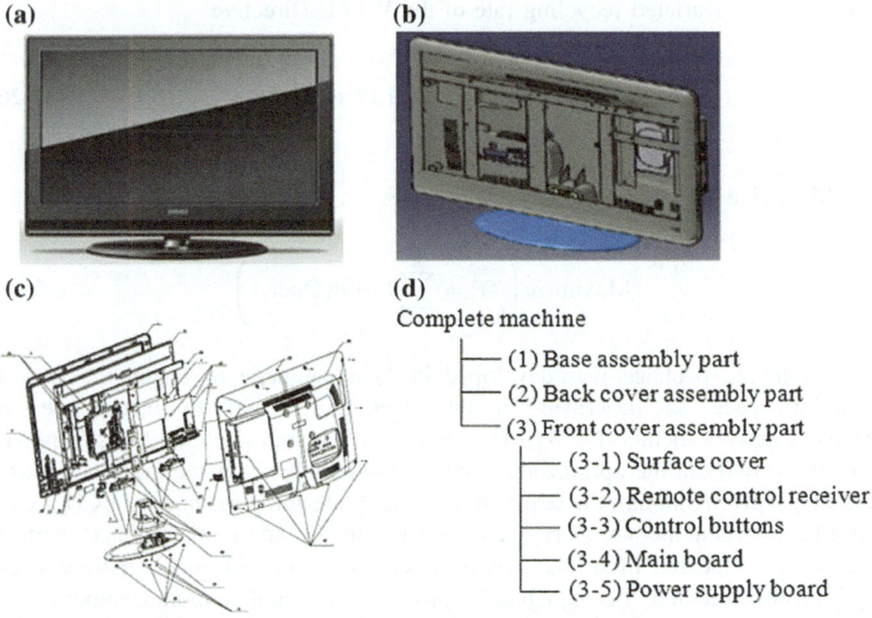

Fig. 12.7 The LCD-TVs and its structures **a** LCD-TV; **b** LCD-TV CAD model; **c** exploded view of LCD-TV structure and **d** parts of LCD-TV

$$
S_{X^+} = \begin{array}{c} \\ A \\ B \\ C \\ D \\ E \\ F \\ G \\ H \\ I \end{array}
\begin{array}{c} \begin{matrix} A & B & C & D & E & F & G & H & I \end{matrix} \\
\begin{bmatrix}
0 & 0 & 0 & 0 & 0 & 0 & 1 & 1 & 1 \\
0 & 0 & 0 & 0 & 0 & 0 & 1 & 1 & 1 \\
0 & 0 & 0 & 0 & 0 & 0 & 1 & 1 & 1 \\
0 & 0 & 0 & 0 & 1 & 1 & 0 & 0 & 0 \\
0 & 0 & 0 & 1 & 0 & 1 & 0 & 0 & 0 \\
0 & 0 & 0 & 1 & 1 & 0 & 1 & 1 & 1 \\
1 & 1 & 1 & 0 & 0 & 0 & 0 & 1 & 0 \\
1 & 1 & 1 & 0 & 0 & 0 & 1 & 0 & 1 \\
0 & 0 & 0 & 0 & 0 & 0 & 0 & 1 & 0
\end{bmatrix} \end{array}
$$

$$
S_{X^-} = \begin{array}{c} \\ A \\ B \\ C \\ D \\ E \\ F \\ G \\ H \\ I \end{array}
\begin{array}{c} \begin{matrix} A & B & C & D & E & F & G & H & I \end{matrix} \\
\begin{bmatrix}
0 & 0 & 0 & 0 & 0 & 0 & 1 & 1 & 1 \\
0 & 0 & 0 & 0 & 0 & 0 & 1 & 1 & 1 \\
0 & 0 & 0 & 0 & 0 & 0 & 1 & 1 & 1 \\
0 & 0 & 0 & 0 & 1 & 1 & 0 & 0 & 0 \\
0 & 0 & 0 & 1 & 0 & 1 & 0 & 0 & 0 \\
0 & 0 & 0 & 1 & 1 & 0 & 1 & 1 & 1 \\
1 & 1 & 1 & 0 & 0 & 0 & 0 & 1 & 0 \\
1 & 1 & 1 & 0 & 0 & 0 & 1 & 0 & 1 \\
0 & 0 & 0 & 0 & 0 & 0 & 0 & 1 & 0
\end{bmatrix} \end{array}
$$

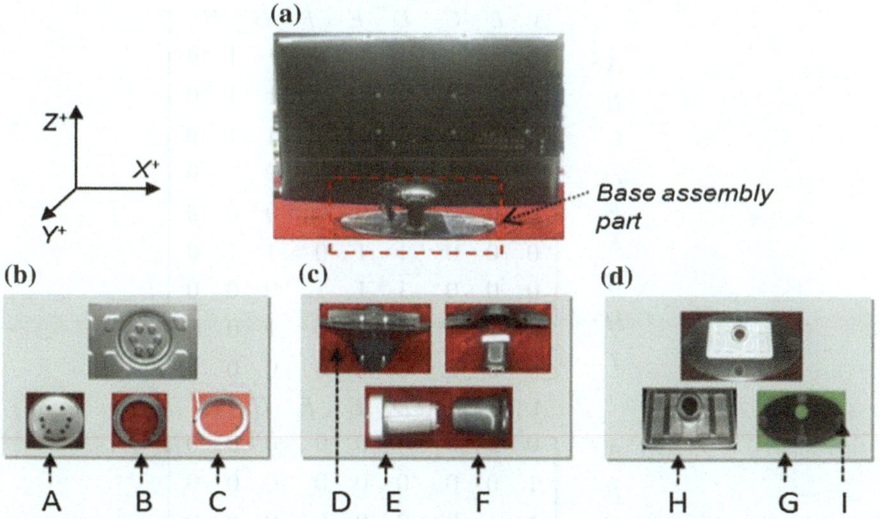

Fig. 12.8 Base assembly part of the LC24F4 LCD-TV: **a** base assembly part, **b** components *A*, *B*, *C*, **c** components *D*, *E*, *F*, and **d** components *G*, *H*, *I*

$$S_{Y^+} = \begin{array}{c} \\ A \\ B \\ C \\ D \\ E \\ F \\ G \\ H \\ I \end{array} \begin{array}{c} \begin{array}{ccccccccc} A & B & C & D & E & F & G & H & I \end{array} \\ \begin{bmatrix} 0 & 0 & 0 & 0 & 0 & 0 & 1 & 1 & 1 \\ 0 & 0 & 0 & 0 & 0 & 0 & 1 & 1 & 1 \\ 0 & 0 & 0 & 0 & 0 & 0 & 1 & 1 & 1 \\ 0 & 0 & 0 & 0 & 1 & 1 & 0 & 0 & 0 \\ 0 & 0 & 0 & 1 & 0 & 1 & 0 & 0 & 0 \\ 0 & 0 & 0 & 1 & 1 & 0 & 1 & 1 & 1 \\ 1 & 1 & 1 & 0 & 0 & 0 & 0 & 1 & 0 \\ 1 & 1 & 1 & 0 & 0 & 0 & 1 & 0 & 1 \\ 0 & 0 & 0 & 0 & 0 & 0 & 0 & 1 & 0 \end{bmatrix} \end{array}$$

$$S_{Y^-} = \begin{array}{c} \\ A \\ B \\ C \\ D \\ E \\ F \\ G \\ H \\ I \end{array} \begin{array}{c} \begin{array}{ccccccccc} A & B & C & D & E & F & G & H & I \end{array} \\ \begin{bmatrix} 0 & 0 & 0 & 0 & 0 & 0 & 1 & 1 & 1 \\ 0 & 0 & 0 & 0 & 0 & 0 & 1 & 1 & 1 \\ 0 & 0 & 0 & 0 & 0 & 0 & 1 & 1 & 1 \\ 0 & 0 & 0 & 0 & 1 & 1 & 0 & 0 & 0 \\ 0 & 0 & 0 & 1 & 0 & 1 & 0 & 0 & 0 \\ 0 & 0 & 0 & 1 & 1 & 0 & 1 & 1 & 1 \\ 1 & 1 & 1 & 0 & 0 & 0 & 0 & 1 & 0 \\ 1 & 1 & 1 & 0 & 0 & 0 & 1 & 0 & 1 \\ 0 & 0 & 0 & 0 & 0 & 0 & 0 & 1 & 0 \end{bmatrix} \end{array}$$

$$S_{Z^+} = \begin{array}{c} \\ A \\ B \\ C \\ D \\ E \\ F \\ G \\ H \\ I \end{array} \begin{array}{c} \begin{array}{ccccccccc} A & B & C & D & E & F & G & H & I \end{array} \\ \begin{bmatrix} 0 & 1 & 1 & 1 & 1 & 1 & 1 & 1 & 0 \\ 0 & 0 & 1 & 1 & 1 & 1 & 1 & 1 & 0 \\ 0 & 0 & 0 & 1 & 1 & 1 & 1 & 1 & 0 \\ 0 & 0 & 0 & 0 & 0 & 0 & 0 & 0 & 0 \\ 0 & 0 & 0 & 1 & 0 & 0 & 0 & 0 & 0 \\ 0 & 0 & 0 & 1 & 1 & 0 & 1 & 1 & 0 \\ 0 & 0 & 0 & 1 & 1 & 1 & 0 & 0 & 0 \\ 0 & 0 & 0 & 1 & 1 & 1 & 1 & 0 & 0 \\ 0 & 0 & 0 & 0 & 0 & 0 & 0 & 0 & 0 \end{bmatrix} \end{array}$$

$$S_{Z^-} = \begin{array}{c} \\ A \\ B \\ C \\ D \\ E \\ F \\ G \\ H \\ I \end{array} \begin{array}{c} \begin{array}{ccccccccc} A & B & C & D & E & F & G & H & I \end{array} \\ \begin{bmatrix} 0 & 0 & 0 & 0 & 0 & 0 & 0 & 0 & 0 \\ 1 & 0 & 0 & 0 & 0 & 0 & 0 & 0 & 0 \\ 1 & 1 & 0 & 0 & 0 & 0 & 0 & 0 & 0 \\ 1 & 1 & 1 & 0 & 1 & 1 & 1 & 1 & 0 \\ 1 & 1 & 1 & 0 & 0 & 1 & 1 & 1 & 0 \\ 1 & 1 & 1 & 0 & 0 & 0 & 0 & 0 & 0 \\ 1 & 1 & 1 & 0 & 0 & 0 & 0 & 1 & 1 \\ 1 & 1 & 1 & 0 & 0 & 0 & 0 & 0 & 0 \\ 0 & 0 & 0 & 0 & 0 & 0 & 0 & 0 & 0 \end{bmatrix} \end{array}$$

The space interference matrices for $X+$, $X-$, $Y+$, $Y-$ directions here are the same, as the base assembly part is a concentric structure along the Z direction so that a component cannot be removed in any direction along the XOY plane if it is surrounded by another component on the same plane. After combining the above six matrices and using Boolean operator 'OR' in rows, the obtained result is as follows:

$S =$

	A	*B*	*C*	*D*	*E*	*F*	*G*	*H*	*I*	Result
A	000000	000010	000010	000010	000010	000010	111110	111110	111100	111110
B	000001	000000	000010	000010	000010	000010	111110	111110	111100	111111
C	000001	000001	000000	000010	000010	000010	111110	111110	111100	111111
D	000001	000001	000001	000000	111101	111101	111101	111101	000000	111101
E	000001	000001	000001	111110	000000	111101	000001	000001	000000	111111
F	000001	000001	000001	111110	111110	000000	111110	111110	111100	111111
G	111101	111101	111101	000010	000010	000010	000000	111101	000001	111111
H	111101	111101	111101	000010	000010	000010	111110	000000	111100	111111
I	000000	000000	000000	000000	000000	000000	000010	111100	000000	111110

Based on the developed matrix analysis algorithm in Sect. 12.3.1, there are totally 918 feasible disassembly sequences for the base assembly part.

Front cover assembly part

The front cover assembly part of the LC24F4 LCD-TV is shown in Fig. 12.9. It is composed of 11 parts: (*J*) control button, (*K*) power switch, (*L*) side loudspeaker, (*M*) control receiver board, (*N*) positive loudspeaker, (*O*) power supply board, (*P*) main board, (*Q*) metal board, (*R*) metal mounting plate, (*S*) surface frame, and (*T*) LCD screen.

The space interference matrices to represent the front cover assembly part in six directions are shown below:

$$S_{X^+} = \begin{array}{c} \\ J \\ K \\ L \\ M \\ N \\ O \\ P \\ Q \\ R \\ S \\ T \end{array} \begin{array}{c} \begin{array}{ccccccccccc} J & K & L & M & N & O & P & Q & R & S & T \end{array} \\ \begin{bmatrix} 0 & 1 & 1 & 1 & 0 & 0 & 0 & 0 & 1 & 0 & 0 \\ 0 & 0 & 1 & 1 & 0 & 0 & 0 & 0 & 1 & 0 & 0 \\ 0 & 0 & 0 & 1 & 0 & 0 & 0 & 0 & 1 & 0 & 0 \\ 0 & 0 & 0 & 0 & 0 & 0 & 0 & 0 & 1 & 0 & 0 \\ 0 & 0 & 0 & 0 & 0 & 0 & 1 & 0 & 1 & 0 & 0 \\ 0 & 0 & 0 & 0 & 0 & 0 & 1 & 1 & 1 & 0 & 0 \\ 0 & 0 & 0 & 0 & 0 & 0 & 0 & 0 & 1 & 0 & 0 \\ 0 & 0 & 0 & 0 & 0 & 0 & 1 & 0 & 1 & 0 & 0 \\ 0 & 1 & 1 & 1 & 1 & 1 & 1 & 1 & 0 & 1 & 1 \\ 0 & 0 & 0 & 0 & 0 & 0 & 0 & 0 & 1 & 0 & 1 \\ 0 & 0 & 0 & 0 & 0 & 0 & 0 & 0 & 1 & 1 & 0 \end{bmatrix} \end{array}$$

$$S_{X^-} = \begin{array}{c} \\ J \\ K \\ L \\ M \\ N \\ O \\ P \\ Q \\ R \\ S \\ T \end{array} \begin{array}{c} \begin{array}{ccccccccccc} J & K & L & M & N & O & P & Q & R & S & T \end{array} \\ \begin{bmatrix} 0 & 0 & 0 & 0 & 0 & 0 & 0 & 0 & 0 & 0 & 0 \\ 1 & 0 & 0 & 0 & 0 & 0 & 0 & 0 & 1 & 0 & 0 \\ 1 & 1 & 0 & 0 & 0 & 0 & 0 & 0 & 1 & 0 & 0 \\ 1 & 1 & 1 & 0 & 0 & 0 & 0 & 0 & 1 & 0 & 0 \\ 0 & 0 & 0 & 0 & 0 & 0 & 0 & 0 & 1 & 0 & 0 \\ 0 & 0 & 0 & 0 & 0 & 0 & 0 & 0 & 1 & 0 & 0 \\ 0 & 0 & 0 & 0 & 1 & 1 & 0 & 1 & 1 & 0 & 0 \\ 0 & 0 & 0 & 0 & 1 & 0 & 0 & 0 & 1 & 0 & 0 \\ 1 & 1 & 1 & 1 & 1 & 1 & 1 & 1 & 0 & 1 & 1 \\ 0 & 0 & 0 & 0 & 0 & 0 & 0 & 0 & 1 & 0 & 1 \\ 0 & 0 & 0 & 0 & 0 & 0 & 0 & 0 & 1 & 1 & 0 \end{bmatrix} \end{array}$$

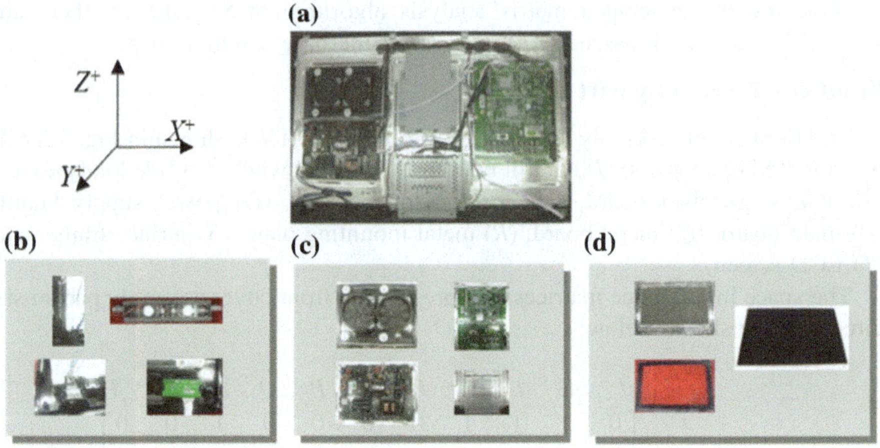

Fig. 12.9 Front assembly part of the LC24F4 LCD-TV: **a** front assembly part, **b** components *J*, *K*, *L*, *M*, **c** components *N*, *O*, *P*, *Q*, and **d** components *R*, *S*, *T*

$$S_{Y^+} = \begin{array}{c} \\ J \\ K \\ L \\ M \\ N \\ O \\ P \\ Q \\ R \\ S \\ T \end{array} \begin{array}{c} \begin{array}{ccccccccccc} J & K & L & M & N & O & P & Q & R & S & T \end{array} \\ \begin{bmatrix} 0 & 0 & 0 & 0 & 0 & 0 & 0 & 0 & 1 & 0 & 0 \\ 0 & 0 & 0 & 0 & 0 & 0 & 0 & 0 & 1 & 0 & 0 \\ 0 & 0 & 0 & 0 & 1 & 1 & 0 & 0 & 1 & 0 & 0 \\ 0 & 0 & 0 & 0 & 0 & 0 & 0 & 0 & 1 & 0 & 0 \\ 0 & 0 & 0 & 0 & 0 & 0 & 0 & 0 & 0 & 0 & 0 \\ 0 & 0 & 0 & 0 & 0 & 0 & 0 & 0 & 0 & 0 & 0 \\ 0 & 0 & 0 & 0 & 0 & 0 & 0 & 0 & 0 & 0 & 0 \\ 0 & 0 & 0 & 0 & 0 & 0 & 1 & 0 & 0 & 0 & 0 \\ 1 & 1 & 1 & 1 & 1 & 1 & 1 & 1 & 0 & 0 & 0 \\ 1 & 1 & 1 & 1 & 1 & 1 & 1 & 1 & 1 & 0 & 1 \\ 1 & 1 & 1 & 1 & 1 & 1 & 1 & 1 & 1 & 0 & 0 \end{bmatrix} \end{array}$$

$$S_{Y^+} = \begin{array}{c} \\ J \\ K \\ L \\ M \\ N \\ O \\ P \\ Q \\ R \\ S \\ T \end{array} \begin{array}{c} \begin{array}{ccccccccccc} J & K & L & M & N & O & P & Q & R & S & T \end{array} \\ \begin{bmatrix} 0 & 0 & 0 & 0 & 0 & 0 & 0 & 0 & 1 & 1 & 1 \\ 0 & 0 & 0 & 0 & 0 & 0 & 0 & 0 & 1 & 1 & 1 \\ 0 & 0 & 0 & 0 & 0 & 0 & 0 & 0 & 1 & 1 & 1 \\ 0 & 0 & 0 & 0 & 0 & 0 & 0 & 0 & 1 & 1 & 1 \\ 0 & 0 & 0 & 0 & 0 & 0 & 0 & 0 & 1 & 1 & 1 \\ 0 & 0 & 1 & 0 & 0 & 0 & 0 & 0 & 1 & 1 & 1 \\ 0 & 0 & 0 & 0 & 0 & 0 & 0 & 1 & 1 & 1 & 1 \\ 0 & 0 & 0 & 0 & 0 & 0 & 0 & 0 & 1 & 1 & 1 \\ 0 & 0 & 0 & 0 & 0 & 0 & 0 & 0 & 0 & 1 & 1 \\ 0 & 0 & 0 & 0 & 0 & 0 & 0 & 0 & 0 & 0 & 0 \\ 0 & 0 & 0 & 0 & 0 & 0 & 0 & 0 & 0 & 1 & 0 \end{bmatrix} \end{array}$$

$$
S_{Z^{+}} = \begin{array}{c} \\ J \\ K \\ L \\ M \\ N \\ O \\ P \\ Q \\ R \\ S \\ T \end{array}
\begin{array}{c} \begin{array}{ccccccccccc} J & K & L & M & N & O & P & Q & R & S & T \end{array} \\
\begin{bmatrix}
0 & 0 & 0 & 0 & 0 & 0 & 0 & 0 & 1 & 0 & 0 \\
0 & 0 & 0 & 0 & 0 & 0 & 0 & 0 & 1 & 0 & 0 \\
0 & 0 & 0 & 0 & 0 & 0 & 0 & 0 & 1 & 0 & 0 \\
0 & 0 & 0 & 0 & 0 & 0 & 0 & 0 & 1 & 0 & 0 \\
0 & 0 & 0 & 0 & 0 & 0 & 0 & 0 & 1 & 0 & 0 \\
0 & 0 & 0 & 0 & 1 & 0 & 0 & 0 & 1 & 0 & 0 \\
0 & 0 & 0 & 0 & 0 & 0 & 0 & 0 & 1 & 0 & 0 \\
0 & 0 & 0 & 0 & 0 & 0 & 1 & 0 & 1 & 0 & 0 \\
1 & 0 & 0 & 0 & 1 & 1 & 1 & 1 & 0 & 1 & 1 \\
0 & 0 & 0 & 0 & 0 & 0 & 0 & 0 & 1 & 0 & 1 \\
0 & 0 & 0 & 0 & 0 & 0 & 0 & 0 & 1 & 1 & 0
\end{bmatrix} \end{array}
$$

$$
S_{Z^{-}} = \begin{array}{c} \\ J \\ K \\ L \\ M \\ N \\ O \\ P \\ Q \\ R \\ S \\ T \end{array}
\begin{array}{c} \begin{array}{ccccccccccc} J & K & L & M & N & O & P & Q & R & S & T \end{array} \\
\begin{bmatrix}
0 & 0 & 0 & 0 & 0 & 0 & 0 & 0 & 1 & 0 & 0 \\
0 & 0 & 0 & 0 & 0 & 0 & 0 & 0 & 0 & 0 & 0 \\
0 & 0 & 0 & 0 & 0 & 0 & 0 & 0 & 0 & 0 & 0 \\
0 & 0 & 0 & 0 & 0 & 0 & 0 & 0 & 0 & 0 & 0 \\
0 & 0 & 0 & 0 & 0 & 1 & 0 & 0 & 1 & 0 & 0 \\
0 & 0 & 0 & 0 & 0 & 0 & 0 & 0 & 1 & 0 & 0 \\
0 & 0 & 0 & 0 & 0 & 0 & 0 & 0 & 1 & 0 & 0 \\
0 & 0 & 0 & 0 & 0 & 0 & 0 & 0 & 1 & 0 & 0 \\
1 & 1 & 1 & 1 & 1 & 1 & 1 & 1 & 0 & 1 & 1 \\
0 & 0 & 0 & 0 & 0 & 0 & 0 & 0 & 1 & 0 & 1 \\
0 & 0 & 0 & 0 & 0 & 0 & 0 & 0 & 1 & 1 & 0
\end{bmatrix} \end{array}
$$

After combining the above six matrices and using Boolean operator 'OR' in rows, the obtained result is shown below:

$S =$

	J	*K*	*L*	*M*	*N*	*O*	*P*	*Q*	*R*	*S*	*T*	Result
J	000000	100000	100000	100000	000000	000000	000000	000000	101111	000100	000100	101111
K	010000	000000	100000	100000	000000	000000	000000	000000	111110	000100	000100	111110
L	010000	010000	000000	100000	000000	001000	001000	000000	111110	000100	000100	111110
M	010000	010000	010000	000000	000000	000000	000000	000000	111110	000100	000100	111110
N	000000	000000	000000	000000	000000	000001	000000	000000	110111	000100	000100	111111
O	000000	000000	000100	000000	000010	000000	100000	100000	110111	000100	000100	110111
P	000000	000000	000000	000000	010000	010000	000000	010000	110111	000100	000100	110111
Q	000000	000000	000000	000000	000000	010000	100000	000000	110111	000100	000100	110111
R	011011	111001	111001	111001	111011	111011	111011	111011	000000	110111	110111	111111
S	001000	001000	001000	001000	001000	001000	001000	001000	111011	000000	111011	111011
T	001000	001000	001000	001000	001000	001000	001000	001000	111011	110111	000000	111111

Based on the developed matrix analysis algorithm, there are a total of 7,096,320 feasible disassembly sequences for the front assembly part.

Back cover assembly part

The back cover assembly part of the LC24F4 LCD-TV is composed of three parts: (*U*) back cover, (*V*) cover plate, and (*W*) support (shown in Fig. 12.10).

The space interference matrices to represent the back cover assembly part in six directions are shown below:

$$S_{X+} = \begin{matrix} U \\ V \\ W \end{matrix}\overset{\begin{matrix} U & V & W \end{matrix}}{\begin{bmatrix} 0 & 1 & 1 \\ 1 & 0 & 1 \\ 0 & 0 & 0 \end{bmatrix}} \quad S_{X-} = \begin{matrix} U \\ V \\ W \end{matrix}\overset{\begin{matrix} U & V & W \end{matrix}}{\begin{bmatrix} 0 & 1 & 0 \\ 1 & 0 & 0 \\ 1 & 1 & 0 \end{bmatrix}}$$

$$S_{Y+} = \begin{matrix} U \\ V \\ W \end{matrix}\overset{\begin{matrix} U & V & W \end{matrix}}{\begin{bmatrix} 0 & 1 & 1 \\ 0 & 0 & 0 \\ 1 & 0 & 0 \end{bmatrix}} \quad S_{Y-} = \begin{matrix} U \\ V \\ W \end{matrix}\overset{\begin{matrix} U & V & W \end{matrix}}{\begin{bmatrix} 0 & 0 & 1 \\ 1 & 0 & 0 \\ 1 & 0 & 0 \end{bmatrix}}$$

$$S_{Z+} = \begin{matrix} U \\ V \\ W \end{matrix}\overset{\begin{matrix} U & V & W \end{matrix}}{\begin{bmatrix} 0 & 1 & 1 \\ 1 & 0 & 0 \\ 1 & 0 & 0 \end{bmatrix}} \quad S_{Z-} = \begin{matrix} U \\ V \\ W \end{matrix}\overset{\begin{matrix} U & V & W \end{matrix}}{\begin{bmatrix} 0 & 1 & 1 \\ 1 & 0 & 0 \\ 1 & 0 & 0 \end{bmatrix}}$$

The combined matrix can be obtained as follows:

$$S = \begin{matrix} U \\ V \\ W \end{matrix}\overset{\begin{matrix} U & \quad V & \quad W \end{matrix}}{\begin{bmatrix} 000000 & 111011 & 101111 \\ 110111 & 000000 & 100000 \\ 011111 & 010000 & 000000 \end{bmatrix}} \quad \begin{matrix} \text{Result} \\ 111111 \\ 110111 \\ 011111 \end{matrix}$$

Based on the developed matrix analysis algorithm, the number of feasible disassembly sequences for the back cover assembly part is 4.

Fig. 12.10 Back cover assembly part of LCD-TV

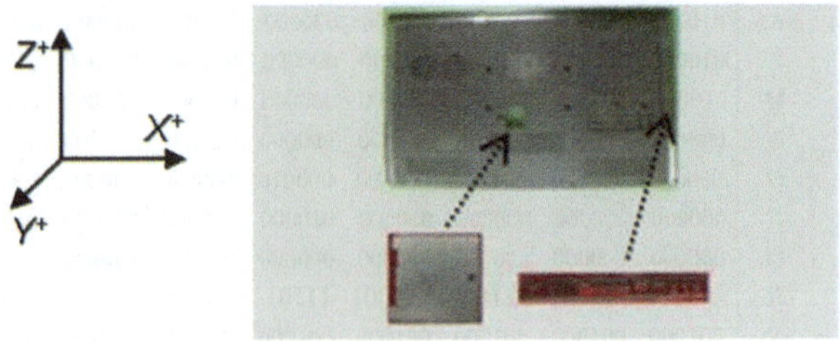

Table 12.8 Comparison between our developed method and full search space

This developed method: (all feasible disassembly sequences)	$918 \times 7{,}096{,}320 \times 4 = 2.6058e^{+10}$
Full search space: (all disassembly sequences)	$23! = 23 \times 22 \ldots 2 \times 1 = 2.5852e^{+22}$
Search range reduction:	$2.5852e^{+22}/2.6058e^{+10} = 9.9209e^{+11}$ times

Based on the above analysis, the number of all the feasible disassembly sequences with geometric constraints of the LC24F4 LCD-TV is $2.6058e^{+10} = 918 \times 7{,}096{,}320 \times 4$ (base assembly part $\times$ front cover assembly part $\times$ back over assembly part). Compared with the theoretical full search space, which could be as large as $23! = 23 \times 22 \ldots 2 \times 1 = 2.5852e^{+22}$, the search range for a disassembly planning algorithm to find the optimized disassembly sequence is reduced by $9.9209e^{+11}$ times (shown in Table 12.8). It is obvious that the developed feasible solution space method can dramatically reduce the search range and obtain all the feasible disassembly sequences of the LC24F4 LCD-TV to alleviate the effort on the search of the optimal disassembly sequence.

Evaluation and decision making on LCD-TVs

The components and some properties of the LC24F4 LCD-TVs provided by Changhong Electronics Company, Ltd. are listed in Table 12.9. The total mass of a LC24F4 LCD-TV is 5648.2 g. Among the component/material composition, the PCBs, which are mainly the main board and power supply board, loudspeaker, and LCD screen are quite complex and are composed of several different materials. Other components of LCD-TVs are usually made of a single material such as: steel, aluminum, copper, ABS.

The calculation of disassembly time, value, cost of the disassembly operations, and the identification of hazardous, heavy and high-value components of the LC24F4 LCD-TVs are described in the following. The results of the calculations are listed in Table 12.10.

- *Disassembly time*: Firstly, the base assembly part, front assembly part, and back cover assembly part are disassembled manually to calculate the disassembly time of each component.
- *Disassembly cost*: The disassembly cost can be calculated in Eqs. (12.15)–(12.17). The cost of labor and factory operation are set as 150 and 3000 Yuan/day separately, there are 100 workers in the disassembly factory.
- *Residual value*: The residual value can be calculated in Eq. (12.13). All the potential values are calculated based on the values of materials of components as there is no component that can be reused and repaired in the LC24F4 LCD-TVs.
- *Hazardous components*: The hazardous components can be identified in Eq. (12.19). In the LC24F4 LCD-TVs, (*O*) power supply board, (*P*) main board, and (*T*) LCD screen are identified as hazardous as discussed in Sect. 12.3.2.2 and they have to be removed for further recycling and processing.

Table 12.9 Components and some properties of the LC24F4 LCD-TVs)

Assembly part	Components	Mass (g)	Materials
Base assembly part	(*A*) Metal fixing plate	15.0	Aluminum
	(*B*) Metal washer 1	10.0	Steel
	(*C*) Metal washer 2	10.0	Steel
	(*D*) Top metal support	25.0	Aluminum
	(*E*) Cylindrical support 1	30.0	PS
	(*F*) Cylindrical support 2	20.0	PS
	(*G*) Toughened glass seat	150.0	Glass
	(*H*) Steel plate	50.0	Steel
	(*I*) Rubber gasket	20.0	Black plastic
Front assembly part	(*J*) Control button	9.2	ABS
	(*K*) Power switch	5.0	TPE, Copper
	(*L*) Side loudspeaker	152.0	Steel, Copper, Plastic, etc.
	(*M*) Control receiver board	3.0	Copper, FP4
	(*N*) Positive loudspeaker	77.8	Steel, Copper, Plastic, etc.
	(*O*) Power supply board	118.0	Copper, Gold, Lead, Cadmium, etc.
	(*P*) Main board	196.0	Copper, Gold, Lead, Cadmium, etc.
	(*Q*) Metal board	183.0	Steel
	(*R*) Metal mounting plate	639.0	Steel
	(*S*) Surface frame	270.8	ABS
	(*T*) LCD screen	2900.0	Silicon, Glass, Polymer, Mercury, etc.
Back cover assembly part	(*U*) Back cover	723.8	PS
	(*V*) Cover plate	25.0	PET
	(*W*) Support	15.6	ABS

- *Heavy components*: The heavy components can be identified in Eq. (12.20). If the component weight is over 2.5% (141.205 g = 5648.2 × 2.5% g) of the whole mass of LC24F4 LCD-TVs, the component is identified as heavy component by the disassembly factory. Components, (*G*) toughened glass seat, (*L*) side loudspeaker, (*P*) main board, (*Q*) metal board, (*R*) metal mounting plate, (*S*) surface frame, (*T*) LCD screen, and (*U*) back cover, are identified as heavy components.

Table 12.10 Results of the calculations for hazardous, heavy and high-value components

Components	Disassembly Time (min)	Disassembly Cost (Yuan)	Residual Value (Yuan)	Hazardous	Heavy	High-Value
(*A*) Metal fixing plate	0.30	0.1125	0.2970			
(*B*) Metal washer 1	0.04	0.0150	0.0660			
(*C*) Metal washer 2	0.04	0.0150	0.0660			
(*D*) Top metal support	0.35	0.1313	0.4950			
(*E*) Cylindrical support 1	0.10	0.0376	0.2400			
(*F*) Cylindrical support 2	0.10	0.0376	0.1600			
(*G*) Toughened glass seat	0.23	0.0863	0.2380		O	
(*H*) Steel plate	0.20	0.0751	0.3300			
(*I*) Rubber gasket	0.10	0.0376	0.0200			
(*J*) Control button	0.08	0.0300	0.0100			
(*K*) Power switch	0.08	0.0300	0.0100			
(*L*) Side loudspeaker	0.35	0.1313	0.6000		O	
(*M*) Control receiver board	0.10	0.0376	0.4000			
(*N*) Positive loudspeaker	0.25	0.0940	0.3071			
(*O*) Power supply board	0.70	0.2626	0.6466	O		
(*P*) Main board	0.70	0.2626	0.7908	O	O	
(*Q*) Metal board	0.59	0.2213	1.2078		O	O
(*R*) Metal mounting plate	1.82	0.6826	4.2174		O	O
(*S*) Surface frame	1.23	0.4613	1.1000		O	
(*T*) LCD screen	1.42	0.4438	9.6684	O	O	O
(*U*) Back cover	1.65	0.5326	1.7904		O	O
(*V*) Cover plate	0.03	0.0113	0.2280			
(*W*) Support	0.04	0.0150	0.0169			

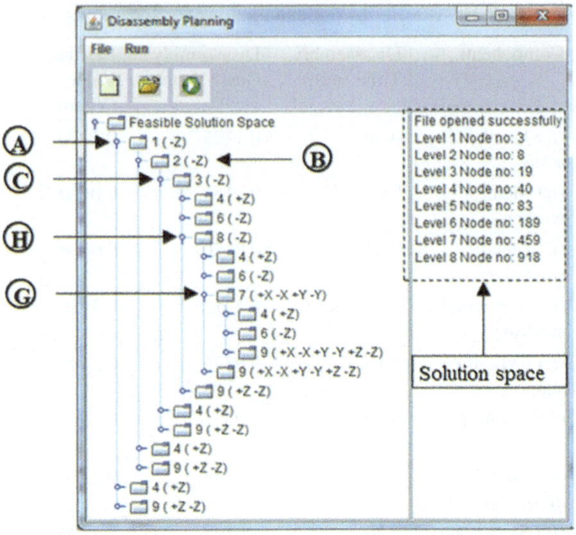

Fig. 12.11 Developed software and the obtained optimized sequence

- *High-value components*: The high-value components can be identified by Eq. (12.21). If the potential value of a component is over 5% (1.1403 Yuan = 22.8054 × 5% Yuan) of the whole residual value of LC24F4 LCD-TV, the component is identified as a high-value component by the disassembly factory. Here, components, (*Q*) metal board, (*R*) metal mounting plate, (*T*) LCD screen, (*U*) back cover, are identified as high-value components.

For the base assembly part, only component (Ⓖ—toughened glass seat) is required to be removed. Based on the obtained feasible solution space for the base assembly part in the previous section, Fig. 12.11 shows the developed software and the obtained optimized sequence (ⒶⒷⒸⒽⒼ) with the maximum profit to dismantle component (Ⓖ) based on computing. The optimized selective disassembly sequences for the front assembly part and back cover assembly part are also obtained using the developed software. Table 12.11 shows the obtained result and the related disassembly cost.

An EOL process flow is then generated with the obtained optimal selective disassembly sequence for the LC24F4 LCD-TV (Shown in Fig. 12.12). If the disassembled component is composed of a single material, the EOL process of the component is Type 3 (remanufacture) as there is no reuse and repair components in the LC24F4 LCD-TV, and the recycling rate of the material is 100% (except glass is 80%); if the disassembled component is composed of several materials, the EOL process of the component is Type 4 (shredding), and the recycling rate of the material is different based on different separation methods. After the EOL disassembly process, the remaining components are recycled for valuable materials with EOL shredding process. In the end, all the worthless materials and components are disposed with landfill and incineration processes.

Table 12.11 Obtained optimal disassembly sequences for LCD-TVs)

Sub-assembly	Selective disassembly sequence	Disassembly cost (Yuan)
Base assembly part	ⒶⒷⒸⒽⒼ	0.3049
Front assembly part	ⓋⓊ	0.5439
Back cover assembly part	ⓁⓃⓄⓅⓆⓈⓉⓇ	2.5971

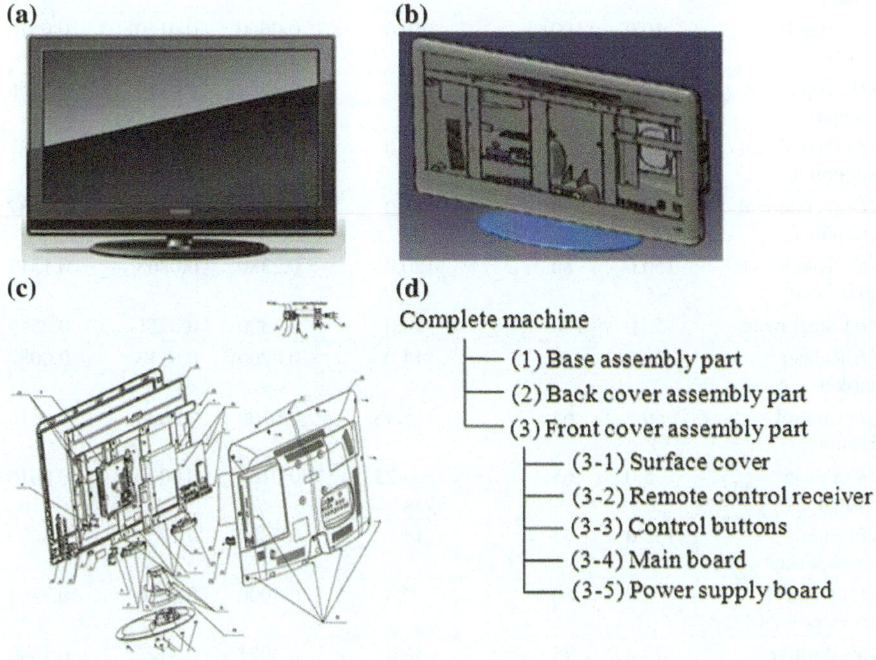

Fig. 12.12 EOL process flow for the LC24F4 LCD-TV

Based on the above analysis, the recycling rate and the total disassembly profit of the LC24F4 LCD-TV can be calculated in the following. The results are listed in Table 12.12.

- *The recycling rate*: Based on the mass and recycling rate of different materials/components, the weights of recycled materials of each component can be calculated, and the recycling rate of the LC24F4 LCD-TV is 86.55% (86.55% = 4888.93/5648.2 100%).
- *The total disassembly profit*: The total disassembly cost can be calculated in Eq. (5.14). The costs of labor and factory operation are set as 150 and 450 Yuan/day separately. There are 10 workers in each cleaning, repair, shredding, separation, landfill, and incineration factories. The total disassembly profit is 12.9616 Yuan (12.9616 = 22.9054−9.9438).

Table 12.12 Recycling rate and profit of the LC24F4 LCD-TV

Components	Mass (g)	Efficiency (%)	Material recycled (g)	Value (Yuan)	Total cost (Yuan)	Profit (Yuan)
(*A*) Metal fixing plate	15.0	100	15.0	0.2970	0.1125	0.1845
(*B*) Metal washer 1	10.0	100	10.0	0.0660	0.0150	0.0510
(*C*) Metal washer 2	10.0	100	10.0	0.0660	0.0150	0.0510
(*D*) Top metal support	25.0	90	22.5	0.4950	0.0394	0.4556
(*E*) Cylindrical support 1	30.0	70	21.0	0.2400	0.0113	0.2287
(*F*) Cylindrical support 2	20.0	70	14.0	0.1600	0.0113	0.1487
(*G*) Toughened glass seat	150.0	80	120.0	0.2380	0.0863	0.1517
(*H*) Steel plate	50.0	100	50.0	0.3300	0.0751	0.2549
(*I*) Rubber gasket	20.0	70	14.0	0.0200	0.0113	0.0087
(*J*) Control button	9.2	65	5.98	0.0100	0.0090	0.0010
(*K*) Power switch	5.0	65	3.25	0.0100	0.0090	0.0010
(*L*) Side loudspeaker	152.0	75	114	0.6000	0.1707	0.4293
(*M*) Control receiver board	3.0	80	2.4	0.4000	0.0436	0.3564
(*N*) Positive loudspeaker	77.8	75	58.4	0.3071	0.1222	0.1849
(*O*) Power supply board	118.0	80	94.4	0.6466	0.4986	0.1480
(*P*) Main board	196.0	80	156.8	0.7908	0.6546	0.1362
(*Q*) Metal board	183.0	100	183.0	1.2078	0.2213	0.9865
(*R*) Metal mounting plate	639.0	100	639.0	4.2174	0.6826	3.5348
(*S*) Surface frame	270.8	100	270.8	1.1000	0.4613	0.6387
(*T*) LCD screen	2900.0	80	2320	9.6684	6.2438	3.4246
(*U*) Back cover	723.8	100	723.8	1.7904	0.5326	1.2578
(*V*) Cover plate	25.0	100	25.0	0.2280	0.0113	0.2167
(*W*) Support	15.6	100	15.6	0.0169	0.0150	0.0019
Total	5648.2	86.55	4888.93	22.9054	9.9438	12.9616

Total recycling material: 4888.93, Total profit: 12.9616
Recycling rate (86.55%) $\geq$ WEEE Directive (75%)

After the above process, the disassembly results of the LC24F4 LCD-TV are achieved as follows:

(1) All the hazardous, heavy and high-value components are removed, including:
   - Hazardous components: (*O*) power supply board, (*P*) main board, and (*T*) LCD screen.
   - Heavy components: (*G*) toughened glass seat, (*L*) side loudspeaker, (*P*) main board, (*Q*) metal board, (*R*) metal mounting plate, (*S*) surface frame, (*T*) LCD screen, and (*U*) back cover.
   - High-value components: (*Q*) metal board, (*R*) metal mounting plate, (*T*) LCD screen, (*U*) back cover.
(2) The recycling rate meets the restricted recycling rate of the WEEE Directive. The recycling rate is 86.55%, which is greater than the restricted recycling rate 75% of the WEEE Directive.
(3) The optimized selective disassembly sequence has been obtained with the maximum profit of the disassembly operation.

12.5 Conclusions

In this paper, a systematic selective disassembly approach is developed to handle WEEE to meet the environmental and economic requirements. The characteristics and contributions of the research include:

(1) Space interference matrix is used to represent the space interference relationship of components in six directions of WEEE. In this manner, all the space interference relationship between components can be digitally recorded and analyzed in the next step;
(2) A matrix analysis algorithm is developed to obtain all the feasible disassembly sequences by analyzing six space interference matrices in a 3D environment. It is capable of obtaining all the feasible disassembly sequences of WEEE, and the result can be used as a solution space to search for an optimized disassembly sequence within an acceptable runtime;
(3) An evaluation and decision-making method is developed to find out the optimized selective disassembly. It is capable of removing all hazardous, high-value and heavy components from WEEE, maximizing disassembly profit and meeting the restricted recycling rate of the WEEE Directive;
(4) An industrial case study on LC24F4 LCD-TVs has been used to demonstrate the performance of the developed approach.

Appendix

Notation	
$C^{\text{clean}}(\text{Oper}_i)$	Clean operation cost
$C^{\text{disassembly}}(\text{Oper}_i)$	Disassembly operation cost
$C^{\text{hazardous}}$	Hazardous components
C^{heavy}	Heavy components
$C^{\text{high-value}}$	High-value components
C_i	Component element in matrix
$C^{\text{landfill/incinerate}}(\text{Oper}_i)$	Disposal operation cost
$C(\text{ Oper}_i)$	Operation cost
Comp_i	Components
$C^{\text{repair}}(\text{Oper}_i)$	Repair operation cost
$C^{\text{shredding}}(\text{Oper}_i)$	Shredding operation cost
$C^{\text{separation}}(\text{Oper}_i)$	Separation operation cost
E^{factory}	Factory expense
N^{labor}	Worker number
$\text{Profit (Oper}_i)$	Operation profit
r_{ij}	Space interference relationship
r_{ji}^{cyc}	Recycling rate of the material
$R_{\text{recycling}}$	Recycling rate
$T^{\text{disassembly}}(\text{Oper}_i)$	Disassembly operation time
TP	Total profit
$V\ (\text{Oper}_i)$	Operation value
w_{ji}	Recycling weight of the material
W^{labor}	Worker wage
W_{total}	Total weight

References

1. Stevels, A., & Huisman, J. (2003). An industry vision on the implementation of WEEE and RoHS. In *Proceedings of EcoDesign2003: Third international symposium on environmentally conscious design and inverse manufacturing* (pp. 8–11). Tokyo, Japan.
2. Li, W. D., Xia, K., Gao, L., & Chao, K. M. (2013). Selective disassembly planning for waste electrical and electronic equipment with case studies on liquid crystal displays. *Robotics and Computer-Integrated Manufacturing, 29*(4), 248–260.
3. Kara, S., Pornprasitpol, P., & Kaebernick, H. (2006). Selective disassembly sequencing: A methodology for the disassembly of end-of-life products. *CIRP—Manufacturing Technology, 55*(1), 37–40.
4. Jin, G. Q., & Li, W. D. (2008). Life Cycle Management of LCD televisions—A case study. In *Handbook of Manufacturing Engineering and Technology* (pp. 1–28). Springer.

5. Ying, T., Zhou, M. C., Zussman, E., & Caudill, R. (2000). Disassembly modelling, planning, and application: A review. In *Proceedings of the 2000 IEEE International Conference on Robotics & Automation* (pp. 2197–2202). San Francisco, USA.
6. Lee, D. H., Kang, J. G., & Xirouchakis, P. (2001). Disassembly planning and scheduling: review and further research. *Proceedings of the Institution of Mechanical Engineers, Part B: Journal of Engineering Manufacture, 215*(5), 695–709.
7. Lambert, A. J. D. (2003). Disassembly sequencing: A survey. *International Journal of Production Research, 41*(16), 3721–3759.
8. Hsieh, D. (2017). 4K TV Boom—Over 30M in 2014? *Analyst Blog 2014*. Available online http://www.displaysearchblog.com/2013/11/4k-tv-boom%E2%80%93-over-30m-in-2014/. Accessed December 8, 2017.
9. Kara, S., Pornprasitpol, P., & Kaebernick, H. (2005). A selective disassembly methodology for end-of-life products. *Assembly Automation, 25*(2), 124–134.
10. Nevins, J. L., & Whitney, D. E. (1989). *Concurrent design of products & processes: A strategy for the next generation in manufacturing*. New York, USA: McGraw-Hill.
11. Garcia, M. A., Larre, A., Lopez, B., & Oller, A. (2000). Reducing the complexity of geometric selective disassembly. In *Proceedings of the 2000 IEEE/RSJ International Conference on Intelligent Robots and Systems* (pp. 1474–1479), Takamatsu, Japan.
12. Behdad, S., Kwak, M., Kim, H., & Thurston, D. (2010). Simultaneous selective disassembly and end-of-life decision making for multiple products that share disassembly operations. *Journal of Mechanical Design, 132*(4), 041002–041009.
13. Elsayed, A., Kongar, E., & Gupta, S. M. (2012). An evolutionary algorithm for selective disassembly of end-of-life products. *International Journal of Swarm Intelligence and Evolutionary Computation, 1,* 1–7.
14. Srinivasan, H., & Gadh, R. (1998). A geometric algorithm for single selective disassembly using the wave propagation abstraction. *Computer-Aided Design, 30*(8), 603–613.
15. Srinivasan, H., Figueroa, R., & Gadh, R. (1999). Selective disassembly for virtual prototyping as applied to de-manufacturing. *Robotics & Computer Integrated Manufacturing, 15*(15), 231–245.
16. Smith, S. S., & Chen, W. H. (2011). Rule-based recursive selective disassembly sequence planning for green design. *Advanced Engineering Informatics, 25*(1), 77–87.
17. Smith, S. S., Smith, G., & Chen, W. H. (2012). Disassembly sequence structure graphs: an optimal approach for multiple-target selective disassembly sequence planning. *Advanced Engineering Informatics, 26*(2), 306–316.
18. Ryan, A., O'Donoghue, L., & Lewis, H. (2011). Characterising components of liquid crystal displays to facilitate disassembly. *Journal of Cleaner Production, 19*(9–10), 1066–1071.
19. Mizuno, T., Kunii, E., Fukushige, S., & Umeda, Y. (2012). Recyclability evaluation of LCD TVs based on end-of-life scenarios. In *Leveraging Technology for a Sustainable World: 19th CIRP international conference on life cycle engineering* (pp. 179–183). Berkeley, USA.
20. Fukushige, S., Mizuno, T., Kunii, E., Matsuyama, Y., & Umeda, Y. (2013). Quantitative design modification for the recyclability of products. In *Re-engineering Manufacturing for Sustainability: 20th CIRP international conference on life cycle engineering* (pp. 27–33). Singapore.
21. Chiodo, J. D., McLaren, J., Billett, E. H., & Harrison, D. J. (2000). Isolating LCD's at end-of-life using active disassembly technology a feasibility study. In *Proceedings of the 2000 IEEE international symposium on electronics and the environment* (pp. 318–323).
22. Li, W. D., Xia, K., Lu, B., Chao, K. M., Gao, L., & Yang, J. X. (2013). A distributed service of selective disassembly planning for waste electrical and electronic equipment with case studies on liquid crystal display. In *Springer Series in Advance Manufacturing: Cloud Manufacturing* (pp. 23–47).
23. Brute-force search. (2017). Available online http://en.wikipedia.org/wiki/Brute-force_search. Accessed December 8, 2017.
24. He, W., & Li, G. (2006). WEEE recovery strategies and the WEEE treatment status in China. *Journal of Hazardous Materials, B136,* 502–512.

25. Abubakar, M. S., & Rahimifard, S. (2008). Ecological and economical assessment of end-of-life waste recycling in the electrical and electronic recovery sector. *International Journal of Sustainable Engineering, 1*(4), 261–277.
26. Cusack, P., & Perrett, T. (2006). The EU ROHS directive and its implications for the plastics industry. *Plastics, Additives and Compounding, 8*(3), 46–49.
27. The Code of Federal Regulations: Hazardous Waste Regulations. (2014). Available online http://www.epa.gov/osw/laws-regs/regs-haz.htm. Accessed May 15, 2014.
28. Cui, J., & Forssberg, E. (2003). Mechanical recycling of waste electric and electronic equipment: a review. *Journal of Hazardous Materials, B99*, 243–263.
29. Sander, K., Schilling, S., Tojo, N., Vanrossem, C., Vernon, J., & George, C. (2014). *The producer responsibility principle of the WEEE Directive*. Final Report (August 19) DG ENV. Study y Contract No. 07010401/2006/ 449269/ MAR/ G4). Available online http://reach-consultants.co.uk/documents/J572_WEEEPRPReportFinal.pdf. Accessed December 8, 2017.

Chapter 13
Disassembly Sequence Planning Using a Simplified Teaching-Learning-Based Optimization Algorithm

Kai Xia, Liang Gao, Weidong Li and Kuo-Ming Chao

Abstract Disassembly sequence planning (DSP) is a challenging NP-hard combinatorial optimization problem. As a new and promising population-based evolutional algorithm, teaching-learning-based optimization (TLBO) algorithm has been successfully applied to various research problems. However, TLBO is not capable or effective in DSP optimization problems with discrete solution spaces and complex disassembly precedence constraints. This chapter presents a simplified teaching-learning-based optimization (STLBO) algorithm to solve DSP problems effectively. The STLBO algorithm inherits the main idea of the teaching-learning-based evolutionary mechanism from the TLBO algorithm, while the realization method of the evolutionary mechanism and the adaptation methods of the algorithm parameters are different. Three new operators are developed and incorporated in the STLBO algorithm to ensure its applicability to DSP problems with complex disassembly precedence constraints: i.e., a feasible solution generator (FSG) used to generate a feasible disassembly sequence, a teaching phase operator (TPO), and a learning phase operator (LPO) used to learn and evolve the solutions toward better ones by applying the method of precedence preservation crossover operation. Numerical experiments and case studies on waste product disassembly planning have been carried out to demonstrate the effectiveness of the designed operators and the results exhibited that the developed algorithm performs better than other relevant algorithms under a set of public benchmarks.

K. Xia
Wuhan Second Ship Design and Research Institute, Wuhan, China

L. Gao (✉)
State Key Laboratory of Digital Manufacturing Equipment and Technology, School of Mechanical Science and Engineering, Huazhong University of Science and Technology, Wuhan, China
e-mail: gaoliang@mail.hust.edu.cn

W. Li · K.-M. Chao
Faculty of Engineering, Environment and Computing, Coventry University, Coventry, UK

W. Li and S. Wang (eds.), *Sustainable Manufacturing and Remanufacturing Management*, https://doi.org/10.1007/978-3-319-73488-0_13

Keywords Disassembly · Disassembly sequence planning · Teaching-learning-based optimization · Simplified teaching-learning-based optimization · Meta-heuristics

13.1 Introduction

Mass-customized productions, technology updating, and shortening life span of products in modern societies have resulted in a generation of enormous amount of waste products such as waste electrical and electronic equipment (WEEE). Developing technical solutions for sustainable recovery of waste products becomes a global trend. End-of-life recovery options include part reuse, remanufacturing, material recycling, energy recovery, and disposal. As shown in Fig. 13.1, disassembly, which is a systematic method for separating a product into its constituent components and subassemblies [1], is a critical stage for end-of-life recovery. Finding an optimum or near-optimum disassembly sequence is crucial to increasing the efficiency of the disassembly process.

Disassembly sequence planning (DSP) determines the order in which components are removed from products aiming at minimizing the disassembly time or cost, while considering the disassembly direction, disassembly method, and other attributes of components. DSP has been proved as a NP-hard problem [2] and has been becoming an important but still a challenging research topic in recent years. In the previous research, heuristics and meta-heuristics were used to find near-optimum or optimum solutions and generate cost-effective and feasible disassembly sequences. Heuristics include rule-based recursive method [3] and graph-based heuristic approach [4]. Meta-heuristics, which have been widely applied for solving such problems as well, include genetic algorithm (GA) [5–12], particle swarm optimization (PSO) [12–18], ant colony optimization (ACO) [19–22], and greedy randomized adaptive search procedure (GRASP) [23–25]. However, the controlling parameters in the above meta-heuristics need to be tuned, such as crossover rate and mutation probability in GA, inertia weight and two acceleration constants in PSO, pheromone weight, balance weight and evaporation

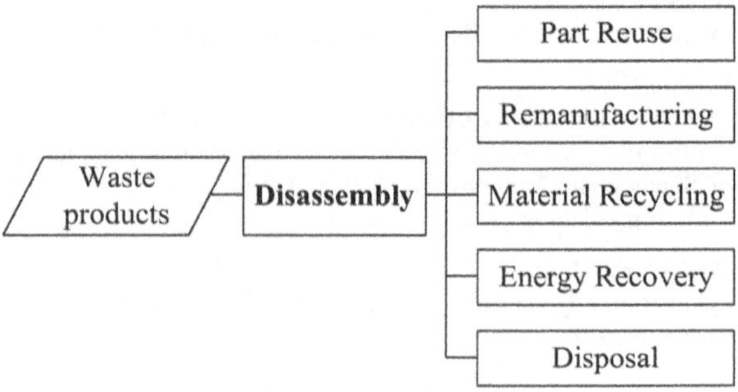

Fig. 13.1 End-of-life recovery options for waste products

rate in ACO, and window size in GRASP. This characteristic makes the above meta-heuristics not adaptive and robust enough for various situations.

A new population-based evolutional algorithm named teaching-learning-based optimization (TLBO) algorithm, which was originally introduced by Rao, et al. in 2012 [26, 27], has been successfully applied to continuous nonlinear large-scale problems [28, 29] including mechanical design optimization [27], parameter optimization of machining processes [30–32], high-dimensional real parameter optimization [33], and economic emission load dispatch [34, 35]. Unlike the above optimization techniques, the TLBO algorithm does not require any algorithm parameters to be tuned and outperforms some of the well-known meta-heuristics regarding constrained benchmark functions, constrained mechanical design, and continuous nonlinear numerical optimization problems.

However, the TLBO algorithm is not suitable for solving the DSP problems directly as it was designed for continuous optimization problems while the DSP problems are discrete combinatorial optimization problems with complex disassembly precedence constraints. In continuous problems, a solution is a vector of design variables, each of which belongs to a continuous rang. Sequencing the components in a product for disassembly planning is a typical DSP problem. In this problem, a solution can be represented as a permutation of integers, which are the serial numbers of components in a product. For this DSP problem, the solution space is not continuous and TLBO is unable to be applied directly. Meanwhile, a directivity discretized TLBO algorithm could not be a good choice for the above DSP problem either. In a DSP problem, the search space for an optimal solution is growing exponentially according to the number of components in the product, while the disassembly constraints could be complex. These characteristics cause that there are few feasible solutions in the population by using a random solution generation method embedded in TLBO. Furthermore, disassembly precedence constraints cannot be preserved simply using an arithmetic operation method during the evolutions toward optimization, which leads to few feasible solutions in the offspring. Hence, the directivity discretized TLBO algorithm can hardly converge and solve the DSP problem effectively.

In order to solve the DSP problems more efficiently, this chapter proposes a new optimization algorithm named simplified teaching-learning-based optimization (STLBO) algorithm. The STLBO algorithm inherits the main idea of the teaching-learning-based evolutionary mechanism so as to take the merits of the TLBO algorithm. Three new operators, including a feasible solution generator (FSG), a teaching phase operator (TPO), and a learning phase operator (LPO), have been designed and incorporated into the algorithm so as to make the algorithm applicable for DSP problems with complex constraints. In the meantime, the complex and multidimensional matrix computation used to modify solutions in the TLBO algorithm is simplified to a precedence preservation crossover operation in the STLBO algorithm. Detecting of feasibilities of new generated solutions would be also avoided. With the designed operators, STLBO can converge faster in the optimization or search process with higher accuracy so as to eventually improve the disassembly efficiency as well as reducing disassembly cost.

The rest of the chapter is organized as follows: In Sect. 13.2, the proposed STLBO algorithm for DSP problems is presented in detail. Section 13.3 demonstrates the performance of STLBO algorithm through numerical experiments and benchmark tests with case studies of waste product disassembly planning. Finally, conclusions are drawn in Sect. 13.4.

13.2 STLBO Algorithm for DSP Problems

This section presents the STLBO algorithm developed for DSP problems in detail. First, the TLBO algorithm is introduced briefly. And then, the framework of the STLBO algorithm is described. Subsequently, three key operators implemented in STLBO algorithm are presented in detail with illustrative examples of disassembly of waste products. Lastly, the comparison analysis of STLBO algorithm with other optimization algorithms is given and the implementation steps of STLBO algorithm for the DSP problems are outlined.

13.2.1 Brief Introduction to the TLBO Algorithm

The TLBO algorithm is a teaching-learning process inspired method simulating two fundamental modes of learning: through teacher (teaching phase) and interacting with the other learners (learning phase), where a group of learner (i.e., class) is considered as a population and the different learning subjects offered to the learners represent the different decision variables of the optimization problem. The grade of a learning subject offered to a learner represents the value of a decision variable in a solution. The result of a learner calculated from the grades of all the offered learning subjects represents the fitness function value of a solution. The learner with the best result in the entire population is considered as the teacher. The TLBO algorithm is explained below with the teaching phase and learning phase.

In the teaching phase, a good teacher conveys knowledge among the learners and tries to improve the mean results of the learners. Suppose there are m learning subjects (decision variables) offered to n learners in a class. At any sequential teaching-learning cycle t, the grades of learner i (i = 1, 2, …, n) are denoted as $X_{t,i} = (x_{t,i,1}, x_{t,i,2}, \ldots, x_{t,i,m})$, where $x_{t,i,j}$ denotes the grade of learner i in subject j (j = 1, 2, …, m) in cycle t. The mean grade of all the learners in subject j is denoted as $M_{t,j}$. The result of learner i considering all the subjects is denoted as $f(X_{t,i})$. The grades of the teacher equal those of the best learner $X_{t,\text{best}}$. The difference between the grade of the teacher and mean grade of the learners in subject j is calculated by the following equation:

$$\text{Difference\_Mean}_{t,j} = r_t \cdot (x_{t,\text{best},j} - \text{round}(1 + r_t) \cdot M), \quad j = (1, 2, \ldots m) \quad (13.1)$$

where $x_{t,\text{best},j}$ is the grade of the teacher (best learner) in subject j,r_t is a random number in the range [0, 1] and round(1 + r_t) can be either 1 or 2 decided by r_t. Based on the Difference_Mean$_{t,j}$, the existing learner i is modified according to the following equation:

$$x_{t,i,j,\text{new}} = +\text{Difference\_Mean}_{t,j}, \quad j = (1, 2, \ldots m), \quad i = (1, 2, \ldots n) \quad (13.2)$$

If $X_{t,i,\text{new}}$ gives a better result (fitness value) than $X_{t,i}$, update $X_{t,i}$ with $X_{t,i,\text{new}}$.

In the learning phase, a learner learns something new if the other learners have better results than his/hers. Another learner is randomly selected such that $i \neq j$. Learner modification is expressed as the following equation:

$$\begin{array}{l} f : f(X_{t,p}) < f(X_{t,q}) \\ x_{t,p,j,\text{new}} = x_{t,p,j} + r_t(x_{t,p,j} - x_{t,q,j}), \quad j = (1, 2, \ldots m) \\ \text{else :} \\ x_{t,p,j,\text{new}} = X_{t,p,j} + r_t(x_{t,q,j} - x_{t,p,j}), \quad j = (1, 2, \ldots m) \end{array} \quad (13.3)$$

If $X_{t,i,\text{new}}$ gives a better result (fitness value) than $X_{t,i}$, update $X_{t,i}$ with $X_{t,i,\text{new}}$.

As analyzed in the introduction section, neither directly implementation nor directly discretization of the TLBO algorithm could solve the DSP problems effectively. The newly proposed STLBO algorithm is presented in detail in the following subsections.

13.2.2 Framework of the STLBO Algorithm

The STLBO algorithm is based on the framework of a discrete and population-based optimization method and inherits the main idea of the teaching-learning-based evolutionary mechanism from the TLBO algorithm. The STLBO algorithm includes two phases: teaching phase and learning phase, and three key parts: FSG, TPO, and LPO. The pseudo-code of the STLBO algorithm is illustrated in Fig. 13.2.

In the teaching phase, the teacher is identified firstly, and then the mean of the fitness function values (mean result) of the learners is calculated. According to the mean result and the result of the teacher, the teaching factor is calculated, which is a self-adaptive parameter determining the probability that a learner learns from the teacher. Each learner will be modified using TPO according to the teaching factor.

In the learning phase, another learner is randomly selected for each learner and a new learner is generated using FSG with its result calculated. According to the results of these three learners, self-learning factor and the factor of learning from another are calculated, which are self-adaptive parameters, respectively, determine

```
STLBO
//Initialization:
For i=1 to class size do
    Generate learner X_i using FSG;
    Calculate the fitness function value of each learner f(X_i);
End for
//Loop:
While (stop criterion is not satisfied & maximum iterations are not exceeded) do
    //Teaching phase:
    Identify the teacher;
    Calculate the mean of the fitness function values of learners;
    Calculate the teaching factor;
    For i=1 to class size do
        Update learner X_i using TPO;
    End for
    //Learning phase:
    For i=1 to class size do
        Randomly select another learner X_j;
        Generate a new learner X_r using FSG;
        Calculate f(X_r);
        Calculate the self-learning factor and factor of learning from another;
        Update learner X_i using LPO;
    End for
End while
```

Fig. 13.2 Pseudo-code of the STLBO algorithm

the probability that a learner learns from himself and the probability that a learner learns from another one.

The detailed explanations of FSG, TPO, and LPO are presented in the following subsections.

13.2.3 Feasible Solution Generator

A feasible solution for a DSP problem is a disassembly sequence that satisfies the given disassembly precedence constraints. The implementation model of FSG depends on the way how disassembly precedence constraints are represented. The general steps for implementing the FSG are as follows:

Step 1 Set the vector of components available for disassembly *C_A* to ∅;
Step 2 Check the current disassembly state and disassembly precedence constraints to find new components available for disassembly, and push them back to *C_A*;
Step 3 Choose one component from *C_A* randomly, push it back to *Solution*, erase it from *C_ A* and suppose that it is disassembled from the product;
Step 4 If all the components are disassembled from the product, the *Solution* is randomly generated; otherwise, go to Step 2.

This chapter studies two types of DSP problems with different ways of representing the disassembly precedence constraints. The aforementioned four steps are general ones for implementing FSG; however, disassembly precedence constraints are represented in different ways, so the methods for finding the available components for disassembly by checking disassembly precedence constraints vary.

Problem 13.1 All the components and joints in the waste product are uniformly considered as components; in other words, the components and joints in the waste product are not discriminated. The disassembly precedence constraints are represented as rules, which are described as pairs of precedent set Pre_i and following set Fol_i. In each rule *i*, the components in Pre_i have disassembly precedence over the components in Fol_i. Figure 13.3 gives an example for Problem 13.1. As illustrated in Fig. 13.3, c_1 and c_2 have the disassembly precedence over c_3, and component c_3 has the disassembly precedence over components c_4 and c_5. The pseudo-code of FSG implemented for Problem 13.1 is outlined in Fig. 13.4.

Problem 13.2 The components and joints in the waste product are considered discriminatorily. In the disassembly process, the components are to be removed while the joints are to be broken. The disassembly precedence constraints are represented in a mathematical formation: two matrices and a list are used to represent the information on interferences between joints and components and among the components themselves.

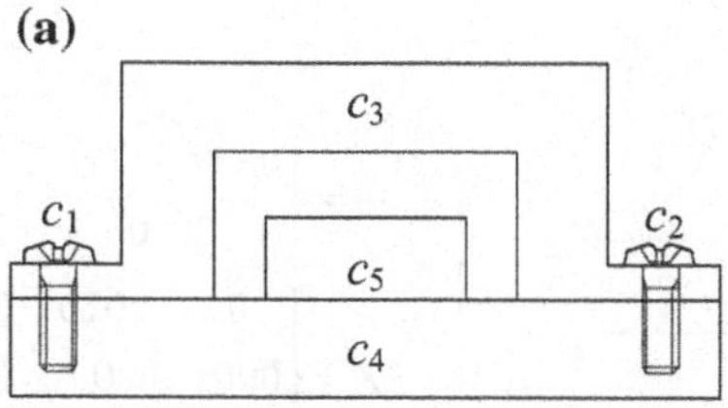

(b)

$$Rule_1 : Pre_1 = \varnothing, Fol_1 = \{c_1, c_2\};$$
$$Rule_2 : Pre_2 = \{c_1, c_2\}, Fol_2 = \{c_3\};$$
$$Rule_3 : Pre_3 = \{c_3\}, Fol_3 = \{c_4, c_5\}.$$

Fig. 13.3 An example for Problem 13.1: **a** 2D product model and **b** disassembly precedence constraints represented by rules

FSG implemented in Problem 1

```
Set the vector of components available for disassembly C_A to Ø;
For i = 0 to length of Solution
    For r = 0 to number of rules
        If Fol_r has not been inserted to C_A and Pre_r ⊆ Solution then
            Insert Fol_r into C_A;
        End if
    End for
    Randomly choose a component from C_A;
    Push the chose component back to Solution;
    Delete the chose component from C_A;
End for
```

Fig. 13.4 Pseudo-code of the FSG implemented in Problem 13.1

- Joint accessibility matrix A. The rows correspond to joints and the columns to components. The element a_{ij} is binary and indicates whether joint i is restricted by the presence of component j.
- Disassembly precedence matrix P. The rows and columns correspond to components. The element p_{ij} indicates, for each component i, which access directions are obstructed by component j. If five possible access directions ($\pm x$, $\pm y$, $+z$) are considered, assuming that the product is fixed on a horizontal table or on the floor, each element p_{ij} can be expressed as five bits, each of which corresponds to one of the five access directions.
- Joints list for each component $L = (L_1, L_2, \ldots, L_n)$. The element L_i indicates a list of all the joints that involve component i.

Figure 13.5 shows an example of Problem 13.2. The j_2 needs to break first before removing c_3. After j_2 is broken and c_3 is removed, j_1 become accessible. After j_1 is broken, c_1 and c_2 can be removed. The pseudo-code of the FSG implemented in Problem 13.2 is outlined in Fig. 13.6.

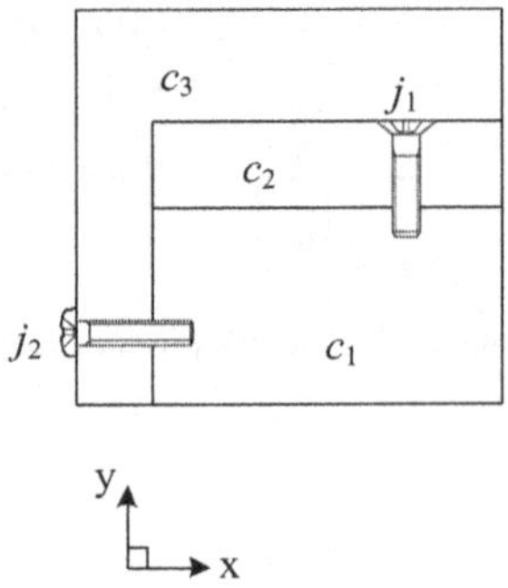

$$A = \begin{Bmatrix} 0 & 0 & 1 \\ 0 & 0 & 0 \end{Bmatrix}$$

$$P = \begin{Bmatrix} 0 & 0010 & 0110 \\ 0001 & 0 & 0110 \\ 1001 & 1001 & 0 \end{Bmatrix};$$

$L_1 \quad \{j_1, j_2\};$

$L_2 \quad \{j_1\};$

$L_3 \quad \{j_2\}.$

Fig. 13.5 An example for Problem 13.2: **a** 2D product model and **b** disassembly precedence constraints represented in mathematical formation

FSG implemented in Problem 2

```
Set the vector of components available for disassembly C_A to ∅;
Set the vector of inaccessible joints J_I to {all the joints}
For i = 0 to length of Solution
        For each component c in the product
                If c∈C_A or c∈Solution then
                        continue;
                End if
                If c can be extracted by checking P then
                        push c back to C_A;
                End if
        End for
        For each joint j in the product
                If j∉J_I then
                        continue;
                End if
                If j is accessible by checking A then
                        erase j from J_I;
                End if
        End for
        For each component c in C_A
                If L_c ∩ J_I ≠ ∅ then
                        erase c from C_A;
                End if
        End for
        Randomly Choose a component c from C_A;
        Push the c back to Solution;
        Delete c from C_A;
        Set all the elements in row c or column c in A and P to 0;
End for
```

Fig. 13.6 Pseudo-code of the FSG implemented in Problem 13.2

13.2.4 Teaching Phase Operator

TPO is used to replace current learners with new better learners in terms of better results (fitness function values) in the teaching phase. The teacher tries to improve the mean of the results of learners by giving his/her experiences and knowledge (e.g., better disassembly sequence) for the learners with expectation of their fitness function value to reach his/her level eventually. In other words, the possible solutions in the teaching phase follow a random process to evolve and the quality of solutions depends on teacher's and population's capability.

New solutions can be randomly generated using TPO according to the teacher and the class in the teaching phase. First, the low bound of the objective function value (f_{lb}) is given according to the best disassembly situation. Then, a self-adaptive parameter named teaching factor (p_t) is used for modifying the existing learners, which can be calculated according to the mean of the objective function value of the population (f_m), the objective function value of the teacher (f_t), and the low bound of the objective function value (f_{lb}):

$$g(x) = \frac{x}{x - f_{lb}} \tag{13.4}$$

$$p_t = \frac{g(f_t)}{g(f_t) + g(f_m)} \tag{13.5}$$

Finally, the precedence preservative operator is applied to preserve the precedence relationship in possible solutions when updating them. The pseudo-code of the TPO is outlined in Fig. 13.7.

In the procedure of TPO, a new solution is generated by a procedure of choosing and setting the elements one by one, from the left to the right of the solution. First, the probability is set randomly, which is used to compare with the teaching factor (p_t). If the probability is less than p_t, the leftmost element of the teacher is chosen and pushed back to the new solution; otherwise, the leftmost element of the learner is chosen and pushed back to the new solution. Then, the chosen element is erased from both the teacher and the learner. Finally, the learner is updated with the newly

Fig. 13.7 Pseudo-code of the TPO

```
TPO
For each element in a new Solution
    Set probability∈[0, 1] randomly;
    If probability < p_t then
        Push Teacher's leftmost element back to Solution;
    Else
        Push Learner's leftmost element back to Solution;
    End if
    erase the chosen element from Teacher;
    erase the chosen element from Learner;
End for
Calculate the objective function value of the new Solution;
If the new Solution is better than the Learner;
    Update the Learner with the new Solution;
End if
```

Table 13.1 An example for using TPO to generate a new solution from a teacher and a learner with 5 elements

Step	p_t	Probability	Teacher	Learner	Solution
1	0.6	0.5	[2 3 1 4 5]	[4 3 5 1 2]	[2]
2	0.6	0.2	[3 1 4 5]	[4 3 5 1]	[2 3]
3	0.6	0.8	[1 4 5]	[4 5 1]	[2 3 4]
4	0.6	0.7	[1 5]	[5 1]	[2 3 4 5]
5	0.6	0.3	[1]	[1]	[2 3 4 5 1]

Remarks The underlined figure indicates the chose element in each step

generated solution if the new one is better. In this way, the disassembly precedence in solutions is preserved.

Table 13.1 gives an example for using TPO to generate a new solution with 5 elements: A new solution ([2 3 4 5 1]) is generated from a teacher ([2 3 1 4 5]) and a learner ([4 3 5 1 2]). Suppose component 5 has precedence over component 3 and the teaching factor (p_t) equals 0.6. It is seen that the precedence is preserved in the new solution.

13.2.5 Learning Phase Operator

LPO is used to modify the existing learners with new better learners in the learning phase, where learners learn from each other and improve their abilities. Learning phase also follows a random process to generate possible solutions, but the qualities of the solutions depend on current learner's ability, other learners' ability and random distribution.

New solutions can be randomly generated using LPO according to the learners themselves, other learners, and randomly generated learners in the learner phase. Two self-adaptive parameters named self-learning factor (p_s) and factor of learning from another (p_a) are used for modifying the existing learners, which can be calculated according to the objective function value of the learner itself (f_s), the objective function value of another learner (f_a), the objective function value of a randomly generated learner (f_r), and the low bound of the objective function value (f_{lb}):

$$p_s = \frac{g(f_s)}{g(f_s) + g(f_a) + g(f_r)} \tag{13.6}$$

$$p_a = \frac{g(f_s) + g(f_a)}{g(f_s) + g(f_a) + g(f_r)} \tag{13.7}$$

The pseudo-code of LPO is similar to that of TPO outlined in Fig. 13.8 A new solution is generated from three parent solutions with two self-adaptive parameters

```
LPO
For each element in a new Solution
    Set probability∈[0, 1] randomly;
    If probability < p_s then
        Push Teacher's leftmost element back to Solution;
    Else if probability < p_a then
        Push Another's leftmost element back to Solution;
    Else
        Push Generated's leftmost element back to Solution;
    End if
    erase the chosen element from Learner;
    erase the chosen element from Another;
    erase the chosen element from Generated;
End for
Calculate the objective function value of the new Solution;
If the new Solution is better than the Learner then
    Update the Learner with the new Solution;
End if
```

Fig. 13.8 Pseudo-code of the LPO

in LPO, while a new solution is generated from two parent solutions with one self-adaptive parameter in TPO. In the learner phase, the learning factors for different learners are different, and they need to be calculated before implementing LPO, while the teaching factor just needs to be calculated once in the teaching phase.

13.2.6 Comparison of STLBO Algorithm with Other Optimization Algorithms

Like GA, PSO, TLBO, etc., the STLBO algorithm is also a population-based technique which implements a group of solutions to evolve and learns to find a near-optimum or optimum solution. The STLBO algorithm inherits the main idea of the teaching-learning-based evolutionary mechanism so as to take the merits of the TLBO algorithm: All the algorithm parameters are self-adapted and need not to be tuned, and the teacher of each iteration is used to modify the learners, thereby to increase search efficiency and effectiveness.

The STLBO algorithm is different from the TLBO algorithm in the following aspects. (a) The realization methods for evolutionary mechanism are different. In the STLBO algorithm, learners are modified using the TPO and LPO, applying the method of precedence preservation crossover operation. While in the TLBO algorithm, learners are modified using multidimensional matrix computation. (b) The adaptation methods for the algorithm parameters are different. In the STLBO algorithm, the teaching factor (p_t), self-learning factor (p_s), and factor of learning from another (p_a) are calculated according to the results of learners (objective function values of solutions). While in the TLBO algorithm, the difference mean is calculated according to the grades of learners in each learning subject, in which the learning factor (r_t) is a random number in the range [0, 1] and the teaching factor [round(1 + r_t)] can be either 1 or 2 decided by r_t.

Compared with GA and discrete PSO applied in solving DSP problems in the previous research works, the teaching-leaning-based evolutionary mechanism in the STLBO algorithm is novel and different from the crossover and mutation mechanisms of GA and particle position update mechanism of discrete PSO. In the STLBO algorithm, all the algorithm parameters are self-adapted and need not to be tuned. While GA requires the crossover rate and mutation probability, the discrete PSO requires inertia weight and two acceleration constants.

13.2.7 Implementation of STLBO Algorithm for DSP Problems

The steps using STLBO algorithm to solve the DSP problems can be summarized as follows:

Step 1 Define and formulate a DSP problem in a mathematical model and represent the disassembly precedence constraints;
Step 2 Initialize the optimization parameters: solution length (number of components), population size (number of learners), and iteration times;
Step 3 Generate a random population using FSG according to the population size and the number of components for disassembly;
Step 4 Calculate the teaching factor and modify all the learners using TPO;
Step 5 Calculate the self-learning factor and factor of learning from another for each learner and modify all of them using LPO;
Step 6 Stop if the stop criteria are satisfied or maximum iteration times are achieved; otherwise repeat from Step 4.

13.3 Experimental Studies

In order to test the performance of the STLBO algorithm and conduct a further comparative study, two kinds of disassembly sequence planning problems are described and used for testing. In both problems, the case of complete disassembly of wasted products is considered and it assumes that the structure of waste products and disassembly precedence constraints are known. Two types of representation methods of disassembly precedence constraints and the corresponding methods for generating feasible disassembly sequence are presented in Sect. 13.2.3, respectively. All the algorithms tested were coded in the C++ programming language and carried out on a personal computer with 2 GHz Intel Core2 Duo CPU T5750 and 2 GB memory.

13.3.1 Problem 1

In the first DSP problem, all the components and joints in a waste product are uniformly considered as components. Four types of attributes of the disassembly process are considered: disassembly direction, disassembly method, demand after disassembly, and material contained in component [5, 16]. A brief description of these attributes is presented as follows, where $x \in (0, \ldots, n-1)$ is the index of component and n is the number of components.

- Disassembly direction: dir(x) $\in$ ($\pm x$, $\pm y$, $\pm z$). Each component must be disassembled in a particular direction. The $\pm x$, $\pm y$, and $\pm z$ indicate six possible directions, respectively.
- Disassembly method: met(x) $\in$ (D, N). The letter D indicates destructive disassembly, which focuses on recycling of materials, while the letter N indicates nondestructive disassembly, which focuses on components reuse.
- Demand after disassembling: dem(x) $\in$ (0, 1, 2). The values 0, 1, and 2 indicate no demand, demand of reuse, and demand of recycling, respectively.
- Material contained in a component: mat(x) $\in$ (A, P, S). The letters A, P, and S indicate three types of materials contained in components: aluminum, plastic, and steel, respectively.

The purpose of this optimization problem is to plan a feasible disassembly sequence X to minimize the total disassembly time $f(X)$, which depends on the basic disassembly time $bt(x_i)$, the penalty of direction change $dt(x_i)$, and the penalty of method change $mt(x_i)$ for each component x_i, where i is the index of sequence. The mathematical formula for Problem 13.1 is as follows:

Find:

$$X = [x_0, x_1, \ldots, x_{n-1}] \tag{13.8}$$

Minimize:

$$f(X) = \sum_{i=0}^{n-1} bt(x_i) + \sum_{i=0}^{n-2} dt(x_i) + \sum_{i=0}^{n-2} mt(x_i) \tag{13.9}$$

where:
If

$$\text{dem}(x_i) = \text{dem}(x_{i+1}) = 2 \quad \text{and} \quad \text{mat}(x_i) = \text{mat}(x_{i+1}) \tag{13.10}$$

then

$$bt(x_i) = \text{bas}(x_i), \quad dt(x_i) = 0, \quad mt(x_i) = 0 \tag{13.11}$$

otherwise

$$dt(x_i) = \begin{cases} 0, & \text{direction is not changed} \\ 1, & \text{direction is changed by } 90^\circ \\ 2, & \text{direction is changed by } 180^\circ \end{cases} \tag{13.12}$$

$$mt(x_i) = \begin{cases} 0, & \text{method is not changed} \\ 1, & \text{method is changed} \end{cases} \tag{13.13}$$

In (13.10), dem(x_i) = dem(x_{i+1}) = 2 and mat(x_i) = mat(x_{i+1}) represents a special situation: Two components in a consecutive sequence, for example, i and i + 1, are made by the same material, and both are demanded for recycling. These two components will not be separated from each other and will be disassembled as a combination. In (13.11), bas(x_i) indicates the basic disassembly time for component x_i.

For instance, an example for Problem 13.1 is developed as shown in Fig. 13.9, where the waste product is modeled by a tree and the disassembly precedence constraints are defined among the nodes. In this tree, the dashed-line arrow indicates that the component/subassembly at the start point has priority for disassembly over the component/subassembly at the end point. The tree is generated based on the bill of materials (BOM) and the disassembly precedence constraints can be acquired from the geometric information and connector information using a connector-knowledge-based approach [36]. The waste product has 13 components indexed by numbers 0–12. The input data of the example are given in Table 13.2.

According to the given example, the disassembly precedence constraints are represented using the method described in Sect. 13.2.3 as follows:

$$\begin{aligned} &\text{Rule}_1 : \text{Pre}_1 = \varnothing, \text{Fol}_1 = \{1, 2\}; \\ &\text{Rule}_2 : \text{Pre}_2 = \{1, 2\}, \text{Fol}_2 = \{0, 3, 6, 8, 11\}; \\ &\text{Rule}_3 : \text{Pre}_3 = \{3\}, \text{Fol}_3 = \{4, 5, 7\}; \\ &\text{Rule}_4 : \text{Pre}_4 = \{5\}, \text{Fol}_3 = \{9, 10, 12\}. \end{aligned} \tag{13.14}$$

Two experiments based on the example for Problem 13.1 are taken: role test of TPO or LPO and comparison of STLBO with GA [5] and simplified swarm optimization (SSO) algorithm [16].

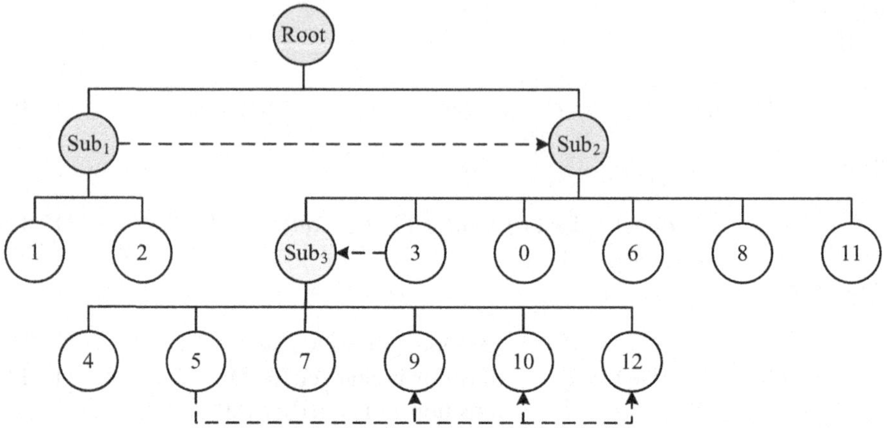

Fig. 13.9 Example for Problem 13.1 with waste product structure and disassembly precedence constraints

Table 13.2 Input data of the example for Problem 13.1

x	0	1	2	3	4	5	6	7	8	9	10	11	12
bas(x)	3	4	1	1	2	2	2	1	2	1	2	1	3
dir(x)	Z	$-X$	X	X	$-Y$	Z	$-Y$	Z	$-Y$	X	$-Y$	$-X$	$-Z$
met(x)	D	D	D	D	N	N	N	N	D	N	D	N	D
dem(x)	1	1	1	0	0	1	2	1	2	0	1	2	1
mat(x)	S	S	P	S	P	P	S	A	A	A	S	A	S

(1) Role test of TPO and LPO

To demonstrate the effectiveness of the proposed TPO or LPO, the following three cases are implemented:

Case 1: STLBO with TPO while without LPO;
Case 2: STLBO with LPO while without TPO; and
Case 3: STLBO with both TPO and LPO.

The following parameters are initialized as:

- Solution length = 13;
- Population size = 10;
- Iteration times = 100.

The numerical results are shown in Fig. 13.10

As shown in Fig. 13.10a, the algorithm in Case 1 converges fast to the local optimum at its early stage, and the objective function values $f(X)$ obtained in Case 1 are located in a large range in the first half of the iterations. Then all the learners

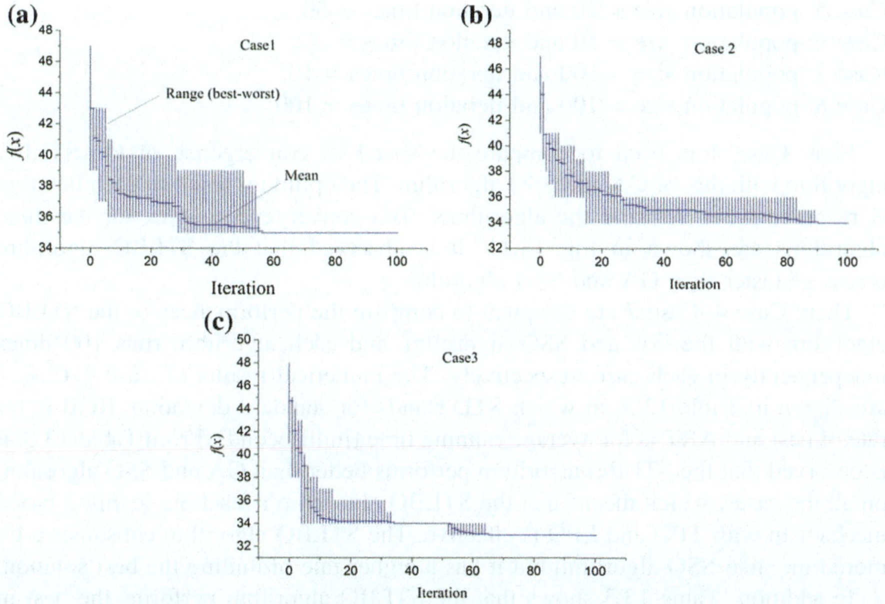

Fig. 13.10 Range (best-worst-mean) of the obtained objective function values in: **a** Case 1, **b** Case 2, and **c** Case 3

converge fast to a certain level for that the teacher's ability and population's capability are limited.

As shown in Fig. 13.10b, the algorithm in Case 2 converges slow with low optimization accuracy, and the objective function values $f(X)$ obtained in Case 2 are located in a range during almost the whole iterations. When modifying the learners using LPO, a new solution is generated according to two existed learners and one randomly generated solution. As thus, there is a chance to find better solutions in Case 2, thus making the speed of convergence of the algorithm slow.

As shown in Fig. 13.10c, the complete the STLBO algorithm in Case 3 converges faster than the algorithms in Case 1 and Case 2 with higher optimization accuracy. The complete STLBO takes the advantages of both TPO and LPO.

From the numerical results, it can be concluded that learning from teacher using TPO makes the speed of convergence fast, and self-learning and learning from each other using LPO make additional chances to find better solutions. Combining TPO and LPO together makes the STLBO algorithm converges fast and enhances the search accuracy of it.

(2) Comparison of STLBO algorithm with GA and SSO

The following cases were implemented to test and compare the STLBO algorithm with GA [5] and SSO algorithm [16]:

Case 4: population size = 10 and iteration times = 100;

Case 5: population size = 20 and iteration times = 50;
Case 6: population size = 50 and iteration times = 20;
Case 7: population size = 100 and iteration times = 10;
Case 8: population size = 100 and iteration times = 100.

First, Case 4 is used to compare the speed of convergence of the STLBO algorithm with that of GA and SSO algorithm. The optimum result of each iteration is recorded when running the algorithms. The convergence curves for the three algorithms are shown in Fig. 13.11. It is observed that the STLBO algorithm converge faster than GA and SSO algorithm.

Then, Case 4–Case 7 are designed to compare the performances of the STLBO algorithm with the GA and SSO algorithm, and each algorithm runs 100 times independently in each case, respectively. The numerical results of Case 4–Case 7 are shown in Table 13.3, in which STD stands for standard deviation, ROB is for rate of best and ART is for average running time (millisecond). From Table 13.3, it is observed that the STLBO algorithm performs better than GA and SSO algorithm on all the cases, which means that the STLBO algorithm's teaching-learning-based mechanism with TPO and LPO is effective. The STLBO algorithm consumes a bit more time than SSO algorithm but it has a higher rate of finding the best solution.

In addition, Table 13.3 shows that the STLBO algorithm performs the best in Case 6, GA has the best performance in Case 7, and SSO algorithm has produced the best result in Case 7. So it can be inferred that a bigger population size makes STLBO, GA, and SSO to perform better even with less iteration times. As the value of iteration times in Case 7 is quite small, the performance of the algorithms can be improved by increasing the iteration times. Therefore, Case 8 is implemented and the numerical results are shown in Table 13.4. It is observed that all the three algorithms' performance is improved by increasing the iteration times to 100. The STLBO algorithm still performs better than the other two algorithms. One of the optimal solutions found by the STLBO algorithm for the example of Problem 13.1 is given in Table 13.5 and the total disassembly time is 33.

Fig. 13.11 Convergence curves for STLBO, GA, and SSO in Case 4

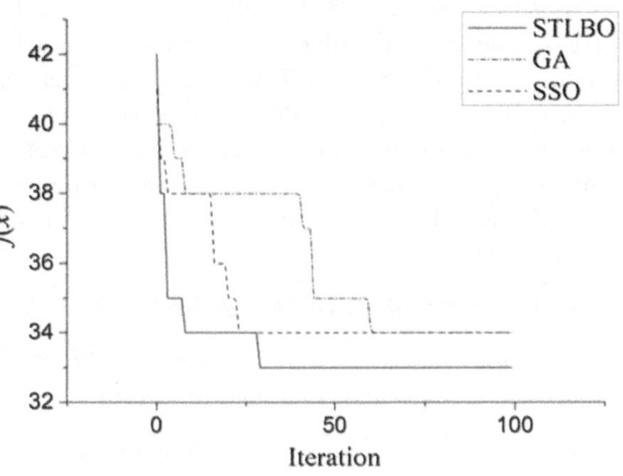

Table 13.3 Numerical results for Case 4–Case 7

Algorithm	Case	Best	Worst	Mean	STD	ROB (%)	ART
STLBO	Case 4	33	36	33.72	0.60	35	1127
	Case 5	33	35	33.59	0.65	50	1201
	Case 6	33	35	33.51	0.52	50	1098
	Case 7	33	35	33.53	0.54	49	1078
GA	Case 4	33	38	35.48	1.02	2	184
	Case 5	34	38	35.18	0.95	0	198
	Case 6	33	37	34.51	0.94	8	291
	Case 7	33	36	34.23	0.68	8	569
SSO	Case 4	33	36	33.75	0.56	30	966
	Case 5	33	36	33.77	0.60	31	993
	Case 6	33	36	33.78	0.59	30	964
	Case 7	33	36	33.73	0.63	36	965

Table 13.4 Numerical results for Case 8

Algorithm	Best	Worst	Mean	STD	ROB (%)	ART
STLBO	33	34	33.14	0.35	86	11,732
GA	33	35	33.8	0.61	31	5971
SSO	33	34	33.16	0.37	83	9732

Table 13.5 One of the optimal solutions fund by SLTBO for the example of Problem 13.1

Sequence	1	2	3	0	8	11	5	7	9	6	4	10	12
Direction	−1	1	1	3	−2	−1	3	3	1	−2	−2	−2	−3
Method	D	D	D	D	D	N	N	N	N	N	N	D	D
Demand	1	1	0	1	2	2	1	1	0	2	0	1	1
Material	S	P	S	S	A	A	P	A	A	S	P	S	S

13.3.2 Problem 2

In the second DSP problem, the components and joints in a waste product are considered, respectively. Three types of attributes of the disassembly process are considered as follows:

- Direction changes of breaking joints. More direction changes lead to the higher cost of breaking joints.
- Tool changes of breaking joints. More tool changes lead to the higher cost of breaking joints.

- The number of directions is obstructed by other components when extracting a component. More obstructed directions lead to the higher cost of extracting process.

The total disassembly cost is taken as the objective function value. The mathematical model of Problem 13.1 is presented in the benchmarks published by Adenso-Díaz and García-Carbajal [37, 38] as follows:

Find:

$$X = [x_0, x_1, \ldots, x_{n-1}] \tag{13.15}$$

Minimize:

$$f(X) = \sum_{i=0}^{n-1} (\text{cost\_joints}(x_i) + \text{cost\_component}(x_i)) \tag{13.16}$$

where

$$\text{cost\_joints}(x_i) = \sum_{j \in \text{Br\_joints}(x_i)} c(j) \cdot c_1(j) \cdot c_2(j) \tag{13.17}$$

$$\text{cost\_component}(x_i) = 100 - (4 - \text{num\_obstr}(x_i)) \times 24.75 \tag{13.18}$$

$$c_1(j) = \begin{cases} 1.1, & \text{broke a joint just before and the tool is changed} \\ 1.1, & \text{extracted a component ust before} \\ 1, & \text{otherwise} \end{cases} \tag{13.19}$$

$$c_2(j) = \begin{cases} 1.15, & \text{broke a joint just before and the direction is changed} \\ 1.15, & \text{extracted a component just before} \\ 1, & \text{otherwise} \end{cases} \tag{13.20}$$

In (13.15)–(13.20), n indicates the number of components, x_i indicates the index of component, cost_joints(x_i) indicates the cost of breaking all the joints that involve component x_i after x_0, x_1, …, x_{i-1} are all extracted, cost_component(x_i) indicates the cost of extracting component x_i after x_0, x_1, …, x_{i-1} are all extracted, Br_joints(x_i) indicates the list of joints that involve component x_i after x_0, x_1, …, x_{i-1} are all extracted, num_obstr(x_i) indicates the number of directions obstructed by other components when extracting component x_i after x_0, x_1, …, x_{i-1} are all extracted, j indicates the index of joints, $c(j)$ indicates the based cost of breaking joint j, $c_1(j)$ indicates the cost penalty of tool change when breaking joint j and $c_2(j)$ indicates the cost penalty of direction change when breaking joint j.

The published disassembly benchmarks of 24 instances were taken for experimental studies. The best solutions and corresponding objective function values for these instances are kept up-to-date online (http://coruxa.epsig.uniovi.es/~adenso/file_d.html). The test problems (of size n = 25 and 50) were generated using the

Fig. 13.12 Example of an instances (100-025-005-05) with n = 25 components in the published benchmarks

GEN-PRODIS tool [39], which is able to generate an unlimited number of random test cases with specified characteristics (number of components and degree of compactness). All the components are right-angled parallelepipeds, and a component may not be involved in more than eight joints and the total number of disassembly tools was set to five. Figure 13.12 gives an example of an instances (100-025-005-05) with n = 25 components in the published benchmarks.

To test the benchmarks, the following parameters are initialized as:

- Solution length = the number of components in each instance;
- Population size = 20;
- Iteration times = 200.

The best results obtained by the STLBO algorithm are shown in Table 13.6. The STLBO algorithm is able to find top 8 best solutions which are better than the previous best solutions and 5 of them are equal to the previous ones. Table 13.7 outlines the 8 new best solutions.

Table 13.6 Objective function values of the previous best solutions in the published benchmarks, and of the solutions found by the STLBO algorithm

No.	Instance	Number of components	Previous best cost	Best found by STLBO
1	100-025-005-05	25	2649.0699	2649.2542
2	101-025-009-20	25	2229.7465	*2229.7465*
3	102-025-012-40	25	2670.7680	[a]2642.0680
4	200-025-025-40	25	2503.6491	[a]2486.1487
5	201-025-037-40	25	2536.6414	[a]2519.2001
6	202-025-050-40	25	2350.0500	[a]2343.5000

(continued)

Table 13.6 (continued)

No.	Instance	Number of components	Previous best cost	Best found by STLBO
7	800-025-005-40	25	3144.3953	3210.2789
8	301-025-009-20	25	2411.9750	*2411.9750*
9	302-025-012-05	25	2381.0032	*2381.0032*
10	400-025-025-05	25	2245.2589	*2245.2589*
11	401-025-037-05	25	2347.4863	2351.2366
12	402-025-050-05	25	2199.3607	*2199.3607*
13	103-050-010-05	50	5505.0209	[a]5492.0913
14	104-050-017-20	50	6069.4630	6141.9410
15	105-050-025-40	50	5793.1697	6018.2916
16	203-050-050-40	50	5374.3406	5456.9973
17	204-050-075-40	50	5592.0996	[a]5489.2619
18	905-050-100-40	50	5621.3563	[a]5618.9372
19	303-050-010-40	50	6178.4352	6324.8494
20	304-050-017-20	50	5422.0031	[a]5339.4085
21	305-050-025-05	50	4720.4362	4725.2437
22	403-050-050-05	50	4613.7832	4668.9782
23	404-050-075-05	50	4523.0915	4539.8000
24	405-050-100-05	50	4759.9078	4786.0266

Remarks The value with [a] in the left indicates it is better than the previous best one, and the italicized value represents it is equal to the previous best one

Table 13.7 New best solutions found by the STLBO algorithm

No.	Instance	Best solution
1	102-025-012-40	[10 15 2 20 19 13 24 16 23 5 1 3 6 8 11 4 18 22 17 9 21 0 7 12 14]
2	200-025-025-40	[23 21 0 24 15 22 12 20 13 5 6 17 10 9 16 14 1 3 18 8 19 4 7 11 2]
3	201-025-037-40	[6 20 23 17 21 8 7 13 24 1 15 5 10 11 9 19 22 3 18 16 14 12 0 2 4]
4	202-025-050-40	[16 11 17 23 13 1 18 22 2 24 14 12 8 20 19 15 21 7 10 5 4 3 9 6 0]
5	103-050-010-05	[47 21 18 45 30 20 15 48 7 49 39 27 32 44 46 42 31 35 23 16 33 38 37 36 41 2 22 28 26 43 9 12 19 25 40 17 10 24 13 29 3 14 8 6 11 5 34 0 1 4]
6	204-050-075-40	[46 41 38 43 49 31 45 21 6 23 48 40 9 44 12 34 22 36 47 30 17 39 33 28 2 5 42 37 15 29 25 10 7 35 26 3 20 19 32 14 11 8 13 0 24 27 16 1 4 18]
7	905-050-100-40	[39 36 43 33 10 14 44 30 46 47 38 21 37 23 45 49 26 22 40 4 29 41 42 24 19 12 32 35 18 48 3 20 6 11 28 15 27 34 5 31 16 2 17 0 7 1 8 25 9 13]
8	304-050-017-20	[4 32 41 45 33 26 40 34 30 46 28 8 19 39 1 36 20 16 44 23 0 13 2 11 17 22 49 25 47 18 37 48 12 27 42 21 38 10 14 3 5 7 43 35 9 6 31 15 24 29]

In summary, the experimental results illustrate that the developed STLBO algorithm performs well on DSP since it outperforms other methods such as GA and SSO and find 8 new best solutions out of 24 instances in the published benchmarks. The success of the STLBO algorithm could come from a better balance of global exploration and local exploitation, resulting from a powerful teaching-learning-based mechanism. The teaching phase is more like a kind of global exploration and learner phase is more like a kind of local exploitation. Moreover, all the parameters in the STLBO algorithm are self-adapted and need not to be tuned, thus making the implementation simpler. And the teacher of each iteration is used to modify the learners, thereby increasing the convergence rate.

13.4 Conclusions

This chapter proposes a novel STLBO algorithm to solve the DSP problems. The STLBO algorithm is divided into two phase: teaching phase and learning phase. And three new key operators are presented: FSG, TPO, and LPO. The developed STLBO algorithm is a discrete and population-based optimization algorithm with a new teaching-learning-based evolutionary mechanism. The characteristics and advantages of the developed STLBO are summarized as follows:

- The STLBO algorithm is specially designed for DSP problems and adapts to different DSP problems with different ways of representing the disassembly precedence constraints.
- The implementation of the STLBO algorithm is simple for that all the algorithm parameters are self-adapted and need not to be tuned. Thus, the performance of the algorithm is stable.
- And STLBO can converge fast with high optimization accuracy by implementing the new teaching-learning-based evolutionary mechanism.

Two types of DSP problems are described and formulated in a mathematical model. The effectiveness and good performance of STLBO to solve the DSP problems are proved by experimental studies and benchmark test.

In the future, more complex case studies of DSP problems will be introduced. The performance of the STLBO algorithm can be enhanced by combining the TPO and LPO with other advanced search technique. The STLBO algorithm can also be used to solve complex combinatorial optimization problems. Moreover, the STLBO algorithm can be extended as a multiobjective optimization algorithm to solve multiobjective disassembly sequence planning problems.

References

1. Gungor, A., & Gupta, S. M. (1999). Issues in environmentally conscious manufacturing and product recovery: a survey. *Computers & Industrial Engineering, 36,* 811–853.
2. Lambert, A. J. D. (2003). Disassembly sequencing: A survey. *International Journal of Production Research, 41,* 3721–3759.
3. Smith, S. S., & Chen, W. H. (2011). Rule-based recursive selective disassembly sequence planning for green design. *Advanced Engineering Informatics, 25,* 77–87.
4. Kuo, T. C., Zhang, H. C., & Huang, S. H. (2000). Disassembly analysis for electromechanical products: A graph-based heuristic approach. *International Journal of Production Research, 38,* 993–1007.
5. Kongar, E., & Gupta, S. M. (2006). Disassembly sequencing using genetic algorithm. *International Journal of Advanced Manufacturing Technology, 30,* 497–506.
6. Giudice, F., & Fargione, G. (2007). Disassembly planning of mechanical systems for service and recovery: A genetic algorithms based approach. *Journal of Intelligent Manufacturing, 18,* 313–329.
7. Go, T. F., Wahab, D. A., Ab Rahman, M. N., Ramli, R., & Hussain, A. (2012). Genetically optimised disassembly sequence for automotive component reuse. *Expert Systems with Applications, 39,* 5409–5417.
8. Wang, H., Xian, D. G., & Duan, G. H. (2008). A genetic algorithm for product disassembly sequence planning. *Neurocomputing, 71,* 2720–2726.
9. Wu, H., & Zuo, H. F. (2009). Using genetic/simulated annealing algorithm to solve dis-assembly sequence planning. *Journal of System Engineering Electron, 20,* 906–912.
10. Li, J. R., Khoo, L. P., & Tor, S. B. (2002). A novel representation scheme for disassembly sequence planning. *International Journal of Advanced Manufacturing Technology, 20,* 621–630.
11. Seo, K. K., Park, J. H., & Jang, D. S. (2001). Optimal disassembly sequence using genetic algorithms considering economic and environmental aspects. *International Journal of Advanced Manufacturing Technology, 18,* 371–380.
12. Li, J. R., Khoo, L. P., & Tor, S. B. (2005). An object-oriented intelligent disassembly se-quence planner for maintenance. *Computers in Industry, 56,* 699–718.
13. Tseng, Y. J., Yu, F. Y., & Huang, F. Y. (2011). A green assembly sequence planning model with a closed-loop assembly and disassembly sequence planning using a particle swarm optimization method. *International Journal of Advanced Manufacturing Technology, 57,* 1183–1197.
14. Wang, Y., & Liu, J. H. (2010). Chaotic particle swarm optimization for assembly sequence planning. *Robotics and Computer-Integrated Manufacturing, 26,* 212–222.
15. Lv, H. G., & Lu, C. (2010). An assembly sequence planning approach with a discrete particle swarm optimization algorithm. *International Journal of Advanced Manufacturing Technology, 50,* 761–770.
16. Yeh, W. C. (2012). Optimization of the disassembly sequencing problem on the ba-sis of self-adaptive simplified swarm optimization. *IEEE Transactions on System, Man, Cybernetics, Part A, 42,* 250–261.
17. Yeh, W. C. (2012). Simplified swarm optimization in disassembly sequencing problems with learning effects. *Computer Operation Research, 39,* 2168–2177.
18. Li, W. D., Xia, K., Gao, L., & Chao, K. M. (2013). Selective disassembly planning for waste electrical and electronic equipment with case studies on liquid crystal displays. *Robotics and Computer-Integrated Manufacturing, 29,* 248–260.
19. Wang, J. F., Liu, J. H., & Zhong, Y. F. (2005). A novel ant colony algorithm for assembly sequence planning. *International Journal of Advanced Manufacturing Technology, 25,* 1137–1143.

20. Agrawal, S., & Tiwari, M. K. (2008). A collaborative ant colony algorithm to stochastic mixed-model U-shaped disassembly line balancing and sequencing problem. *International Journal of Production Research, 46,* 1405–1429.
21. McGovern, S. M., & Gupta, S. M. (2006). Ant colony optimization for disassembly sequencing with multiple objectives. *International Journal of Advanced Manufacturing Technology, 30,* 481–496.
22. Wang, J. F., Liu, J. H., Li, S. Q., & Zhong, Y. F. (2003). Intelligent selective disassembly using the ant colony algorithm. *Ai Edam, 17,* 325–333.
23. Adenso-Diaz, B., Garcia-Carbajal, S., & Lozano, S. (2007). An efficient GRASP algorithm for disassembly sequence planning. *OR Spectrum, 29,* 535–549.
24. Adenso-Díaz, B., García-Carbajal, S., & Gupta, S. M. (2008). A path-relinking approach for a bi-criteria disassembly sequencing problem. *Computer Operation Research, 35,* 3989–3997.
25. Andres, C., Lozano, S., & Adenso-Diaz, B. (2007). Disassembly sequence planning in a disassembly cell context. *Robotics and Computer-Integrated Manufacturing, 23,* 690–695.
26. Rao, R. V., Savsani, V. J., & Vakharia, D. P. (2012). Teaching–Learning-Based Optimiza-tion: An optimization method for continuous non-linear large scale prob-lems. *Information Sciences, 183,* 1–15.
27. Rao, R. V., Savsani, V. J., & Vakharia, D. P. (2011). Teaching-learning-based optimiza-tion: A novel method for constrained mechanical design optimization prob-lems. *Computer Aided Design, 43,* 303–315.
28. Rao, R., & Patel, V. (2012). An elitist teaching-learning-based optimization algorithm for solving complex constrained optimization problems. *International Journal of Industrial Engineering Computations, 3.*
29. Rao, R., & Patel, V. (2013). Comparative performance of an elitist teach-ing-learning-based optimization algorithm for solving unconstrained opti-mization problems. *International Journal of Industrial Engineering Computations, 4.*
30. Pawar, P. J., & Rao, R. V. (2012). Parameter optimization of machining processes using teaching–learning-based optimization algorithm. *The International Journal of Advanced Manufacturing Technology, 67,* 995–1006.
31. Venkata, Rao R., & Kalyankar, V. D. (2013). Parameter optimization of modern ma-chining processes using teaching–learning-based optimization algorithm. *Engineering Applications on Artificial Intelligence, 26,* 524–531.
32. Rao, R. V., & Kalyankar, V. D. (2012). Parameter optimization of machining processes using a new optimization algorithm. *Material Manufacturing Processing, 27,* 978–985.
33. Chandra, S. S., Naik, A., & Parvathi, K. (2012). High dimensional real parameter optimization with teaching learning based optimization. *International Journal of Industrial Engineering Computations, 3,* 807–816.
34. Jadhav, H. T., Chawla, D., & Roy, R. (2012). Modified teaching learning based algorithm for economic load dispatch incorporating wind power. In *Proceedings of 2012 11th International Conference on Environment and Electrical Engineering (EEEIC),* (pp. 397–402).
35. Niknam, T., Golestaneh, F., & Sadeghi, M. S. (2012). Multiobjective teaching–learning-based optimization for dynamic economic emission dispatch. *IEEE Systems Journal, 6,* 341–352.
36. Li, J. R., Wang, Q. H., Huang, P., & Shen, H. Z. (2010). A novel connect-or-knowledge-based approach for disassembly precedence constraint gener-ation. *International Journal of Advanced Manufacturing Technology, 49,* 293–304.
37. González, B., & Adenso-Díaz, B. (2005). A bill of materials-based approach for end-of-life decision making in design for the environment. *International Journal of Production Research, 43,* 2071–2099.
38. Gonzalez, B., & Adenso-Diaz, B. (2006). A scatter search approach to the optimum dis-assembly sequence problem. *Computer Operation Research, 33,* 1776–1793.
39. Adenso-Diaz, B., Moure, F., & Rendueles, M. (2002). Automatic disassembly plans: Application to the continuous process industries. *Journal of Manufacturing Systems, 21,* 276–286.

Index

W. Li and S. Wang (eds.), *Sustainable Manufacturing and Remanufacturing Management*, https://doi.org/10.1007/978-3-319-73488-0